THE MASTER HANDBOOK OF ACOUSTICS

F. Alton Everest

FOURTH EDITION

McGraw-Hill

New York   San Francisco   Washington, D.C.   Auckland   Bogotá
Caracas   Lisbon   London   Madrid   Mexico City   Milan
Montreal   New Delhi   San Juan   Singapore
Sydney   Tokyo   Toronto

Cataloging-in-Publication Data is on file with the Library of Congress

Notices

DFR, ETC, QRD, RFZ, RPG, Abffusor, Diffractal, Diffusorblox, Flutterfree, Korner Killer, Omniffusor, Triffusor
RPG Diffusor System, Inc.
651-C Commerce Drive
Upper Marlboro, MD 20772

Quick Sound Field, Tube Traps, Snap Traps
Acoustic Sciences Corporation
P.O. Box 1189
Eugene, OR 97440

Techron, Sound Lab, AcoustaEQ, TEF, EFC, ETC, FTC, EFTE
Techron Division of Crown International
1718 W. Mishwaka Rd.
Elkhart, IN 46517

McGraw-Hill

A Division of The McGraw·Hill Companies

4 5 6 7 8 9 0   DOC/DOC   0 9 8 7 6 5 4 3 2

ISBN 0-07-136097-2

The sponsoring editor for this book was Scott Grillo, the editing supervisor was Steven Melvin, and the production supervisor was Pamela Pelton. It was set in Melior per the CMS design specs by Michele Pridmore of McGraw-Hill's Professional Book Group Hightstown composition unit.

Printed and bound by R. R. Donnelley & Sons Company.

 This book is printed on recycled, acid-free paper containing a minimum of 50% recycled de-inked fiber.

McGraw-Hill books are available at special quantity discounts to use as premiums and sales promotions, or for use in corporate training programs. For more information, please write to the Director of Special Sales, Professional Publishing, McGraw-Hill, Two Penn Plaza, New York, NY 10121-2298. Or contact your local bookstore.

THE MASTER HANDBOOK
OF ACOUSTICS

To Bonnie Gail, whose love of art, nature, and the author now embraces acoustics.

CONTENTS

Directly or indirectly, all questions connected with this subject must come for decision to the ear, as the organ of hearing; and from it there can be no appeal. But we are not therefore to infer that all acoustical investigations are conducted with the unassisted ear. When once we have discovered the physical phenomena which constitute the foundation of sound, our explorations are in great measure transferred to another field lying within the dominion of the principles of Mechanics. Important laws are in this way arrived at, to which the sensations of the ear cannot but conform.

Lord Raleigh in *The Theory of Sound*,
First Edition 1877.
(Also in first American edition, 1945,
courtesy of Dover Publications Inc.)

INTRODUCTION

Excerpts from the introduction to the third edition.

In 1981, the copyright year of the first edition of this book, Manfred Schroeder was publishing his early ideas on applying number theory to the diffusion of sound. In the third edition a new chapter has been added to cover numerous applications of diffraction-grating diffusors to auditoriums, control rooms, studios and home listening rooms.

Introduction to the fourth edition.

The science of acoustics made great strides in the 20th century, during which the first three editions of this book appeared. This fourth edition, however, points the reader to new horizons of the 21st century. A newly appreciated concept of distortion of sound in the medium itself (Chap. 25), a program for acoustic measurements (Chap. 26), and the optimization of placement of loudspeakers and listener (Chap. 27), all based on the home computer, point forward to amazing developments in acoustics yet to come.

As in the previous three editions, this fourth edition balances treatment of the fundamentals of acoustics with the general application of fundamentals to practical problems.

F. Alton Everest
Santa Barbara

Fundamentals of Sound

*S*ound can be defined as a wave motion in air or other elastic media (stimulus) or as that excitation of the hearing mechanism that results in the perception of sound (sensation). Which definition applies depends on whether the approach is physical or psychophysical. The type of problem dictates the approach to sound. If the interest is in the disturbance in air created by a loudspeaker, it is a problem in physics. If the interest is how it sounds to a person near the loudspeaker, psychophysical methods must be used. Because this book addresses acoustics in relation to people, both aspects of sound will be treated.

These two views of sound are presented in terms familiar to those interested in audio and music. *Frequency* is a characteristic of periodic waves measured in hertz (cycles per second), readily observable on a cathode-ray oscilloscope or countable by a frequency counter. The ear perceives a different *pitch* for a soft 100 Hz tone than a loud one. The pitch of a low-frequency tone goes down, while the pitch of a high-frequency tone goes up as intensity increases. A famous acoustician, Harvey Fletcher, found that playing pure tones of 168 and 318 Hz at a modest level produces a very discordant sound. At a high intensity, however, the ear hears the pure tones in the 150-300 Hz octave relationship as a pleasant sound. We cannot equate frequency and pitch, but they are analogous.

The same situation exists between *intensity* and *loudness.* The relationship between the two is not linear. This is considered later in more detail because it is of great importance in high fidelity work.

Similarly, the relationship between *waveform* (or spectrum) and perceived *quality* (or timbre) is complicated by the functioning of the hearing mechanism. As a complex waveform can be described in terms of a fundamental and a train of harmonics (or partials) of various amplitudes and phases (more on this later), the frequency-pitch interaction is involved as well as other factors.

The Simple Sinusoid

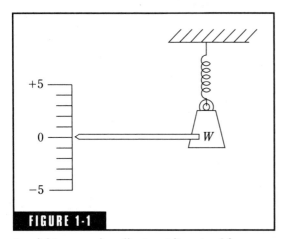

FIGURE 1-1

A weight on a spring vibrates at its natural frequency because of the elasticity of the spring and the inertia of the weight.

The *sine wave* is a basic waveform closely related to simple harmonic motion. The weight (mass) on the spring shown in Fig. 1-1 is a vibrating system. If the weight is pulled down to the −5 mark and released, the spring pulls the weight back toward 0. The weight will not, however, stop at zero; its inertia will carry it beyond 0 almost to +5. The weight will continue to vibrate, or oscillate, at an amplitude that will slowly decrease due to frictional losses in the spring, the air, etc.

The weight in Fig. 1-1 moves in what is called simple harmonic motion. The piston in an automobile engine is connected to the crankshaft by a connecting rod. The rotation of the crankshaft and the up-and-down motion of the pistons beautifully illustrate the relationship between rotary motion and linear simple harmonic motion. The piston position plotted against time produces a sine wave. It is a very basic type of mechanical motion, and it yields an equally basic waveshape in sound and electronics.

If a ballpoint pen is fastened to the pointer of Fig. 1-2, and a strip of paper is moved past it at a uniform speed, the resulting trace is a sine wave.

In the arrangement of Fig. 1-1, vibration or oscillation is possible because of the *elasticity* of the spring and the *inertia* of the weight.

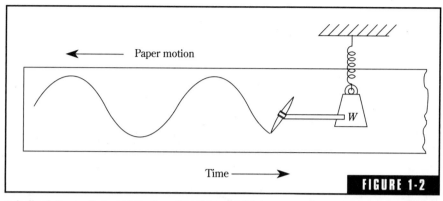

A ballpoint pen fastened to the vibrating weight traces a sine wave on a paper strip moving at uniform speed. This shows the basic relationship between simple harmonic motion and the sine wave.

Elasticity and inertia are two things all media must possess to be capable of conducting sound.

Sine-Wave Language

The sine wave is a specific kind of alternating signal and is described by its own set of specific terms. Viewed on an oscilloscope, the easiest value to read is the *peak-to-peak* value (of voltage, current, sound pressure, or whatever the sine wave represents), the meaning of which is obvious in Fig. 1-3. If the wave is symmetrical, the peak-to-peak value is twice the peak value.

The common ac voltmeter is, in reality, a dc instrument fitted with a rectifier that changes the alternating sine current to pulsating unidirectional current. The dc meter then responds to the *average* value as indicated in Fig. 1-3. Such meters are, however, almost universely calibrated in terms of *rms* (described in the next paragraph). For pure sine waves, this is quite an acceptable fiction, but for nonsinusoidal waveshapes the reading will be in error.

An alternating current of one ampere rms (or effective) is exactly equivalent in heating power to 1 ampere of direct current as it flows through a resistance of known value. After all, alternating current can heat up a resistor or do work no matter which direction it flows, it is just a matter of evaluating it. In the right-hand positive loop of Fig. 1-3 the ordinates (height of lines to the curve) are read off for each marked

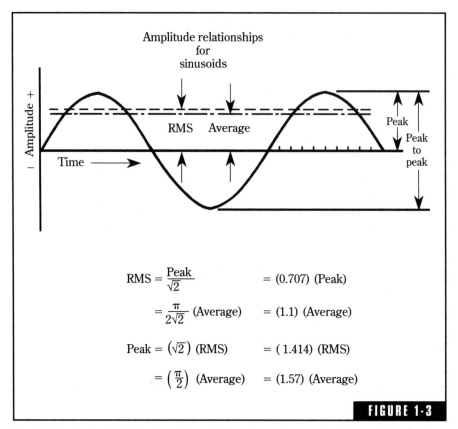

Amplitude relationships
for
sinusoids

$$RMS = \frac{Peak}{\sqrt{2}} \qquad\qquad = (0.707)\ (Peak)$$

$$= \frac{\pi}{2\sqrt{2}}\ (Average) \qquad = (1.1)\ (Average)$$

$$Peak = \left(\sqrt{2}\right)\ (RMS) \qquad = (1.414)\ (RMS)$$

$$= \left(\frac{\pi}{2}\right)\ (Average) \qquad = (1.57)\ (Average)$$

FIGURE 1-3

Amplitude relationships for sinusoids, which apply to sine waves of electrical voltage or current, as well as to acoustical parameters such as sound pressure. Another term which is widely used in the audio field is *crest factor,* or peak divided by rms.

increment of time. Then (a) each of these ordinate values is squared, (b) the squared values are added together, (c) the average is found, and (d) the square root is taken of the average (or mean). Taking the square root of this average gives the root-mean-square or rms value of the positive loop of Fig. 1-3. The same can be done for the negative loop (squaring a negative ordinate gives a positive value), but simply doubling the positive loop of a symmetrical wave is easier. In this way the root-mean-square or "heating power" value of any alternating or periodic waves can be determined whether the wave is for voltage, current, or sound pressure. Such computations will help you understand the meaning of rms, but fortunately reading meters is far easier. Figure 1-3 is a useful summary of relationships pertaining only to the sine wave.

Propagation of Sound

If an air particle is displaced from its original position, elastic forces of the air tend to restore it to its original position. Because of the inertia of the particle, it overshoots the resting position, bringing into play elastic forces in the opposite direction, and so on.

Sound is readily conducted in gases, liquids, and solids such as air, water, steel, concrete, etc., which are all elastic media. As a child, perhaps you heard two sounds of a rock striking a railroad rail in the distance, one sound coming through the air and one through the rail. The sound through the rail arrives first because the speed of sound in the dense steel is greater than that of tenuous air. Sound has been detected after it has traveled thousands of miles through the ocean.

Without a medium, sound cannot be propagated. In the laboratory, an electric buzzer is suspended in a heavy glass bell jar. As the button is pushed, the sound of the buzzer is readily heard through the glass. As the air is pumped out of the bell jar, the sound becomes fainter and fainter until it is no longer audible. The sound-conducting medium, air, has been removed between the source and the ear. Because air is such a common agent for the conduction of sound, it is easy to forget that other gases as well as solids and liquids are also conductors of sound. Outer space is an almost perfect vacuum; no sound can be conducted except in the tiny island of air (oxygen) within a spaceship or a spacesuit.

The Dance of the Particles

Waves created by the wind travel across a field of grain, yet the individual stalks remain firmly rooted as the wave travels on. In a similar manner, particles of air propagating a sound wave do not move far from their undisplaced positions as shown in Fig. 1-4. The disturbance travels on, but the propagating particles do their little dance close to home.

There are three distinct forms of particle motion. If a stone is dropped on a calm water surface, concentric waves travel out from the point of impact, and the water particles trace *circular* orbits (for deep water, at least) as in Fig. 1-5(A). Another type of wave motion is illustrated by a violin string, Fig. 1-5(B). The tiny elements of the string move *transversely*, or at right angles to the direction of travel of the waves along the string. For sound traveling in a gaseous medium such as air, the particles move in the direction the sound is traveling. These are called *longitudinal waves*, Fig. 1-5C.

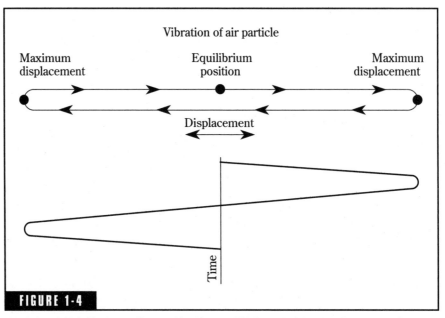

FIGURE 1-4

An air particle is made to vibrate about its equilibrium position by the energy of a passing sound wave because of the interaction of the elastic forces of the air and the inertia of the air particle.

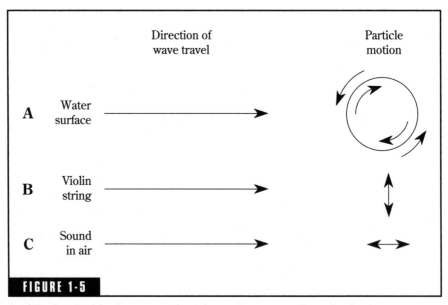

FIGURE 1-5

Particles involved in the propagation of sound waves can dance in circular, transverse, or longitudinal motions.

How a Sound Wave Is Propagated

How are air particles jiggling back and forth able to carry beautiful music from the loudspeaker to our ears at the speed of a rifle bullet? The little dots of Fig. 1-6 represent air molecules. There are more than a million molecules in a cubic inch of air; hence this sketch is greatly exaggerated. The molecules crowded together represent areas of *compression* in which the air pressure is slightly greater than the prevailing atmospheric pressure. The sparse areas represent *rarefactions* in which the pressure is slightly less than atmospheric. The small arrows indicate that, on the average, the molecules are moving to the right of the compression crests and to the left in the rarefaction troughs between the crests. Any given molecule will move a certain distance to the right and then the same distance to the left of its undisplaced position as the sound wave progresses uniformly to the right.

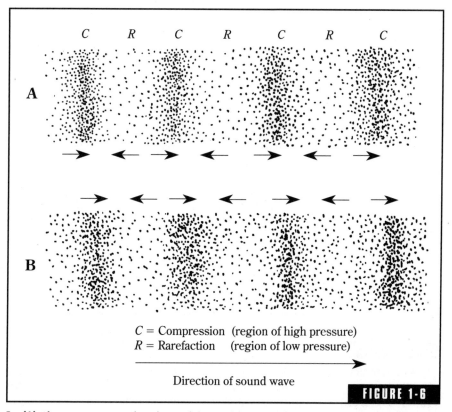

C = Compression (region of high pressure)
R = Rarefaction (region of low pressure)

Direction of sound wave

FIGURE 1-6

In (A) the wave causes the air particles to be pressed together in some regions and spread out in others. An instant later (B) the wave has moved slightly to the right.

Why does the sound wave move to the right? The answer is revealed by a closer look at the arrows of Fig. 1-6. The molecules tend to bunch up where two arrows are pointing toward each other, and this occurs a bit to the right of each compression. When the arrows point away from each other the density of molecules will decrease. Thus, the movement of the higher pressure crest and the lower pressure trough accounts for the small progression of the wave to the right.

As mentioned previously, the pressure at the crests is higher than the prevailing atmospheric barometric pressure and the troughs lower than the atmospheric pressure, as shown in the sine wave of Fig. 1-7. These fluctuations of pressure are very small indeed. The faintest sound the ear can hear (20 μPascal) is some 5,000 million times smaller than atmospheric pressure. Normal speech and music signals are represented by correspondingly small ripples superimposed on the atmospheric pressure.

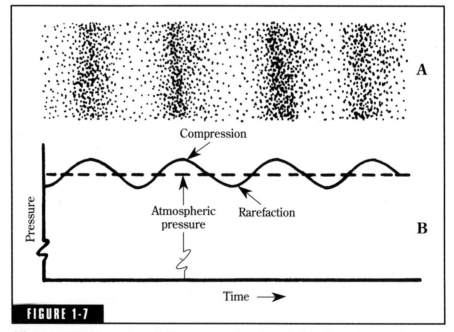

FIGURE 1-7

(A) An instantaneous view of the compressed and rarefied regions of a sound wave in air. (B) The compressed regions are very slightly above and the rarefied regions very slightly below atmospheric pressure. Pressure variations of sound waves are thus superimposed on prevailing barometric pressure.

Sound in Free Space

The intensity of sound decreases as the distance to the source is increased. In free space, far from the influence of surrounding objects, sound from a point source is propagated uniformly in all directions. The intensity of sound decreases as shown in Fig. 1-8. The same sound power flows out through A1, A2, A3, and A4, but the areas increase as the square of the radius, r. This means that the sound power per unit area (intensity) decreases as the square of the radius. Doubling the distance reduces the intensity to one-fourth the initial value, tripling the distance yields $\frac{1}{9}$, and increasing the distance four times yields $\frac{1}{16}$ of

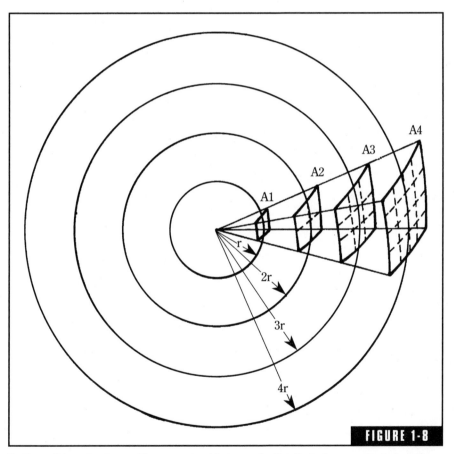

FIGURE 1-8

In the solid angle shown, the same sound energy is distributed over spherical surfaces of increasing area as r is increased. The intensity of the sound is inversely proportional to the square of the distance from the point source.

the initial intensity. The *inverse square law* states that the *intensity of sound in a free field is inversely proportional to the square of the distance from the source.* This law provides the basis of estimating the sound level in many practical circumstances and is discussed in a later chapter.

Wavelength and Frequency

A simple sine wave is illustrated in Fig. 1-9. The *wavelength* is the distance a wave travels in the time it takes to complete one cycle. A wavelength can be measured between successive peaks or between any two corresponding points on the cycle. This holds for periodic waves other than the sine wave as well. The *frequency* is the number of cycles per second (or hertz). Frequency and wavelength are related as follows:

$$\text{Wavelength (ft)} = \frac{\text{Speed of sound (ft/sec)}}{\text{Frequency (hertz)}} \qquad \textbf{(1-1)}$$

which can be written as:

$$\text{Frequency} = \frac{\text{Speed of sound}}{\text{Wavelength}} \qquad \textbf{(1-2)}$$

The speed of sound in air is about 1,130 feet per second (770 miles per hour) at normal temperature and pressure. For sound traveling in air, Equation 1-1 becomes:

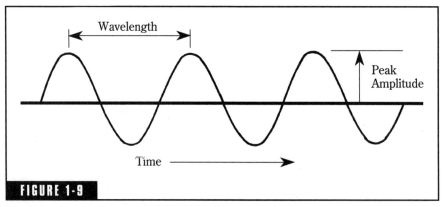

Wavelength

Peak Amplitude

Time

FIGURE 1-9

Wavelength is the distance a wave travels in the time it takes to complete one cycle. It can also be expressed as the distance from one point on a periodic wave to the corresponding point on the next cycle of the wave.

$$\text{Wavelength} = \frac{1,130}{\text{Frequency}} \qquad \text{(1-3)}$$

This relationship is used frequently in following sections. Figure 1-10 gives two graphical approaches for an easy solution to Equation 1-3.

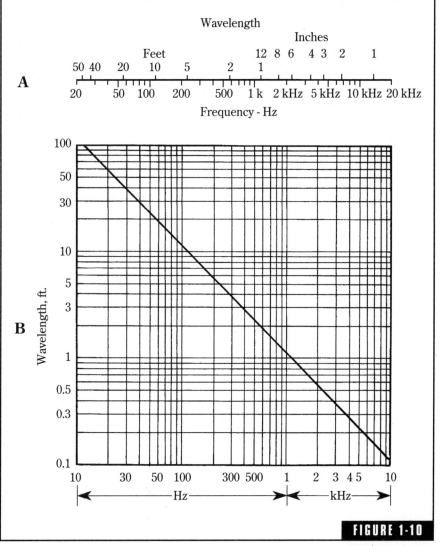

(A) Convenient scales for rough determination of wavelength of sound in air from known frequency, or vice versa. (B) A chart for easy determination of the wavelength in air of sound waves of different frequencies. (Both based on speed of sound of 1,139 ft per second.)

Complex Waves

Speech and music waveshapes depart radically from the simple sine form. A very interesting fact, however, is that no matter how complex the wave, as long as it is periodic, it can be reduced to sine components. The obverse of this is that, theoretically, any complex periodic wave can be synthesized from sine waves of different frequencies, different amplitudes, and different time relationships (phase). A friend of Napoleon, named Joseph Fourier, was the first to develop this surprising idea. This idea can be viewed as either a simplification or complication of the situation. Certainly it is a great simplification in regard to concept, but sometimes complex in its application to specific speech or musical sounds. As we are interested primarily in the basic concept, let us see how even a very complex wave can be reduced to simple sinusoidal components.

Harmonics

A simple sine wave of a given amplitude and frequency, f_1, is shown in Fig. 1-11A. Figure 1-11B shows another sine wave half the amplitude and twice the frequency (f_2). Combining A and B at each point in time the waveshape of Fig. 1-11C is obtained. In Fig. 1-11D, another sine wave half the amplitude of A and three times its frequency (f_3) is shown. Adding this to the $f_1 + f_2$ wave of C, Fig. 1-11E is obtained. The simple sine wave of Fig. 1-11A has been progressively distorted as other sine waves have been added to it. Whether these are acoustic waves or electronic signals, the process can be reversed. The distorted wave of Fig. 1-11E can be disassembled, as it were, to the simple $f_1, f_2,$ and f_3 sine components by either acoustical or electronic filters. For example, passing the wave of Fig. 1-11E through a filter permitting only f_1 and rejecting f_2 and f_3, the original f_1 sine wave of Fig. 1-11A emerges in pristine condition.

Applying names, the sine wave with the lowest frequency (f_1) of Fig. 1-11A is called the *fundamental,* the one with twice the frequency (f_2) of Fig. 1-11B is called the *second harmonic,* and the one three times the frequency (f_3) of Fig. 1-11D is the *third harmonic.* The fourth harmonic, the fifth harmonic, etc., are four and five times the frequency of the fundamental, and so on.

Phase

In Fig. 1-11, all three components, $f_1, f_2,$ and f_3, start from zero together. This is called an *in-phase condition.* In some cases, the time relation-

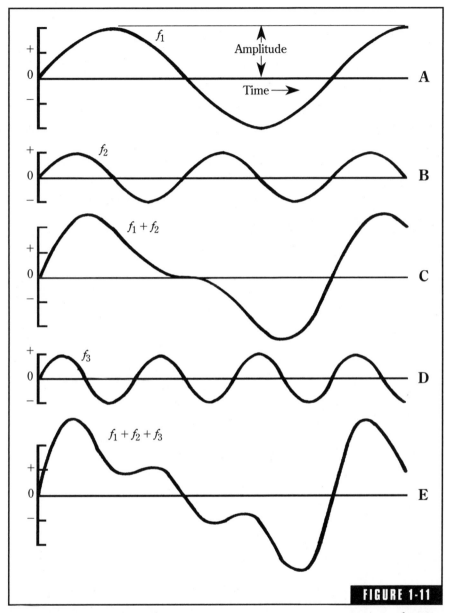

FIGURE 1-11

A study in the combination of sine waves. (A) The fundamental of frequency f_1. (B) A second harmonic of frequency $f_2 = 2f_1$ and half the amplitude of f_1. (C) The sum of f_1 and f_2 obtained by adding ordinates point by point. (D) A third harmonic of frequency $f_3 = 3f_1$ and half the amplitude of f_1. (E) The waveshape resulting from the addition of f_1, f_2, and f_3. All three components are "in phase," that is, they all start from zero at the same instant.

ships between harmonics or between harmonics and the fundamental are quite different from this. Remember how one revolution of the crankshaft of the automobile engine (360°) was equated with one cycle of simple harmonic motion of the piston? The up-and-down travel of the piston spread out in time traces a sine wave such as that in Fig. 1-12. One complete sine-wave cycle represents 360° of rotation. If another sine wave of identical frequency is delayed 90°, its time relationship to the first one is a quarter wave late (time increasing to the right). A half-wave delay would be 180°, etc. For the 360° delay, the wave at the bot-

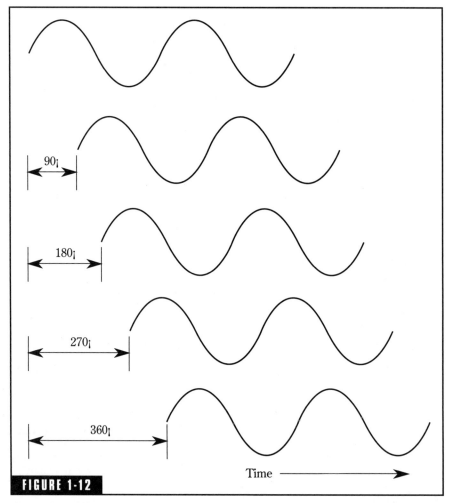

90¡

180¡

270¡

360¡

Time

FIGURE 1-12

Illustration of the phase relationships between waves with the same amplitude and frequency. A rotation of 360 degrees is analogous to one complete sine cycle.

tom of Fig. 1-12 falls in step with the top one, reaching positive peaks and negative peaks simultaneously and producing the in-phase condition.

In Fig. 1-11, all three components of the complex wave of Fig. 1-11E are in phase. That is, the f_1 fundamental, the f_2 second harmonic, and the f_3 third harmonic all start at zero at the same time. What happens if the harmonics are out of phase with the fundamental? Figure 1-13 illustrates this case. The second harmonic f_2 is now advanced 90°, and the third harmonic f_3 is retarded 90°. By combining f_1, f_2, and f_3 for each instant of time, with due regard to positive and negative signs, the contorted wave of Fig. 1-13E is obtained.

The only difference between Figs. 1-11E and 1-13E is that a phase shift has been introduced between harmonics f_2 and f_3, and the fundamental f_1. That is all that is needed to produce drastic changes in the resulting waveshape. Curiously, even though the shape of the wave is dramatically changed by shifting the time relationships of the components, the ear is relatively insensitive to such changes. In other words, waves E of Figs. 1-11 and 1-13 would sound very much alike to us.

A common error is confusing polarity with phase. Phase is the time relationship between two signals while polarity is the $+/-$ or the $-/+$ relationship of a given pair of signal leads.

Partials

A musician is inclined to use the term *partial* instead of harmonic, but it is best that a distinction be made between the two terms because the partials of many musical instruments are not harmonically related to the fundamental. That is, partials might not be exact multiples of the fundamental frequency, yet richness of tone can still be imparted by such deviations from the true harmonic relationship. For example, the partials of bells, chimes, and piano tones are often in a nonharmonic relationship to the fundamental.

Octaves

Audio and electronics engineers and acousticians frequently use the integral multiple concept of harmonics, closely allied as it is to the physical aspect of sound. The musician often refers to the octave, a logarithmic concept that is firmly embedded in musical scales and terminology because of its relationship to the ear's characteristics. Audio people are also involved with the human ear, hence their common use of logarithmic scales for frequency, logarithmic measuring units, and

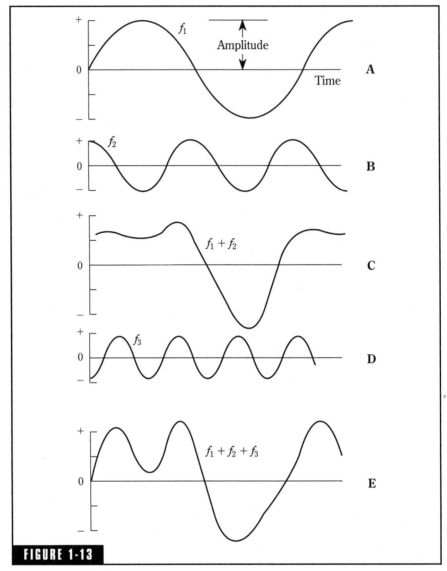

FIGURE 1-13

A study of the combination of sine waves that are not in phase. (A) The fundamental of frequency f_1. (B) The second harmonic f_2 with twice the frequency and half the amplitude of f_1 advanced 90 degrees with respect to f_1. (C) The combination of f_1 and f_2 obtained by adding ordinates point by point. (D) The third harmonic f_3 with phase 90 degrees behind f_1, and with half the amplitude of f_1. (E) The sum of f_1, f_2, and f_3. Compare this waveshape with that of Fig. 1-11(E). The difference in waveshapes is due entirely to the shifting of the phase of the harmonics with respect to the fundamental.

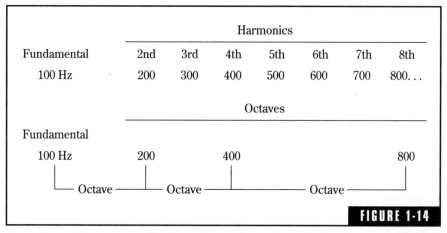

	Harmonics						
Fundamental	2nd	3rd	4th	5th	6th	7th	8th
100 Hz	200	300	400	500	600	700	800...

	Octaves		
Fundamental			
100 Hz	200	400	800

Octave — Octave — Octave

FIGURE 1-14

Comparison of harmonics and octaves. Harmonics are linearly related; octaves are logarithmically related.

various devices based on octaves, which are more fully discussed later. Harmonics and octaves are compared in Fig. 1-14.

The Concept of Spectrum

Chapter 3 relates the commonly accepted scope of the audible spectrum, 20 Hz to 20 kHz, to specific characteristics of the human ear. Here, in the context of sine waves, harmonics, etc., we need to establish the spectrum concept. The visible spectrum of light has its counterpart in sound in the audible spectrum, the range of frequencies that fall within the perceptual limits of the human ear. We cannot see far-ultraviolet light because the frequency of its electromagnetic energy is too high for the eye to perceive. Nor can we see the far-infrared light because its frequency is too low. There are likewise sounds of too low (infrasound) and too high frequency (ultrasound) for the ear to hear.

Figure 1-15 shows several waveforms that typify the infinite number of different waveforms commonly encountered in audio. These waveforms have been photographed directly from the screen of a cathode-ray oscilloscope. To the right of each photograph is the spectrum of that particular signal. The spectrum tells how the energy of the signal is distributed in frequency. In all but the bottom signal of Fig. 1-15, the audible range of the spectrum was searched with a wave analyzer having a very sharp filter with a passband only 5 Hz wide. In this way,

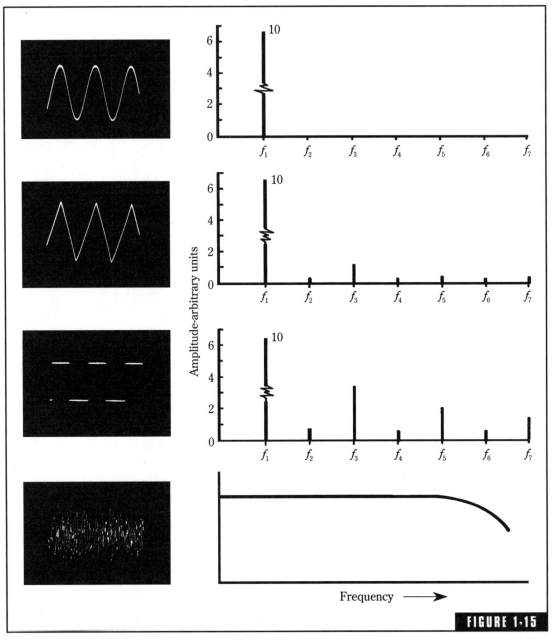

FIGURE 1·15

The spectral energy of a pure sinusoid is contained entirely at a single frequency. The triangular and square waves each have a prominent fundamental and numerous harmonics at integral multiples of the fundamental frequency. Random noise (white noise) has energy distributed uniformly throughout the spectrum up to some point at which energy begins to fall off due to generator limitations. Random noise may be considered a mixture of sine waves with continuously shifting frequencies, amplitudes, and phases.

concentrations of energy were located and measured with an electronic voltmeter.

For an ideal sine wave, all the energy is concentrated at one frequency. The sine wave produced by this particular signal generator is not really a pure sine wave. No oscillator is perfect and all have some harmonic content, but in scanning the spectrum of this sine wave, the harmonics measured were too low to show on the graph scale of Fig. 1-15.

The triangular wave of this signal generator has a major fundamental component of 10 units magnitude. The wave analyzer detected a significant second harmonic component at f_2, twice the frequency of the fundamental with a magnitude of 0.21 units. The third harmonic showed an amplitude of 1.13 units, the fourth of 0.13 units, etc. The seventh harmonic still had an amplitude of 0.19 units and the fourteenth harmonic (about 15 kHz in this case) an amplitude of only 0.03 units, but still easily detectable. So we see that this triangular wave has both odd and even components of modest amplitude down through the audible spectrum. If you know the amplitude and phases of each of these, the original triangular wave shape can be synthesized by combining them.

A comparable analysis reveals the spectrum of the square wave shown in Fig. 1-15. It has harmonics of far greater amplitude than the triangular wave with a distinct tendency toward more prominent odd than even harmonics. The third harmonic shows an amplitude 34 percent of the fundamental! The fifteenth harmonic of the square wave is still 0.52 units! If the synthesis of a square wave stops with the fifteenth harmonic, the wave of Fig. 1-16C results.

A glance at the spectra of sine, triangular, and square waves reveals energy concentrated at harmonic frequencies, but nothing between. These are all so-called *periodic waves,* which repeat themselves cycle after cycle. The fourth example in Fig. 1-15 is a random noise. The spectrum of this signal cannot be measured satisfactorily by a wave analyzer with a 5-Hz passband because the fluctuations are so great that it is impossible to get a decent reading on the electronic voltmeter. Analyzed by a wider passband of fixed bandwidth and with the help of various integrating devices to get a steady indication, the spectral shape shown is obtained. This spectrum tells us that the energy of the random-noise signal is equally distributed throughout the spectrum until the drooping at high frequencies indicates that the upper frequency limit of the random noise generator has been reached.

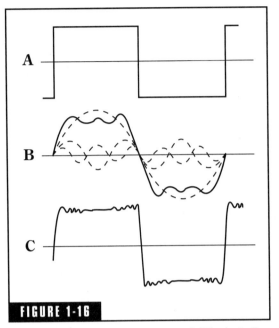

FIGURE 1-16

In synthesizing the square wave of (A), including only the fundamental and two harmonics yields (B). Including fifteen components yields (C). It would take many more than fifteen harmonics to smooth the ripples and produce the square corners of (A).

There is little visual similarity between the sine and the random-noise signals as revealed by the cathode-ray oscilloscope, yet there is a hidden relationship. Even random noise can be considered as being made up of sine-wave components constantly shifting in frequency, amplitude, and phase. If you pass random noise through a narrow filter and observe the filter output on a cathode-ray oscilloscope, you will see a restless, sinelike wave that constantly shifts in amplitude. Theoretically, an infinitely narrow filter would sift out a pure, but nervous, sine wave. (See chapter 5.)

Electrical, Mechanical, and Acoustical Analogs

An acoustical system such as a loudspeaker can be represented in terms of an equivalent electrical or mechanical system. The physicist freely uses these equivalents to set up his mathematical approach for analyzing a given system. Although such approaches are far outside the scope of this book, it is useful to develop some appreciation for these methods. For example, the effect of a cabinet on the functioning of a loudspeaker is clarified by thinking of the air in the enclosed space as acting like a capacitor in an electrical circuit, absorbing and giving up the energy imparted by the cone movement.

Figure 1-17 shows the graphical representation of the three basic elements in electrical, mechanical, and acoustical systems. Inductance in an electrical circuit is equivalent to mass in a mechanical system and inertance in an acoustical system. Capacitance in an electrical circuit is analogous to compliance in a mechanical system and capacitance in an acoustical system. Resistance is resistance in all three systems, whether it be the frictional losses offered to air-particle movement in glass fiber, frictional losses in a wheel bearing, or resistance to the flow of current in an electrical circuit.

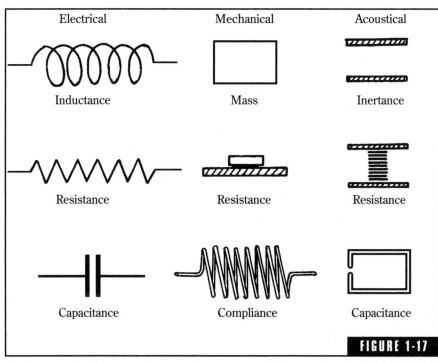

The three basic elements of electrical systems and their analogs in mechanical and acoustical systems.

CHAPTER 2

Sound Levels and the Decibel

The decibel is as commonly used in audio circles as the minute or the mile is in general usage. A superficial understanding of the decibel can hinder the study of the science of sound and be a barrier in the proper use and development of its many applications. The goal of this chapter is to show the need for the decibel concept and how decibels can be applied in many different ways.

Levels in decibels make it easy to handle the extremely wide range of sensitivity in human hearing. The threshold of hearing matches the ultimate lower limit of perceptible sound in air, the noise of air molecules beating a tattoo on the eardrum. The sensitivity of normal human eyes also matches the ultimate limit by responding to one or a very few photons of light. From these threshold responses to the most feeble stimuli, the ear and eye are also capable of handling high intensities of sound and light. A level in decibels is a convenient way of handling the billion-fold range of sound pressures to which the ear is sensitive without getting bogged down in long strings of zeros.

Ratios vs. Differences

Imagine a sound source set up in a room completely protected from interfering noise. (The term *sound-proof* is avoided because there will be much sound in it.) The sound source is adjusted for a weak sound with a sound pressure of 1 unit, and its loudness is carefully noted.

When the sound pressure is increased until it sounds twice as loud, the level dial reads 10 units. This completes observation A. For observation B, the source pressure is increased to 10,000 units. To double the loudness, you find that the sound pressure must be increased from 10,000 to 100,000 units. The results of this experiment can now be summarized as follows:

Observations	Two Pressures	Ratio of Two Pressures
A	10 – 1	10: 1
B	100,000 – 10,000	10: 1

Observations A and B accomplish the same doubling of perceived loudness. In observation A, this was accomplished by an increase in sound pressure of only 9 units, where in observation B it took 90,000 units. Ratios of pressures seem to describe loudness changes better than differences in pressure. Ernst Weber (1834), Gustaf Fechner (1860), Hermann von Helmholtz (1873), and other early researchers pointed out the importance of ratios, which we know apply equally well to sensations of vision, hearing, vibration, or even electric shock.

Many years ago, a friend working in a university research laboratory demonstrated his experiment on the hearing of cats, which in many ways is similar to that of humans. A tone of 250 Hz, radiated from a nearby loudspeaker, was picked up by the ears of an anesthetized cat, a portion of whose brain was temporarily exposed. A delicate probe picked up the 250-Hz signal at a highly localized spot on the auditory cortex, displaying it on a cathode-ray oscilloscope. When the tone was shifted to 500 Hz, the signal was picked up at another spot on the cortex. Tones of 1,000 and 2,000 Hz were detected at other specific spots. The fascinating point here is that changing the tone an octave resulted in the signal appearing on the auditory cortex at discrete, equally spaced points. Frequencies in the ratio of 2 to 1 (an octave) seem to have a linear positional relationship in the cat's brain. This indicates a logarithmic response to frequency. Ratios of stimuli come closer to matching up with human perception than do differences of stimuli. This matching is not perfect, but close enough to make a strong case for the use of levels in decibels.

Ratios of powers or ratios of intensities, or ratios of sound pressure, voltage, current, or anything else are dimensionless. For instance, the ratio of 1 watt to 100 watts is 1 watt/100 watts, and the watt unit upstairs and the watt unit downstairs cancel, leaving $\frac{1}{100} = 0.01$, a pure number without dimension. This is important because logarithms can be taken only of nondimensional numbers.

Handling Numbers

Table 2-1 illustrates three different ways numbers can be expressed. The decimal and arithmetic forms are familiar in everyday activity. The exponential form, while not as commonly used, has the charm of simplifying things once the fear of the unknown or little understood is conquered. In writing *one hundred thousand*, there is a choice between 100,000 watts and 10^5 watts, but how about *a millionth of a millionth* of a watt? All those zeros behind the decimal point make it impractical even to reproduce here, but 10^{-12} is easy. And the prefix that means 10^{-12} is *pico*; so the power is 1 picowatt (shown later in Table 2-4). Engineering-type calculators take care of the exponential form in what is called *scientific notation*, by which very large or very small numbers can be entered.

Table 2-1. Ways of expressing numbers.

Decimal form	Arithmetic form	Exponential form
100,000	$10 \times 10 \times 10 \times 10 \times 10$	10^5
10,000	$10 \times 10 \times 10 \times 10$	10^4
1,000	$10 \times 10 \times 10$	10^3
100	10×10	10^2
10	10×1	10^1
1	$10/10$	10^0
0.1	$1/10$	10^{-1}
0.01	$1/(10 \times 10)$	10^{-2}
0.001	$1/(10 \times 10 \times 10)$	10^{-3}
0.0001	$1/(10 \times 10 \times 10 \times 10)$	10^{-4}

Table 2-1. Ways of expressing numbers *(Continued)*.

Decimal form	Arithmetic form	Exponential form
100,000	(100)(1,000)	$10^2 + 10^3 = 10^{2+3} = 10^5$
100	10,000/100	$10^4/10^2 = 10^{4-2} = 10^2$
10	100,000/10,000	$10^5/10^4 = 10^{5-4} = 10^{-1} = 10$
10	$\sqrt{100} = \sqrt[2]{100}$	$100^{1/2} = 100^{0.5}$
4.6416	$\sqrt[3]{100}$	$100^{1/2} = 100^{0.333}$
31.6228	$\sqrt[4]{100^3}$	$100^{3/4} = 100^{0.75}$

Logarithms

Representing 100 as 10^2 simply means that $10 \times 10 = 100$ and that 10^3 means $10 \times 10 \times 10 = 1,000$. But how about 267? That is where logarithms come in. It is agreed that 100 equals 10^2. By definition you can say that the logarithm of 100 to the base 10 = 2, commonly written $\log_{10} 100 = 2$, or simply log 100 = 2, because common logarithms are to the base 10. Now that number 267 needn't scare us; it is simply expressed as 10 to some other power between 2 and 3. The old fashioned way was to go to a book of log tables, but with a simple hand-held calculator punch in 267, push the "log" button, and 2.4265 appears. Thus, $267 = 10^{2.4265}$, and log 267 = 2.4265. Logs are so handy because, as Table 2-1 demonstrates, they reduce multiplication to addition, and division to subtraction. This is exactly how the now-extinct slide rule worked, by positioning engraved logarithmic scales.

Logs should be the friend of every audio worker because they are the solid foundation of our levels in decibels. A level is a logarithm of a ratio. A level in decibels is ten times the logarithm to the base 10 of the ratio of two power like quantities.

Decibels

A power level of a power W_1 can be expressed in terms of a reference power W_2 as follows:

$$L_1 = \log_{10} \frac{W_1}{W_2} \text{ bels} \qquad\qquad \textbf{(2-1)}$$

Because the decibel, from its very name, is ⅒ bel (from Alexander Graham Bell), the level in decibels of a power ratio becomes:

$$L_1 = 10 \log_{10} \frac{W_1}{W_2} \text{ decibels} \qquad\qquad \textbf{(2-2)}$$

Equation 2-2 applies equally to acoustic power, electric power, or any other kind of power. A question often arises when levels other than power need to be expressed in decibels. For example, acoustic intensity is acoustic power per unit area in a specified direction, hence Equation 2-2 is appropriate. Acoustic power is proportional to the square of the acoustic pressure, p, hence the power level is:

$$L_p = 10 \log \frac{p^2_1}{p^2_2}$$

$$= 20 \log \frac{p_1}{p_2} \text{ in decibels} \qquad\qquad \textbf{(2-3)}$$

The tabulation of Table 2-2 will help you decide whether the Equation 2-2 or Equation 2-3 form applies.

Table 2-2. Use of 10 log and 20 log.

Parameter	Eq(2-2) $10 \log_{10} \frac{a_1}{a_2}$	Eq(2-3) $20 \log_{10} \frac{b_1}{b_2}$
Acoustic		
Power	X	
Intensity	X	
Air particle velocity		X
Pressure		X

Parameter	Eq(2-2) $10 \log_{10} \dfrac{a_1}{a_2}$	Eq(2-3) $20 \log_{10} \dfrac{b_1}{b_2}$
Electric		
Power	X	
Current		X
Voltage		X
Distance		
(From source-SPL; inverse square)		X

Table 2-2. Use of 10 log and 20 log (*Continued*).

Sound pressure is usually the most accessible parameter to measure in acoustics, even as voltage is for electronic circuits. For this reason, the Equation 2-3 form is more often encountered in day-to-day technical work.

Reference Levels

A sound-level meter is used to read a certain sound-pressure level. If the corresponding sound pressure is expressed in normal pressure units, a great range of very large and very small numbers results. Ratios are more closely related to human senses than linear numbers, and "the level decibels approach" compresses the large and small ratios into a more convenient and comprehensible range. Basically, our sound-level meter reading is a certain sound-pressure level, 20 log (p_1/p_2), as in Equation 2-3. Some standard reference sound pressure for p_2 is needed. The reference p_2 selected must be the same as that used by others, so that ready comparisons can be made worldwide. Several such reference pressures have been used over the years, but for sound in air the standard reference pressure is 20 µPa (micropascal). This might seem quite different from the reference pressure of 0.0002 microbar or 0.0002 dyne/cm², but it is the same standard merely written in different units. This is a very minute sound pressure and corresponds closely to the threshold of human hearing. The relationship between sound pressure in Pascals, pounds per square inch, and sound pressure level is shown in the graph of Fig. 2-1.

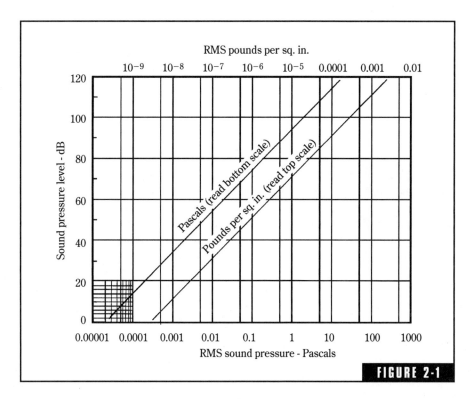

FIGURE 2-1

The relationship between sound pressure in Pascals or pounds per square inch and sound-pressure level (referred to 20 uPa) is shown in this graph. This is a graphical approach to the solution of Eq. 2-2.

When a statement is encountered such as, "The sound pressure *level* is 82 dB," the 82-dB sound-pressure level is normally used in direct comparison with other levels. However, if the sound pressure were needed, it can be computed readily by working backward from Equation 2-3 as follows:

$$82 = 20 \log \frac{p_1}{20 \ \mu Pa}$$

$$\log \frac{p_1}{20 \ \mu Pa} = \frac{82}{20}$$

$$\frac{p_1}{20 \ \mu Pa} = 10^{\frac{82}{20}}$$

The y^x button on the calculator (the Hewlett-Packard 41-C is assumed) helps us to evaluate $10^{4.1}$. Press 10, enter 4.1 press y^x power button, and the answer 12,589 appears.

$$p_1 = (20 \text{ μPa})(12{,}589)$$
$$p_1 = 251{,}785 \text{ μPa}$$

There is another lesson here. The 82 has what is called two *significant figures*. The 251,785 has six significant figures and implies a precision that is not there. Just because the calculator says so doesn't make it so! A better answer is 252,000 μPa, or 0.252 Pa.

Logarithmic and Exponential Forms Compared

The logarithmic and exponential forms are equivalent as can be seen by glancing again at Table 2-1. In working with decibels it is imperative that a familiarity with this equivalence be firmly grasped.

Let's say we have a power ratio of 5:

$$10 \log_{10} 5 = 6.99 \text{ is exactly equivalent to}$$

$$5 = 10^{\frac{6.99}{10}}$$

(2-4)

There are two tens in the exponential statement but they come from different sources as indicated by the arrows. Now let us treat a sound pressure ratio of 5:

$$20 \log_{10} 5 = 13.98$$

$$5 = 10^{\frac{13.98}{20}}$$

(2-5)

Remember that sound-pressure *level* in air means that the reference pressure downstairs (p_2) in the pressure ratio is 20 μPa. There are other reference quantities; some of the more commonly used ones are listed in Table 2-3. In dealing with very large and very small numbers, you should become familiar with the prefixes of Table 2-4. These prefixes are nothing more than Greek names for the powers exponents of 10.

Table 2-3. Reference quantities in common use.

Level in decibels	Reference quantity
Acoustic	
Sound pressure level in air (SPL, dB)	20 micropascal
Power level (Lp,dB)	1 picowatt(10–12 watt)
Electric	
Power level re 1 mW	10–3 watt (1 milliwatt)
Voltage level re 1 V	1 volt
Volume level, VU	10–3 watt

Acoustic Power

It doesn't take many watts of acoustic power to produce some very loud sounds, as anyone who lives downstairs from a dedicated audiophile will testify. We are conditioned by megawatt electrical generating plants, 350-horsepower (261 kilowatt) automobile engines, and 1,500-watt flatirons that eclipse the puny watt or so the hi-fi loudspeakers might radiate as acoustic power. Even though a hundred-watt amplifier may be driving the loudspeakers, loudspeaker efficiency (output for a given input) is very low, perhaps on the order of 10 percent, and head-room must be reserved for the occasional peaks of music. Increasing power to achieve greater results is often frustrating. Doubling power from 1 to 2 watts is a 3-dB increase in power level (10 log 2 = 3.01), a very small increase in loudness; however, the same 3-dB increase in level is represented by an increase in power from 100 to 200 watts or 1,000 to 2,000 watts.

Table 2-5 lists sound pressure and sound-pressure levels of some common sounds. In the sound-pressure column, it is a long stretch

Table 2-4. The Greeks had a word for it.

Prefix	Symbol	Multiple
tera	T	1012
giga	G	109
mega	M	106
kilo	k	103
milli	m	10-3
micro	μ	10-6
nano	n	10-9
pico	p	10-12

Table 2-5. Some common sound-pressure levels and sound pressures.

Sound Source	Sound pressure (Pa)	Sound level* (decibels, A-weighted)
Saturn rocket	100,000. (one atmosphere)	194
Ram jet	2,000.	160
Propeller aircraft	200.	140
Threshold of pain		135
Riveter	20.	120
Heavy truck	2.	100
Noisy office, } Heavy traffic	0.2	80
Conversational speech	0.02	60
Private office		50
Quiet residence	0.0002	40
Recording studio		30
Leaves rustling	0.0002	20
Hearing threshold, good ears at frequency of maximum sensitivity		10
Hearing threshold, excellent ears at frequency maximum response	0.00002	0

* Reference pressure (take your pick, these are identical):
20 micropascal (μPa)
0.00002 pascal
2×10^{-5} newton/meter^2
0.0002 dyne/cm^2 or microbar

from 100,000 Pa (100 kPa), which is atmospheric pressure to, 0.00002 Pa (20 μPa), but this range is reduced to quite a convenient form in the level column. The same information is present in graphical form in Fig. 2-2.

Another way to generate a 194-dB sound-pressure level, besides launching a Saturn rocket, is to detonate 50 pounds of TNT 10 feet away. Common sound waves are but tiny ripples on the steady-state atmospheric pressure. A 194-dB sound-pressure level approaches the atmospheric and, hence, is a ripple of the same order of magnitude as

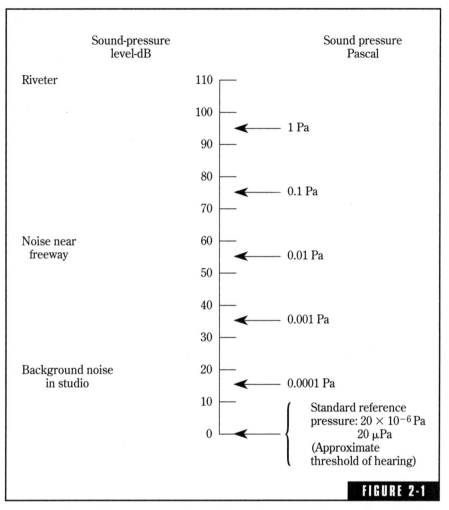

An appreciation of the relative magnitude of a sound pressure of 1 Pascal can be gained by comparison to known sounds. The standard reference pressure for sound in air is 20μPa, which corresponds closely to the minimum audible pressure.

atmospheric pressure. The 194-dB sound pressure is an rms (root mean square) value. A peak sound pressure 1.4 times as great would modulate the atmospheric pressure completely.

Using Decibels

A level is a logarithm of a ratio of two powerlike quantities. When levels are computed from other than power ratios, certain conventions are observed. The convention for Equation 2-3 is that sound power is proportional to (sound pressure)^2. The voltage-level gain of an amplifier in decibels is 20 log (output voltage/input voltage), which holds true regardless of the input and output impedances. However, for power-level gain, the impedances must be considered if they are different. If it is a line amplifier with 600-ohm input and output impedances, well and good. Otherwise, a correction is required. The important lesson is to clearly indicate what type of level is intended, or else label the gain in level as "relative gain, dB." The following examples illustrate the use of the decibel.

Example: Sound-Pressure Level

A sound-pressure level (SPL) is 78 dB. What is the sound pressure, p?

$$78 \text{ dB} = 20 \log p/(20 \times 10^{-6})$$
$$\log p/(20 \times 10^{-6}) = 78/20$$
$$p/(20 \times 10^{-6}) = 10^{3.9}$$
$$p = (20 \times 10^{-6}) (7{,}943.3)$$
$$p = 0.159 \text{ pascals}$$

Remember that the reference level in SPL measurements is 20 μPa.

Example: Loudspeaker SPL

An input of 1 watt produces a SPL of 115 dB at 1 meter. What is the SPL at 20 ft (6.1 meters)?

$$SPL = 115 - 20 \log (6.1/1)$$
$$= 115 - 15.7$$
$$= 99.3 \text{ dB}$$

The assumption made in the 20 log 6.1 factor is that the loudspeaker is operating in a free field and that the inverse square law is operating. This is a reasonable assumption for a 20-foot distance if the loudspeaker is remote from reflecting surfaces.

An Electro-Voice "constant directivity" horn Model HP9040 is rated at a sound pressure level of 115 dB on axis at 1 meter with 1 watt into 8 ohms. If the input were decreased from 1 watt to 0.22 watts, what would be the sound-pressure level at 1 meter distance?

$$SPL = 115 - 10 \log (0.22/1)$$
$$= 115 - 6.6$$
$$= 108.4 \text{ dB}$$

Note that 10 log is used because two powers are being compared.

Example: Microphone Specifications

A Shure Model 578 omnidirectional dynamic microphone open-circuit voltage is specified as −80 dB for the 150-ohm case. They also specify that 0 dB = 1 volt per μbar. What would be the open-circuit voltage, v, be in volts?

$$-80 \text{ dB} = 20 \log v/1$$
$$\log v/1 = -80/20$$
$$v = 0.0001 \text{ volt}$$
$$= 0.1 \text{ millivolt}$$

Example: Line Amplifier

A line amplifier (600 ohms in, 600 ohms out) has a gain of 37 dB. With an input of 0.2 volts, what is the output voltage?

$$37 \text{ dB} = 20 \log (v/0.2)$$
$$\log (v/0.2) = 37/20$$
$$= 1.85$$
$$v/0.2 = 10^{1.85}$$
$$v = (0.2)(70.79)$$
$$v = 14.16 \text{ volts}$$

Example: General-Purpose Amplifier

This amplifier has a bridging input of 10,000 Ω impedance and an output impedance of 600 Ω. With a 50 mv input, an output of 1.5 V is observed. What is the gain of the amplifier? The so-called *voltage gain* is:

$$\text{voltage gain} = 20 \log (1.5/0.05)$$
$$= 29.5 \text{ dB}$$

It must be emphasized that this is not a power level gain because of the differences in impedance. However, voltage gain may serve a practical purpose in certain cases.

Example: Concert Hall

Seat x in a concert hall is 84 feet from the tympani drums. The tympanist strikes a single, mighty note. The sound-pressure level of the direct sound of the note at seat x is measured to be 55 dB. The first reflection from the nearest sidewall arrives at seat x 105 milliseconds after the arrival of the direct sound. (A) How far does the reflection travel to reach seat x? (B) What is the SPL of the reflection at seat x, assuming perfect reflection at the wall? (C) How long will the reflection be delayed after arrival of the direct sound at seat x?

(A)   Distance = (1,130 ft/sec) (0.105 sec)
$$= 118.7 \text{ ft}$$

(B) First, the level, L, 1 foot from the tympani drum must be estimated:
$$55 = L - 20 \log (84/1)$$
$$L = 55 + 38.5$$
$$L = 93.5 \text{ dB}$$

The SPL of the reflection at seat x is:
$$dB = 93.5 - 20 \log (118.7/1)$$
$$= 93.5 - 41.5$$
$$= 52 \text{ dB}$$

(C) The reflection will arrive after the direct sound at seat x after:
$$\text{Delay} = (118.7 - 84)/1{,}130 \text{ ft/sec}$$
$$= 30.7 \text{ milliseconds}$$

A free field is also assumed here. In chapter 3, the 30.7 ms reflection might be called an *incipient* echo.

Example: Combining Decibels

Let's say it is warm in our studio and a fan is brought in to augment the air conditioning (A/C) system. If both fan and the A/C are turned off, a very low noise level prevails, low enough to be neglected in the calculation. If the A/C alone is running, the sound-pressure level at a given position is 55 dB. If the fan alone is running, the sound-pressure level is 60 dB. What will be the sound-pressure level if both are running at the same time?

$$\text{Combined dB} = 10 \log \left(10^{\frac{55}{10}} + 10^{\frac{60}{10}} \right)$$

$$= 61.19 \text{ dB}$$

If the combined level of two noise sources is 80 dB and the level with one of the sources turned off is 75 dB, what is the level of the remaining source?

$$\text{Difference dB} = 10 \log \left(10^{\frac{80}{10}} - 10^{\frac{75}{10}}\right)$$

$$= 78.3 \text{ dB}$$

In other words, combining the 78.3 dB level with the 75 dB level gives the combined level of 80 dB.

Ratios and Octaves

An *octave* is defined as a 2:1 ratio of two frequencies. For example, middle C (C4) on the piano has a frequency close to 261 Hz. The next highest C (C5) has a frequency of about 522 Hz. Ratios of frequencies are very much a part of the musical scale. The frequency ratio 2:1 is the *octave*; the ratio 3:2 is the *fifth*; 4:3 is the *fourth*, etc. Because the octave is very important in acoustical work, it is well to consider the mathematics of the octave.

As the ratio of 2:1 is defined as the octave, its mathematical expression is:

$$\frac{f_2}{f_1} = 2^n \tag{2-6}$$

in which:

f_2 = the frequency of the upper edge of the octave interval.
f_1 = the frequency of the lower edge of the octave interval.
n = the number of octaves.

For 1 octave, $n = 1$ and Equation (2-6) becomes $f_2/f_1 = 2$, which is the definition of the octave. Other applications of Equation (2-6) are now explored:

Example

The low-frequency edge of a band is 20 Hz, what is the high-frequency edge of a band 10 octaves wide?

$$\frac{f_2}{20 \text{ Hz}} = 2^{10}$$

$$f_2 = (20)(2^{10})$$
$$f_2 = (20)(1{,}024)$$
$$f_2 = 20{,}480 \text{ Hz}$$

Example

If 446 Hz is the lower edge of a ⅓ octave band, what is the frequency of the upper edge?

$$\frac{f_2}{446} = 2^{1/3}$$

$$f_2 = (446)(2^{1/3})$$
$$f_2 = (446)(1.2599)$$
$$f_2 = 561.9 \text{ Hz}$$

Example

What is the lower edge of a ⅓ octave band centered on 1,000 Hz? The f_1 is 1,000 Hz but the lower edge would be ⅙ octave lower than the ⅓ octave, so $n = ⅙$:

$$\frac{f_2}{f_1} = \frac{1,000}{f_1} = 2^{1/6}$$

$$f_1 = \frac{1,000}{2^{1/6}}$$

$$f_1 = \frac{1,000}{1.12246}$$

$$f_1 = 890.9 \text{ Hz}$$

Example

At what frequency is the lower edge of an octave band centered on 2,500 Hz?

$$\frac{2,500}{f_1} = 2^{1/2}$$

$$f_1 = \frac{2,500}{2^{1/2}}$$

$$f_1 = \frac{2,500}{1.4142}$$

$$f_1 = 1,767.8 \text{ Hz}$$

What is the upper edge?

$$\frac{f_2}{2,500} = 2^{1/2}$$

$$f_2 = (2,500) \, (2^{1/2})$$
$$f_2 = (2,500) \, (1.4142)$$
$$f_2 = 3,535.5 \text{ Hz}$$

Measuring Sound-Pressure Level

A sound level meter is designed to give readings of sound-pressure level; sound pressure in decibels referred to the standard reference level, 20 μPa. Sound level meters usually offer a selection of weighting networks designated A, B, and C having frequency responses shown in Fig. 2-3. Network selection is based on the general level of sounds to be measured (background noise? jet engines?), such as:

- For sound-pressure levels of 20–55 dB...use network A.
- For sound-pressure levels of 55–85 dB...use network B.
- For sound-pressure levels of 85–140 dB...use network C.

These network response shapes are designed to bring the sound level meter readings into closer conformance to the relative loudness of sounds.

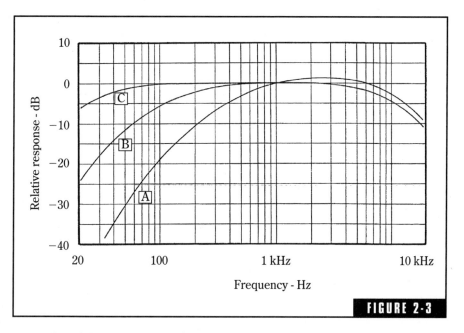

FIGURE 2-3

A, B, and C weighting response characteristics for sound level meters. (ANSI S1 .4-1971.)

The Ear and the Perception of Sound

The study of the structure of the ear is a study in physiology. The study of human perception of sound comes under the general heading of *psychology*. *Psychoacoustics* is an inclusive term embracing the physical structure of the ear, the sound pathways, the perception of sound, and their interrelationships. Psychoacoustics, quite a recent term, is especially pertinent to this study because it emphasizes both structure and function of the human ear.

The stimulus sound wave striking the ear sets in motion mechanical movements that result in neuron discharges that find their way to the brain and create a sensation. Then comes the question, "How are these sounds recognized and interpreted?" In spite of vigorous research activities on all aspects of human hearing, our knowledge is still woefully incomplete.

Sensitivity of the Ear

The delicate and sensitive nature of our hearing can be underscored dramatically by a little experiment. A bulky door of an anechoic chamber is slowly opened, revealing extremely thick walls, and three-foot wedges of glass fiber, points inward, lining all walls, ceiling, and what could be called the floor, except that you walk on an open steel grillwork.

A chair is brought in, and you sit down. This experiment takes time, and as a result of prior briefing, you lean back, patiently counting the glass fiber wedges to pass the time. It is very eerie in here. The sea of sound and noises of life and activity in which we are normally immersed and of which we are ordinarily scarcely conscious is now conspicuous by its absence.

The silence presses down on you in the tomblike silence, 10 minutes, then a half hour pass. New sounds are discovered, sounds that come from within your own body. First, the loud pounding of your heart, still recovering from the novelty of the situation. An hour goes by. The blood coursing through the vessels becomes audible. At last, if your ears are keen, your patience is rewarded by a strange hissing sound between the "ker-bumps" of the heart and the slushing of blood. What is it? It is the sound of air particles pounding against your eardrums. The eardrum motion resulting from this hissing sound is unbelievably small—only $\frac{1}{100}$ of a millionth of a centimeter—or $\frac{1}{10}$ the diameter of a hydrogen molecule!

The human ear cannot detect sounds softer than the rain of air particles on the eardrum. This is the threshold of hearing. There would be no reason to have ears more sensitive, because any lower-level sound would be drowned by the air-particle noise. This means that the ultimate sensitivity of our hearing just matches the softest sounds possible in an air medium. Accident? Adaptation? Design?

At the other extreme, our ears can respond to the roar of a cannon, the noise of a rocket blastoff, or a jet aircraft under full power. Special protective features of the ear protect the sensitive mechanism from damage from all but the most intense noises.

A Primer of Ear Anatomy

The three principal parts of the human auditory system, shown in Fig. 3-1, are the outer ear, the middle ear, and the inner ear. The outer ear is composed of the pinna and the auditory canal or auditory meatus. The auditory canal is terminated by the tympanic membrane or the eardrum. The middle ear is an air-filled cavity spanned by the three tiny bones, the ossicles, called the malleus, the incus, and the stapes. The malleus is attached to the eardrum and the stapes is attached to the oval window of the inner ear. Together these three bones form a mechanical, lever-action connection between the air-actuated eardrum and the fluid-filled cochlea of the inner ear. The inner ear is terminated in the auditory nerve, which sends impulses to the brain.

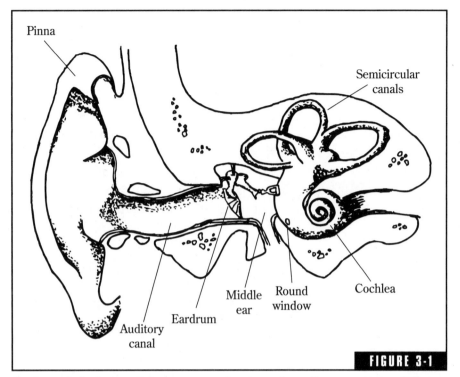

Pinna

Semicircular
canals

Middle
ear

Round
window

Cochlea

Eardrum

Auditory
canal

FIGURE 3-1

The four principal parts of the human ear: the pinna, the auditory canal, the middle ear, and the inner ear.

The Pinna: Directional Encoder of Sound

In ancient times, the pinna was regarded as either a vestigial organ or a simple sound-gathering device. True, it is a sound-gathering device. The pinna offers a certain differentiation of sounds from the front as compared to sound from the rear. Cupping your hand behind the ear increases the effective size of the pinna and thus the apparent loudness by an amount varying with frequency. For the important speech frequencies (2,000 to 3,000 Hz), sound pressure at the eardrum is increased about 5 dB. This front-back differentiation is the more modest contribution of the pinna.

Recent research has revealed that the pinna performs a very crucial function in imprinting directional information on all sounds picked up by the ear. This means that information concerning the direction to the source is superimposed on the sound content itself so that the resultant sound pressure on the eardrum enables the brain to interpret both the content of the sound and the direction from which it comes.

Directional Cues: An Experiment

If the equipment is available, a simple psychoacoustical experiment can illustrate how subjective directional impressions result from simple changes in sounds falling on the ear. Listen with a headphone on one ear to an octave bandwidth of random noise centered on 8 kHz arranged with an adjustable notch filter. Adjusting the filter to 7.2 kHz will cause the noise to seem to come from a source on the level of the observer. With the notch adjusted to 8 kHz the sound seems to come from above. With the notch at 6.3 kHz the sound seems to come from below. This experiment[1] demonstrates that the human hearing system extracts directional information from the shape of the sound spectra at the eardrum.

The Ear Canal

The ear canal also increases the loudness of the sounds traversing it. In Fig. 3-2 the ear canal, with an average diameter of about 0.7 cm and length of about 3 cm, is idealized by straightening and giving it a uniform diameter throughout its length. Acoustically, this is a reasonable approximation. It is a pipe-like duct, closed at the inner end by the eardrum.

Organ pipes were studied intensely by early investigators when the science of acoustics was in its infancy. The acoustical similarity of this ear canal to an organ pipe was not lost on early workers in the field. The resonance effect of the ear canal increases sound pressure at the eardrum at certain frequencies. The maximum is near the frequency at which the 3-cm pipe is one-quarter wavelength—about 3,000 Hz.

Figure 3-3 shows the increase in sound pressure at the eardrum over that at the opening of the ear canal. A primary peak is noted around 3,000 Hz caused by the quarter-wave pipe resonance effect. The primary pipe resonance amplifies the sound pressure at the eardrum approximately 12 dB at the major resonance at about

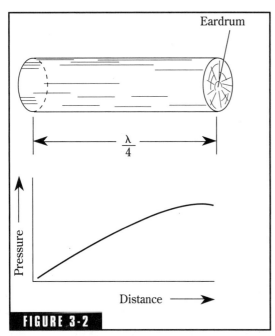

Eardrum

$\frac{\lambda}{4}$

Pressure

Distance

FIGURE 3-2

The auditory canal, closed at one end by the eardrum, acts as a quarter-wavelength "organ pipe." Resonance provides acoustic amplification for the important voice frequencies.

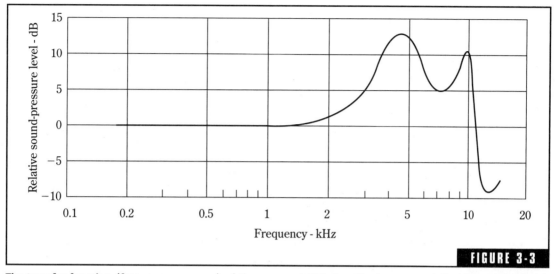

The transfer function (frequency response) of the ear canal. This is a fixed component that is combined with every directionally-encoded sound reaching the eardrum. See also Figs. 3-15 and 3-16. (After Mehrgardt and Mellart.[2])

4,000 Hz. There is a secondary resonance nearer 9,000 Hz of lower peak pressure.[2]

The Middle Ear

Transmitting sound energy from a tenuous medium such as air into a dense medium like water is a serious problem. Without some very special equipment, sound originating in air bounces off water like light off a mirror. It boils down to a matter of matching impedances, and in this case the impedance ratio is something like 4,000:1. Consider how satisfactory it would be to drive the 1-ohm voice coil of a loudspeaker with an amplifier having an output impedance of 4,000 ohms! Clearly not much power would be transferred.

The object is to get the feeble energy represented by the vibratory motion of a rather flimsy diaphragm, transferred with maximum efficiency to the fluid of the inner ear. The two-fold solution is suggested in Fig. 3-4. The three ossicles (hammer, anvil, and stirrup) form a mechanical linkage between the eardrum and the oval window, which is in intimate contact with the fluid of the inner ear. The first of the three bones, the malleus, is fastened to the eardrum. The third, the stapes, is actually a part of the oval window. There is a lever action in this linkage with a

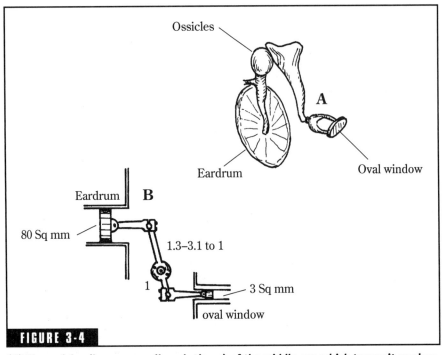

Ossicles

A

Eardrum

Oval window

Eardrum **B**

80 Sq mm

1.3–3.1 to 1

1

3 Sq mm

oval window

FIGURE 3-4

(A) The ossicles (hammer, anvil, and stirrup) of the middle ear, which transmit mechanical vibrations of the eardrum to the oval window of the cochlea. (B) A mechanical analog of the impedance-matching function of the middle ear. The difference in area between the eardrum and the oval window, coupled with the step-down mechanical linkage, match the motion of the air-actuated eardrum to the fluid-loaded oval window.

ratio leverage ranging from 1.3:1 to 3.1:1. That is, the eardrum motion is reduced by this amount at the oval window of the inner ear.

This is only part of this fascinating mechanical-impedance-matching device. The area of the eardrum is about 80 sq mm, and the area of the oval window is only 3 sq mm. Hence, a given force on the eardrum is reduced in the ratio of 80/3, or about 27-fold.

In Fig. 3-4B, the action of the middle ear is likened to two pistons with area ratios of 27:1 connected by an articulated connecting rod having a lever arm ranging from 1.3:1 to 3.1:1, making a total mechanical force increase of between 35 and 80 times. The acoustical impedance ratio between air and water being on the order of 4,000:1, the pressure ratio required to match two media would be $\sqrt{4,000}$, or about 63.2, and we note that this falls within the 35 to 80 range obtained from the mechanics of the middle ear illustrated in Fig. 3-4B.

The problem of matching sound in air to sound in the fluid of the inner ear is beautifully solved by the mechanics of the middle ear. The evidence that the impedance matching plus the resonance amplification of Fig. 3-3 really work is that a diaphragm motion comparable to molecular dimensions gives a threshold perception.

A schematic of the ear is given in Fig. 3-5. The conical eardrum at the inner end of the auditory canal forms one side of the air-filled middle ear. The middle ear is vented to the upper throat behind the nasal cavity by the Eustachian tube. The eardrum operates as an "acoustic suspension" system, acting against the compliance of the trapped air in the middle ear. The Eustachian tube is suitably small and constricted so as not to destroy this compliance. The round window separates the air-filled middle ear from the practically incompressible fluid of the inner ear.

The Eustachian tube fulfills a second function by equalizing the static air pressure of the middle ear with the outside atmospheric pressure so that the eardrum and the delicate membranes of the inner ear can function properly. Whenever we swallow, the Eustachian tubes

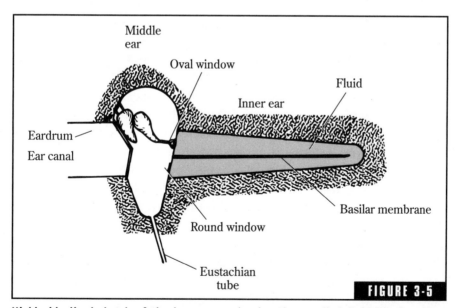

FIGURE 3-5

Highly idealized sketch of the human ear showing the unrolled fluid-filled cochlea. Sound entering the ear canal causes the eardrum to vibrate. This vibration is transmitted to the cochlea through the mechanical linkage of the middle ear. The sound is analyzed through standing waves set up on the basilar membrane.

open, equalizing the middle ear pressure. When an aircraft (at least those without pressurized cabins) undergoes rapid changes in altitude, the occupants might experience momentary deafness or pain until the middle ear pressure is equalized by swallowing. Actually, the Eustachian tube has a third emergency function of drainage if the middle ear becomes infected.

The Inner Ear

Only the acoustical amplifiers and the mechanical impedance matching features of the middle ear have been discussed so far. These are relatively well understood. The intricate operation of the cochlea is still clouded in mystery, but extensive research is steadily adding to our knowledge.

Figure 3-1 shows the close proximity of the three mutually-perpendicular, semicircular canals of the vestibular mechanism, the balancing organ, and the cochlea, the sound-analyzing organ. The same fluid permeates all, but their functions are independent.

The cochlea, about the size of a pea, is encased in solid bone. It is coiled up like a cockleshell from which it gets its name. For the purposes of illustration, this 2¾-turn coil has been stretched out its full length, about one inch, as shown in Fig. 3-5. The fluid-filled inner ear is divided lengthwise by two membranes, *Reissner's membrane* and the *basilar membrane*. Of immediate interest is the basilar membrane and its response to sound vibrations in the fluid.

Vibration of the eardrum activates the ossicles. The motion of the stapes, attached to the oval window, causes the fluid of the inner ear to vibrate. An inward movement of the oval window results in a flow of fluid around the distant end of the basilar membrane, causing an outward movement of the membrane of the round window. Sound actuating the oval window results in standing waves being set up on the basilar membrane. The position of the amplitude peak of the standing wave on the basilar membrane changes as the frequency of the exciting sound is changed.

Low-frequency sound results in maximum amplitude near the distant end of the basilar membrane; high-frequency sound produces peaks near the oval window. For a complex signal such as music or speech, many momentary peaks are produced, constantly shifting in amplitude and position along the basilar membrane. These resonant peaks on the basilar membrane were originally thought to be so broad as to be unable to explain the sharpness of frequency discrimination displayed by the human ear. Recent research is showing that at low sound intensities, the basilar membrane tuning curves are very sharp, broadening only for intense sound. It now appears that the sharpness of the basilar membrane's mechanical tuning curves is comparable to the sharpness of single auditory nerve fibers, which innervate it.

Stereocilia

Waves set up on the basilar membrane in the fluid-filled duct of the inner ear stimulate hairlike nerve terminals that convey signals to the brain in the form of neuron discharges, about 15,000 outer hair cells with about 140 tiny hairs called *stereocilia* jutting from each one. In addition, there are about 3,500 inner hair cells, each having about 40 *stereocilia* attached. These stereocilia are the true transducers of sound energy to electrical discharges. There are two types of hair cells, inner and outer, so-called by their placement and arrangement. As sound causes the cochlear fluid and the basilar membrane to move, the stereocilia on the hair cells are bent, initiating neural discharges to the auditory cortex.

When sound excites the fluid of the inner ear, membrane and hair cells are stimulated, sending an electrical wave through the surrounding tissue. These so-called *microphonic potentials* (analog) can be picked up and amplified, reproducing the sound falling on the ear, which acts as a biological microphone. These potentials are proportional to the sound pressure and linear in their response over an 80-dB range. While interesting, this microphonic potential must not be confused with the *action potentials* of the auditory nerve, which convey information to the brain.

Bending the stereocilia triggers the nerve impulses that are carried by the auditory nerve to the brain. While the microphonic signals are analog, the impulses sent to the acoustic cortex are impulses generated by neuron discharges. A single nerve fiber is either firing or not firing (binary!). When it fires, it causes an adjoining one to fire, and so on. Physiologists liken the process to a burning gunpowder fuse. The rate of travel bears no relationship to how the fuse was lighted. Presumably the loudness of the sound is related to the number of nerve fibers excited and the repetition rates of such excitation. When all the nerve fibers (some 15,000 of them) are excited, this is the maximum loudness that can be perceived. The threshold sensitivity would be represented by a single fiber firing. An overall, well-accepted theory of how the inner ear and the brain really function has not yet been formulated.[3-6]

This has been a highly simplified presentation of a very complex mechanism to which much current research is being devoted. Some of the numbers used and theories discussed are not universally accepted. Popularization of a subject such as the ear is an occupation that might be hazardous to my health, but audio workers must surely be amazed at the delicate and effective workings of the human ear. It is hoped that a new awareness of, and respect for, this delicate organism will be engendered and that damaging high levels of sound be avoided.

Loudness vs. Frequency

The seminal work on loudness was done at Bell Laboratories by Fletcher and Munson and reported in 1933,[7] and refinements have been added by others since that time. The family of equal-loudness contours of Fig. 3-6, the work of Robinson and Dadson,[8] has been adopted as an international standard (I.S.O. 226).

Each equal-loudness contour is identified by its value at 1,000 Hz, and the term *loudness level* in phons is thus defined. For example, the equal-loudness contour passing through 40-dB sound-pressure level at 1,000 Hz is called the 40-phon contour. Loudness is a subjective term; sound-pressure level is strictly a physical term. Loudness level is also a physical term that is useful in estimating the loudness of a sound (in units of sones) from sound-level measurements. The shapes of the equal-loudness contours contain subjective information because they were obtained by a subjective comparison of the loudness of a tone to its loudness at 1,000 Hz.

The surprising thing about the curves of Fig. 3-6 is that they reveal that perceived loudness varies greatly with frequency and sound-pressure level. For example, a sound-pressure level of 30 dB yields a loudness level of 30 phons at 1,000 Hz, but it requires a sound-pressure level of 58 dB more to sound equally loud at 20 Hz as shown in Fig. 3-7. The curves tend to flatten at the higher sound levels. The 90-phon curve rises only 32 dB between 1,000 Hz and 20 Hz. Note that inverting the curves of Fig. 3-7 gives the frequency response of the ear in

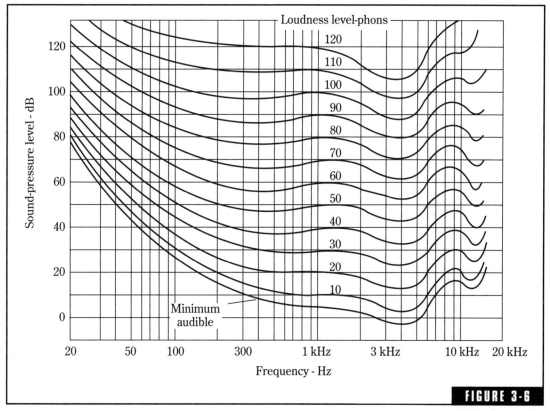

Equal-loudness contours of the human ear. These contours reveal the relative lack of sensitivity of the ear to bass tones, especially at lower sound levels. Inverting these curves give the frequency response of the ear in terms of loudness level. (After Robinson and Dadson.[8])

terms of loudness level. The ear is less sensitive to bass notes than midband notes at low levels. There are wiggles in the ear's high-frequency response that are relatively less noticeable. This bass problem of the ear means that the quality of reproduced music depends on the volume-control setting. Listening to background music at low levels requires a different frequency response than listening at higher levels.

Loudness Control

Let us assume that the high fidelity enthusiast adjusts the volume control on his or her amplifier so that the level of recorded symphony music is pleasing as a background to conversation (assumed to be

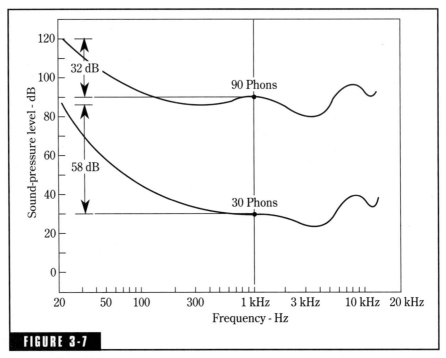

FIGURE 3-7

A comparison of the ear's response at 20 Hz compared to that at 1,000 Hz. At a loudness level of 30 phons, the sound-pressure level of a 20-Hz tone must be 58 dB higher than that at 1,000 Hz to have the same loudness. At 90 phons loudness level, an increase of only 32 dB is required. The ear's response is somewhat flatter at high loudness levels. Loudness level is only an intermediate step to true subjective loudness as explained in the text.

about 60 phons). As the passage was played at something like an 80-phon loudness level in the concert hall, something needs to be done to give the bass and treble of the music the proper balance at the lower-than-concert-hall level. Our enthusiast would find it necessary to increase both bass and treble for good balance.

The loudness control found on many amplifiers adjusts electrical networks to compensate for the change in frequency response of the ear for different loudness levels. But the curve corresponding to a given setting of the loudness control applies only to a specific loudness level of reproduced sound. The loudness control is far from a complete solution to the problem. Think of all the things that affect the volume-control setting in a particular situation. The loudspeakers vary in acoustic output for a given input power. The gain of preamplifiers, power amplifiers, tuners, and phono pickups differs from brand to brand and circuit to circuit. Listening-room conditions vary from dead to highly reverberant. With all of these variables, how can a manufacturer design a loudness control truly geared to the sound-pressure level at the ear of listener x with the particular variables of x's equipment and x's listening environment? For a loudness control to function properly, x's system must be calibrated and the loudness control fitted to it.[9]

Area of Audibility

Curves A and B of Fig. 3-8 were obtained from groups of trained listeners. In this case, the listeners face the sound source and judge whether a tone of a given frequency is barely audible (curve A) or

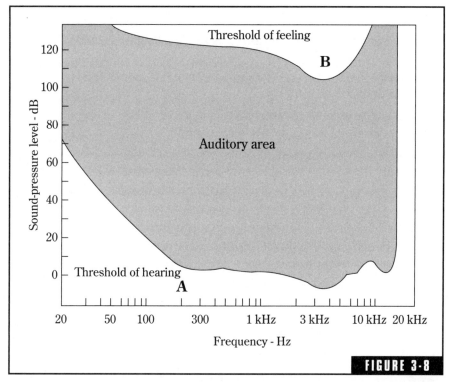

FIGURE 3-8

The auditory area of the human ear is bounded by two threshold curves, (A) the threshold of hearing delineating the lowest level sounds the ear can detect, and (B) the threshold of feeling at the upper extreme. All of our auditory experiences occur within this area.

beginning to be painful (curve B). These two curves represent the extremes of our perception of loudness.

Curve A of Fig. 3-8, the threshold of hearing, tells us that human ears are most sensitive around 3 kHz. Another way to state this is that around 3 kHz a lower-level sound elicits a greater threshold response than higher or lower frequencies. At this most sensitive region, a sound-pressure level of 0 dB can just barely be heard by a person of average hearing acuity. Is it fortuitous that this threshold is at a nice, round, 0-dB level? No, the reference level of pressure of 20 mPa (20 micropascals) was selected for this reason. It is both instructive and comforting to know that a sound-pressure level of 60 dB turns out to be approximately 60 dB above our threshold of hearing.

Curve B of Fig. 3-8 represents the level at each frequency at which a tickling sensation is felt in the ears. This occurs at a sound-pressure level of about 120 or 130 dB. Further increase in level results in an increase in feeling until a sensation of pain is produced. The threshold tickling is a warning that the sound is becoming dangerously loud and that ear damage is either imminent or has already taken place.

In between the threshold of hearing (curve A of Fig. 3-8) and the threshold of feeling (curve B) is the area of audibility. This is an area with two dimensions: the vertical dimension of sound-pressure level and the horizontal range of frequencies that the ear can perceive. All the sounds that humans experience must be of such a frequency and level as to fall within this auditory area. Chapter 5 details more specifically how much of this area is used for common music and speech sounds.

The area of audibility for humans is quite different from that of many animals. The bat specializes in sonar cries that are far above the upper frequency limit of our ears. The hearing of dogs extends higher than ours, hence the usefulness of ultrasonic dog whistles. Sound in the infrasonic and ultrasonic regions, as related to the hearing of humans, is no less true sound in the physical sense, but it does not result in human perception.

Loudness vs. Sound-Pressure Level

The *phon* is the unit of loudness level that is tied to sound-pressure level at 1,000 Hz as we have seen in Figs. 3-6, 3-7, and 3-8. This is useful up to a point, but it tells us little about human reaction to loudness of sound. We need some sort of subjective unit of loudness. Many experiments conducted with hundreds of subjects and many types of sound have yielded a consensus that for a 10-dB increase in sound-pressure level, the average person reports that loudness is doubled. For a 10-dB decrease in sound level, subjective loudness is cut in half. One researcher says this should be 6 dB, others say 10 dB, so work on the problem continues. However, a unit of subjective loudness has been adopted called the *sone*. One sone is defined as the loudness experienced by a person listening to a tone of 40-phon loudness level. A sound of 2 sones is twice as loud, and 0.5 sone half as loud.

Figure 3-9 shows a graph for translating sound-pressure levels to loudness in sones. One point on the graph is the very definition of the

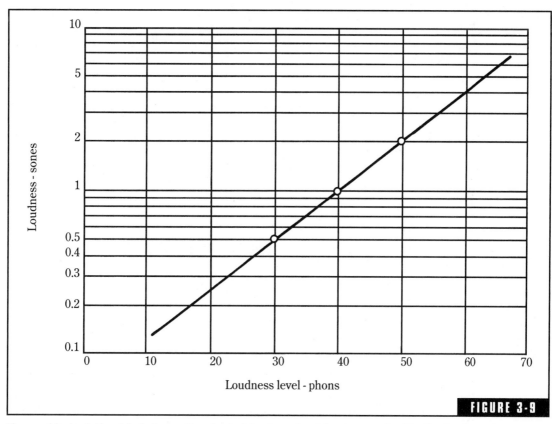

FIGURE 3-9

The graphical relationship between the physical loudness level in phons and subjective loudness in sones.

sone, the loudness experienced by a person hearing a 1,000-Hz tone at 40-dB sound-pressure level, or 40 phons. A loudness of 2 sones is then 10 dB higher; a loudness of 0.5 sones is 10 dB lower. A straight line can be drawn through these three points, which can then be extrapolated for sounds of higher and lower loudness.

As crude as this graph may be, it is a way of getting at the subjective factor of loudness. The value of this line of reasoning is that if a consultant is required by a court to give his or her opinion on the loudness of an industrial noise that bothers neighbors, he or she can make a one-third octave analysis of the noise, translate the sound-pressure levels of each band to sones by the help of a series of graphs such as Fig. 3-9, and by adding together the sones of each band, arrive at an estimate of the loudness of the noise. This idea of being able to add component sones is very nice; adding decibels of sound-pressure levels is a path that leads only to confusion.

Table 3-1 shows the relationship between loudness level in phons to the subjective loudness in sones. Although most audio workers will have little occasion to become involved in phons or sones, it is good to realize that a true subjective unit of loudness (sone) is related to loudness level (phon), which is in turn related by definition to what we can measure with a sound-level meter. There are highly developed empirical methods of calculating the loudness of sound as they would be perceived by humans from purely physical measurements of sound spectra, such as those measured with a sound-level meter and an octave or one-third octave filter.[10]

Table 3-1. Loudness level in phons vs. loudness in sones.

Loudness level (phons)	Subjective loudness (sones)	Typical examples
100	64	Heavy truck passing
80	16	Talking loudly
60	4	Talking softly
40	1	Quiet room
20	0.25	Very quiet studio

Loudness and Bandwidth

In the discussion of loudness we have talked tones up to this point, but single-frequency tones do not give all the information we need to relate subjective loudness to meter readings. The noise of a jet aircraft taking off sounds much louder than a tone of the same sound-pressure level. The bandwidth of the noise affects the loudness of the sound, at least within certain limits.

Figure 3-10A represents three sounds having the same sound-pressure level of 60 dB. Their bandwidths are 100, 160, and 200 Hz, but heights (representing sound intensity per Hz) vary so that areas are equal. In other words, the three sounds have equal intensities. (Sound intensity has a specific meaning in acoustics and is not to be equated to sound pressure. Sound intensity is proportional to the square of sound pressure for a plane progressive wave). The catch is that all three sounds of Fig.3-10A do not have the same loudness. The graph in Fig. 3-10B shows how a bandwidth of noise having a constant 60-dB sound-pressure level and centered on 1,000 Hz is related to loudness as experimentally determined. The 100-Hz noise

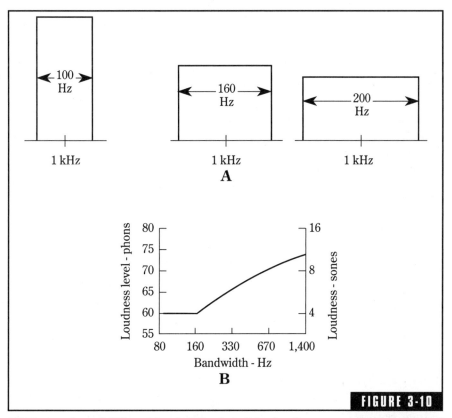

FIGURE 3-10

(A) Three noises of different bandwidths, but all having the same sound-pressure level of 60 dB. (B) The subjective loudness of the 100- and 160-Hz noise is the same, but the 200-Hz band sounds louder because it exceeds the 160-Hz critical band width of the ear at 1,000 Hz.

has a loudness level of 60 phons and a loudness of 4 sones. The 160-Hz bandwidth has the same loudness, but something mysterious happens as the bandwidth is increased beyond 160 Hz. The loudness of the noise of 200-Hz bandwidth is louder, and from 160 Hz or up, increasing bandwidth increases loudness. Why the sharp change at 160 Hz?

It turns out that 160 Hz is the width of the ear's critical band at 1,000 Hz. If a 1,000-Hz tone is presented to a listener along with random noise, only the noise in a band 160 Hz wide is effective in masking the tone. In other words, the ear acts like an analyzer composed of a set of bandpass filters stretching throughout the audible spectrum. This filter set is nothing like that found in the electronics laboratory. The common ⅓-octave filter set may have 28 adjacent filters overlapping at the −3 dB points. The set of critical band filters is continuous; that is, no matter where you might choose to set the signal generator dial, there is a critical band centered on that frequency.

Many years of research on this problem has yielded a modicum of agreement on how the width of the critical-band filters varies with frequency. This classical bandwidth function is shown in the graph of Fig. 3-11. There has been some question as to the accuracy of this graph below about 500 Hz that has led to other methods of measuring the bandwidth. Out of this has come the concept of the *equivalent rectangular bandwidth* (ERB) that applies to young listeners at moderate sound levels.[11] This approach is based on mathematical methods and offers the convenience of being able to calculate the ERB from the equation given in Fig. 3-11.

One-third-octave filter sets have been justified in certain measurements because the filter bandwidths approach those of the critical bands of the ear. For comparison, a plot of one-third-octave bandwidths is included in Fig. 3-11. One-third-octave bands are 23.2 percent of the center frequency. The classical critical-band function is about 17 percent of the center frequency. It is interesting to note that the ERB function (12 percent) is very close to that of one-sixth-octave bands (11.6 percent). This suggests the possibility of one-sixth-octave filter sets playing a larger role in sound measurements of the future.

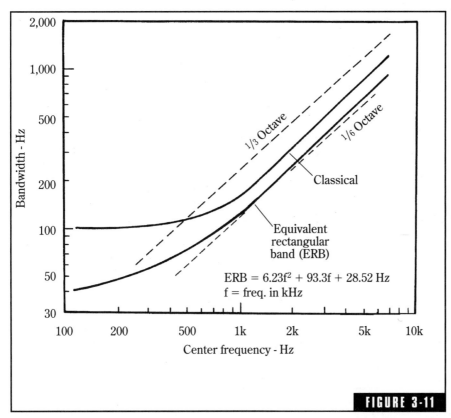

FIGURE 3-11

A comparison of bandwidths of ⅓- and ⅙-octave bands, critical bands of the ear, and equivalent rectangular critical bands (ERB) calculated from the above equation.[11]

Loudness of Impulses

The examples discussed up to this point have been concerned with steady-state tones and noise. How does the ear respond to transients of short duration? This is important because music and speech are essentially made up of transients. To focus attention on this aspect of speech and music, play some tapes backward. The initial transients now appear at the ends of syllables and musical notes and stand out prominently. These transients justify a few words on the ear's response to short-lived sounds.

A 1,000-Hz tone sounds like 1,000 Hz in a 1-second tone burst, but an extremely short burst sounds like a click. The duration of such a burst also influences the perceived loudness. Short bursts do not sound as loud as longer ones. Figure 3-12 shows how much the level of the shorter pulses has to be increased to have the same loudness as a long pulse or steady tone. A pulse 3 milliseconds long must have a level about 15 dB higher to sound as loud as a 0.5-second (500 millisecond) pulse. Tones and random noise follow roughly the same relationship in loudness vs. pulse length.

The 100-msec region is significant in Fig. 3-12. Only when the tones or noise bursts are shorter than this amount must the sound-pressure level be increased to produce a loudness equal to that of long pulses or steady tones or noise. This 100 msec appears to be the integrating time or the time constant of the human ear.

In reality, Fig. 3-12 tells us that our ears are less sensitive to short transients. This has a direct bearing on understanding speech. The consonants of speech determine the meaning of many words. For instance, the only difference between bat, bad, back, bass, ban, and bath are the consonants at the end. The words led, red, shed, bed, fed, and wed have the all-important consonants at the beginning. No matter where they

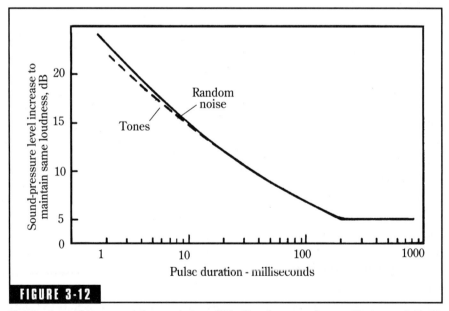

FIGURE 3-12

Short pulses of tones or noise are less audible than longer pulses as these graphs indicate. The discontinuity of the 100- to 200-msec region is related to the integrating time of the ear.

occur, these consonants are genuine transients having durations on the order of 5 to 15 msec. A glance at Fig. 3-12 tells you that transients this short must be louder to be comparable to longer sounds. In the above words, each consonant is not only much shorter than the rest of the word, it is also at a lower level. Thus you need good listening conditions to distinguish between such sets of words. Too much background noise or too much reverberation can seriously impair the understandability of speech because of the consonant problem.[12]

Audibility of Loudness Changes

Modern faders are of the composition type giving gradations in level so small as to be inaudible. Wire-wound faders of early mixing consoles produced discrete steps in level that could be audible. Steps of 5 dB were definitely audible, steps of 0.5 dB were inaudible, but these steps cost too much to produce and 0.5 dB steps were not necessary. Steps of 2 dB, an economic compromise, produced changes in signal level that were barely detectable by an expert ear. Detecting differences in intensity varies somewhat with frequency and also with sound level.

At 1 kHz, for very low levels, a 3-dB change is the least detectable by the ear, but at high levels the ear can detect a 0.25-dB change. A very low level 35-Hz tone requires a 9-dB level change to be detectable. For the important midfrequency range and for commonly used levels, the minimum detectable change in level that the ear can detect is about 2 or 3 dB. Making level changes in increments less than these is usually unnecessary.

Pitch vs. Frequency

Pitch, a subjective term, is chiefly a function of frequency, but it is not linearly related to it. Because pitch is somewhat different from frequency, it requires another subjective unit—the *mel*. Frequency is a physical term measured in cycles per second, now called Hertz. Although a weak 1,000-Hz signal is still 1,000 Hz if you increase its level, the pitch of a sound may depend on sound-pressure level. A reference pitch of 1,000 mels has been defined as the pitch of a 1,000-Hz tone with a sound-pressure level of 60 dB. The relationship between pitch and frequency, determined by experiments with juries of listeners, is shown in Fig. 3-13. Notice that on the experimental curve 1,000 mels coincides with 1,000 Hz, which tells us that the sound-pressure level for this curve is 60 dB. It is interesting to note that the shape of the curve of Fig. 3-13 is quite similar to a plot of position along the basilar membrane of the inner ear as a function of frequency. This suggests that pitch is related to action on this membrane, but much work remains to be done to be certain of this.

Researchers tell us that there are about 280 discernible steps in intensity and some 1,400 discernible steps in pitch that can be

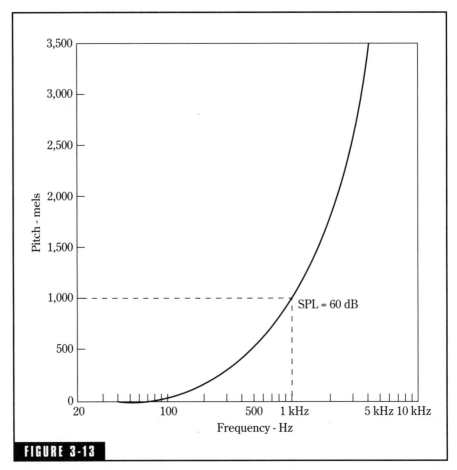

FIGURE 3-13

Pitch (in mels, a subjective unit) is related to frequency (in Hz, a physical unit) according to this curve obtained by juries of listeners. (After Stevens and Volkman.[13])

detected by the human ear. As changes in intensity and pitch are the very stuff of communication, it would be interesting to know how many combinations are possible. Offhand, it might seem that there would be 280 × 1,400 = 392,000 combinations detectable by the ear. This is overly optimistic because the tests were conducted by comparing two simple, single-frequency sounds in rapid succession and bears little resemblance to the complexities of commonly heard sounds. More realistic experiments show that the ear can detect only about 7 degrees of loudness and 7 degrees of pitch or only 49 pitch-loudness combinations. This is not too far from the number of *phonemes* (the smallest unit in a language that distinguishes one utterance from another) which can be detected in a language.

An Experiment

The level of sound affects the perception of pitch. For low frequencies, the pitch goes down as the level of sound is increased. At high frequencies, the reverse takes place—the pitch increases with sound level.

The following is an experiment within the reach of many readers that was suggested by Harvey Fletcher. Two audio oscillators are required, as well as a frequency counter. One oscillator is fed to the input of one channel of a high-fidelity system, the other oscillator to the other channel. After the oscillators have warmed up and stabilized, adjust the frequency of the left channel oscillator to 168 Hz and that of the right channel to 318 Hz. At low level these two tones are quite discordant. Increase the level until the pitches of the 168-Hz and 318-Hz tones decrease to the 150-Hz–300-Hz octave relationship, which gives a pleasant sound. This illustrates the decrease of pitch at the lower frequencies. An interesting follow-up would be to devise a similar test to show that the pitch of higher frequency tones increases with sound level.

Timbre vs. Spectrum

Timbre has to do with our perception of complex sounds. The word is applied chiefly to the sound of various musical instruments. A flute and oboe sound different even though they are both playing A. The tone of each instrument has its own timbre. Timbre is determined by the number and relative strengths of the instrument's partials. Tonal quality comes close to being a synonym for timbre.

Timbre is another subjective term. The analogous physical term is spectrum. A musical instrument produces a fundamental and a set of partials (or harmonics) that can be analyzed with a wave analyzer and plotted as in Fig. 1-15. Suppose the fundamental is 200 Hz, the second harmonic 400 Hz, the third harmonic 600 Hz, etc. The subjective pitch that the ear associates with our measured 200 Hz, for example, varies slightly with the level of the sound. The ear also has its own subjective interpretation of the harmonics. Thus, the ear's perception of the overall timbre of the instrument's note might be considerably different from the measured spectrum in a very complex way.

In listening to an orchestra in a music hall, the timbre you hear is different for different locations in the seating area.[14] The music is composed of a wide range of frequencies, and the amplitude and phase of the various components are affected by reflections from the various surfaces of the room. The only way to get one's analytical hands on studying such differences is to study the sound spectra at different locations. However, these are physical measurements, and the subjective timbre still tends to slip away from us. The important point of this section is to realize that a difference exists between timbre and spectrum.

Localization of Sound Sources

The perception of a direction to a source of a sound is, at least partially, the result of the amazing encoding function of the external ear, the pinna. Sound reflected from the various ridges, convolutions, and surfaces of the pinna combines with the unreflected (direct) sound at the entrance to the auditory canal. This combination, now encoded with directional information, passes down the auditory canal to the eardrum and thence to the middle and inner ear and on to the brain for interpretation.

This directional encoding process of the sound signal is indicated in Fig. 3-14. The sound wavefront can be considered as a multiplicity of sound rays coming from a specific source at a specific horizontal and vertical angle. As these rays strike the pinna they are reflected from the various surfaces, some of the reflections going toward the entrance to the auditory canal. At that point these reflected components combine with the unreflected (direct) component.

For a sound coming directly from the front of the observer (azimuth and vertical angle = 0°), the "frequency response" of the combination at the opening of the ear canal will be that shown in Fig. 3-15. Instead of *frequency response*, a curve of this type is called a *transfer function* because it represents a vector combination involving phase angles.

For the sound at the entrance of the ear canal (Fig. 3-15) to reach the eardrum, the auditory canal must be traversed. As the transfer function at the entrance to the ear canal (Fig. 3-15) and that of the ear canal (Fig. 3-3) are combined, the shape of the resulting

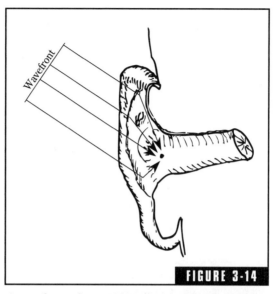

FIGURE 3-14

A wavefront of a sound can be considered as numerous rays perpendicular to that wavefront. Such rays, striking a pinna, are reflected from the various ridges and convolutions. Those reflections directed to the opening of the ear canal combine vectorially (according to relative amplitudes and phases). In this way the pinna encodes all sound falling on the ear with directional information, which the brain decodes as a directional perception.

transfer function impinging on the eardrum is radically changed. Figure 3-3 showed a typical transfer function of the ear canal alone. It is a static, fixed function that does not change with direction of arrival of the sound. The ear canal acts like a quarter-wave pipe closed at one end by the eardrum exhibiting two prominent resonances.

The transfer function representing the specific direction to the source of Fig. 3-15 combining with the fixed transfer function of the ear canal (Fig. 3-3) gives the combined transfer function at the eardrum of Fig. 3-16. The brain translates this to a perception of sound coming from directly in front of the observer.[2]

The transfer function at the entrance to the ear canal (such as Fig. 3-15) is shaped differently for each horizontal and vertical direction. This is how the pinna encodes all arriving sound enabling the brain to yield different perceptions of direction. The sound arriving at the

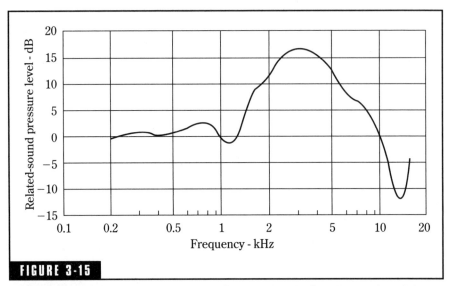

FIGURE 3-15

A measured example of the sound pressure (transfer function) at the opening of the ear canal corresponding to sound arriving from a point immediately in front of the subject. The shapes of such transfer functions vary with the horizontal and vertical angles at which the sound arrives at the pinna. (After Mehrgardt and Mellert.[2])

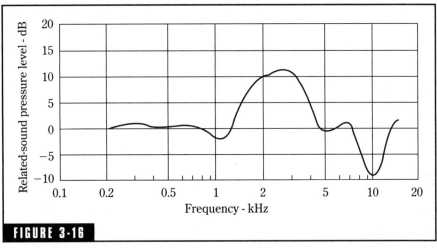

FIGURE 3-16

The transfer function of Fig. 3-15 at the opening of the ear canal is altered to this shape at the eardrum after being combined with the transfer function of the ear canal. In other words, a sound arriving at the opening of the ear canal from a source directly in front of the observer (Fig. 3-15) looks like Fig. 3-16 at the eardrum because it has been combined with the characteristics of the ear canal itself (Fig. 3-3). The brain has no trouble subtracting the fixed influence of the ear canal from every changing arriving sound.

eardrum is the raw material for all directional perceptions. The brain neglects (sees through?) the fixed component of the ear canal and translates the differently shaped transfer functions to directional perceptions.

Another more obvious directional function of the pinna is that of forward-backward discrimination, which does not depend on encoding and decoding. At the higher frequencies (shorter wavelengths), the pinna is an effective barrier. The brain uses this front-back differentiation to convey a general perception of direction.

A crucial question at this juncture is, "How about sounds arriving in the median plane?" The median plane is a vertical plane passing symmetrically through the center of the head and nose. Sources of sound in this plane present identical transfer functions to the two ears. The auditory mechanism uses another system for such localization, that of giving a certain place identity to different frequencies. For example, signal components near 500 and 8,000 Hz are perceived as coming from directly overhead, components near 1,000 and 10,000 Hz as coming from the rear.[15] This is an active area of research that is being continually refined.

The pinna, originally suspected of being only a useless vestigial organ, turns out to be a surprisingly sophisticated sound directional encoding mechanism.

Sound arriving from directly in front of an observer results in a peak in the transfer function at the eardrum in the 2- to 3-kHz region. This is the basis of the successful technique of old-time sound mixers adding "presence" to a recorded voice by adding an equalization boost in this frequency region. A voice can also be made to stand out from a musical background by adding such a peak to the voice response.

Binaural Localization

Stereophonic records and sound systems are a relatively new development. Stereo hearing has been around at least as long as man. Both are concerned with the localization of the source of sound. In early times some people thought that having two ears was like having two lungs or two kidneys, if something went wrong with one the other could still function. Lord Rayleigh laid that idea to rest by a simple experiment on the lawn of Cambridge University. A circle of assistants spoke or struck tuning forks and Lord Rayleigh in the center with his eyes

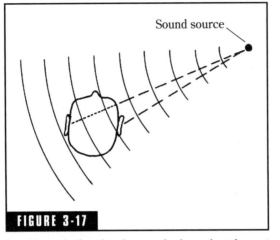

Sound source

FIGURE 3-17

Our binaural directional sense is dependent in part on the difference in intensity and phase of the sound falling on two ears.

closed pointed to the source of sound with great accuracy, confirming the fact that two ears function together in binaural localization.

Two factors are involved, the difference in intensity and the difference in time of arrival (phase) of the sound falling on the two ears. In Fig. 3-17 the ear nearest the source receives a greater intensity than the far ear because the hard skull casts a "sound shadow." Because of the difference of distance to the source, the far ear receives sound somewhat later than the near ear. Below 1 kHz the phase (time) effect dominates while above 1 kHz the intensity effect dominates. There is one localization blind spot. A listener cannot tell whether sounds are coming from directly in front or from directly behind because the intensity of sound arriving at each ear is the same and in the same phase.

Another method of perception of direction comes into play in a relatively small room. The sound reaches the person over a direct path followed by many reflections from many different directions. The sound that arrives first creates in the hearer the main perception of direction. This has been called *the law of the first wavefront.*

Aural Harmonics: Experiment #1

This experiment, suggested by Craig Stark,[16] can be performed easily with your home high-fidelity system and two audio oscillators. Plug one oscillator into the left channel and the other into the right channel, and adjust both channels for an equal and comfortable level at some midband frequency. Set one oscillator to 24 kHz and the other to 23 kHz without changing the level settings. With either oscillator alone, nothing is heard because the signal is outside the range of the ear. (He notes here, however, that the dog might leave the room in disgust!) When both oscillators are feeding their respective channels, one at 24 kHz and the other at 23 kHz, a distinct 1,000-Hz tone is heard if the tweeters are good enough and you are standing in the right place.

The 1,000-Hz tone is the difference between 24,000 and 23,000 Hz. The sum, or 47,000 Hz, which even the dog may not hear even if it were radiated, is another sideband. Such sum and difference sidebands are generated whenever two pure tones are mixed in a nonlinear element. The nonlinear element in the above experiment is the middle and inner ear. In addition to the intermodulation products discussed earlier, the nonlinearity of the ear generates new harmonics that are not present in the sound falling on the eardrum.

Aural Harmonics: Experiment #2

The distortion introduced by the auditory system cannot be measured by ordinary instruments. It is a subjective effect requiring a different approach. Another demonstration of distortion in the ear can be accomplished by the following method with the same equipment used above, with the addition of a pair of headphones.

First, a 150-Hz tone is applied to the left earphone channel. If the hearing mechanism were perfectly linear, no aural harmonics would be heard as the exploratory tone is swept near the frequencies of the second, third, and other harmonics. If it is nonlinear, the presence of aural harmonics is indicated by the generation of beats. When 150 Hz is applied to the left ear, and the exploratory tone of the right ear is slowly varied about 300 Hz, the second harmonic is indicated by the presence of beats between the two. If you change the exploratory oscillator to a frequency around 450 Hz, the presence of a third harmonic will be revealed by beats.

Experts have even estimated the magnitude of the harmonics by the strength of such beats. The amount of distortion produced in the ear is modest at lower levels but becomes appreciable at high levels. Running the above experiment with tones of a higher level will make the presence of aural harmonics even more obvious.

The Missing Fundamental

If tones such as 1,000, 1,200, and 1,400 Hz are reproduced together, a pitch of 200 Hz is heard. This can be interpreted as the fundamental with 1,000 Hz as the 5th harmonic, 1,200 Hz as the 6th harmonic, etc.

At one time this 200 Hz was called "periodicity pitch" but so-called "pattern" theories dominate today. The auditory system is supposed to recognize that the upper tones are harmonics of the 200 Hz and supplies the missing fundamental that would have generated them. This is a very interesting effect but explanations of it are highly controversial.

The Ear as an Analyzer

Listening to a good symphony orchestra in your favorite concert hall, concentrate first on the violins. Now focus your attention on the clarinets, then the percussion section. Next listen to a male quartet and single out the first tenor, the baritone, the bass. This is a very remarkable power of the human ear/brain combination. In the ear canal, all these sounds are mixed together; how does the ear succeed in separating them? The sea surface might be disturbed by many wave systems, one due to local wind, one from a distant storm, and several wakes from passing vessels. The eye cannot separate these, but this is essentially what the ear is constantly doing with complex sound waves. By rigorous training, a keen observer can listen to the sound of a violin and pick out the various overtones apart from the fundamental!

The Ear as a Measuring Instrument

The emphasis on the distinction between physical measurements and subjective sensation would seem to rule out the possibility of using the ear for physical measurements. True, we cannot obtain digital readouts by looking in someone's eyes (or ears), but the ears are very keen at making comparisons. People are able to detect sound-level differences of about 1 dB throughout most of the audible band if the level is reasonable. Under ideal conditions, a change of a third this amount is perceptible. At ordinary levels, and for frequencies less than 1,000 Hz, the ear can tell the difference between tones separated by as little as 0.3%. This would be 0.3 Hz at 100 Hz and 3 Hz at 1,000 Hz.

The eminent Harvey Fletcher[17] has pointed out how the remarkable keenness of the human ear saved the day in many of his researches in synthesizing musical sounds. For example, in his study of piano sounds, he initially postulated that all that is necessary is to measure the frequency and magnitude of fundamental and harmonics and then combine them with the measured values of attack and decay. When this was done, the listening jury unanimously voted that the synthetic sounds did not sound like piano sounds but more like organ tones. Further study revealed the long-known fact that piano strings are very stiff and have properties of both solid rods and stretched strings. The effect of this is that piano partials are *nonharmonic!* By correcting the frequencies of what were assumed to be harmonics in integral multiples, the jury could not distinguish between the synthetic piano sounds and the real thing. The critical faculty of the ears of the jury in comparing sound qualities provided the key.

An Auditory Analyzer: An Experiment

Knowledge of the ear's filterlike critical bands leads to the tantalizing idea of analyzing continuous noises such as traffic noises, underwater background noises, etc., by using the ear instead of heavy and expensive sound-analyzing gear. This must have occurred to Harvey Fletcher, who first proposed the idea of critical bands, and to many investigators in this field who have dealt with critical bands through the years.

The general approach is illustrated in Fig. 3-18.[18] A tape recording of the noise to be analyzed is played back and mixed with a tone from a variable-frequency oscillator. The combination is amplified and listened to with a pair of headphones having a flat frequency response. The oscillator is set, say, at 1,000 Hz and its output

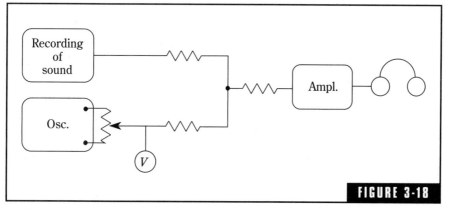

FIGURE 3-18

Equipment arrangement for using the critical bands of the human ear for sound analysis.

adjusted until the tone is just hidden or masked by the noise. Only the noise in the critical band centered on 1,000 Hz is effective in masking the tone. If the noise is expressed in sound-pressure level of a band 1 Hz wide, the voltage of the tone then corresponds to the 1-Hz sound-pressure level of the noise at the masked point. Adjusting the voltage until the tone is just masked should yield one of the points on our noise spectrum graph. For convenience, let us assume that this voltmeter is calibrated in dB referred to some arbitrary base such as 1 volt (dBv). Referring to Fig. 3-11, note that the critical band centered on 1,000 Hz is 160 Hz wide. This can also be expressed in decibels by taking $\log_{10}160 = 22$ dB; this 22 dB, representing the width of the critical band as it does, must be subtracted from the voltmeter reading in dB. This gives one point on the noise spectrum graph. By repeating the process for other frequencies, a series of points is obtained that reveal the shape of the noise spectrum. If the recording and the entire measuring system (including the observer's ears) were calibrated, the absolute levels for the noise spectrum could be obtained.

The important point here is that there is such a set of filters in our head that could be put to such a task, not that this method will ever replace a good sound level meter equipped with octave or one-third octave filters. Surely human variables would far exceed sound-level meter fluctuations from day to day, and what the observer eats for breakfast has no effect on the sound-level meter, although it might affect the dependability of the readings made with physiological equipment.

Meters vs. the Ear

There still remains a great chasm between subjective judgments of sound quality, room acoustics, etc., and objective measurements. Considerable attention is being focused on the problem. Consider the following descriptive words, which are often applied to concert-hall acoustics[19,20]:

warmth	clarity
bassiness	brilliance
definition	resonance
reverberance	balance

fullness of tone   blend

liveness   intimacy

sonority   shimmering

What kind of an instrument measures warmth or brilliance? How would you devise a test for definition? Progress, however, is being made. Take definition for instance. German researchers have adopted the term *deutlichkeit*, which literally means clearness or distinctness, quite close to definition. It can be measured by taking the energy in an echogram during the first 50 to 80 milliseconds and comparing it to the energy of the entire echogram. This compares the direct sound and early reflections, which are integrated by the ear, to the entire reverberant sound. This relatively straightforward measurement of an impulsive sound from a pistol or pricked balloon holds considerable promise for relating the descriptive term definition to an objective measurement. It will be a long time before all of these and a host of other subjective terms can be reduced to objective measurements, but this is a basic problem in acoustics and psychoacoustics.

There comes a time at which meter readings must give way to observations by human subjects. Experiments then take on a new, subjective quality. For example, in a loudness investigation, panels of listeners are presented with various sounds, and each observer is asked to compare the loudness of sound A with the loudness of B or to make judgments in other ways. The data submitted by the jury of listeners are then subjected to statistical analysis, and the dependence of a human sensory factor, such as loudness, upon physical measurements of sound level is assessed. If the test is conducted properly and sufficient observers are involved, the results are trustworthy. It is in this way that we discover that there is no linear relationship between sound level and loudness, pitch and frequency, or between timbre and sound quality.

The Precedence Effect

Our hearing mechanism integrates sound intensities over short intervals and acts somewhat like a ballastic measuring instrument. In simpler terms, in an auditorium situation, the ear and brain have the remarkable ability to gather all reflections arriving within about 50 msec after the direct sound and combine (integrate) them to give the impression that all this sound is from the direction of the original source, even though reflections from other directions are involved. The sound energy integrated over this period also gives an impression of added loudness.

It should not be too surprising that the human ear fuses all sounds arriving during a certain time window. After all, our eyes fuse a series of still pictures at the cinema, giving us the impression of continuous movement. The rate of presentation of the still pictures is important; there must be at least 16 pictures per second (62-millisecond interval) to avoid seeing a series of still pictures or a flicker. Auditory fusion works best during the first 20 or 30 milliseconds; beyond 50 to 80 milliseconds discrete echoes dominate.

Haas[21] set his subjects 3 meters from two loudspeakers arranged so that they subtended an angle of 45 degrees, the observer's line of symmetry splitting this angle. The conditions were approximately anechoic. The observers were called upon to adjust an attenuator until the sound from the "direct" loudspeaker equaled that of the "delayed" loudspeaker. He then proceeded to study the effects of varying the delay.

A number of researchers had previously found that very short delays (less then 1 msec) were involved in our discerning the direction of a source by slightly different times of arrival at our two ears. Delays greater than this do not affect our directional sense.

As shown in Fig. 3-19, Haas found that in the 5 to 35 msec delay range the sound from the delayed loudspeaker has to be increased more than 10 dB over the direct before it sounded like an echo. This is the precedence effect, or Haas effect. In a room, reflected energy arriving at the ear within 35 msec is integrated with the direct sound and is perceived as part of the direct sound as opposed to reverberant sound. These early reflections increase the loudness of the sound, and as Haas said, result in "...a pleasant modification of the sound impression in the sense of broadening of the primary sound source while the echo source is not perceived acoustically."

The transition zone between the integrating effect for delays less than 35 msec and the perception of delayed sound as discrete echo is gradual, and therefore, somewhat indefinite. Some peg the dividing line at a convenient 1/16 second (62 msec), some at 80 msec, and some at 100 msec beyond which there is no question about the discreteness of the echo. In this book we will consider the first 30 msec as in Fig. 3-19, the region of definite integration.

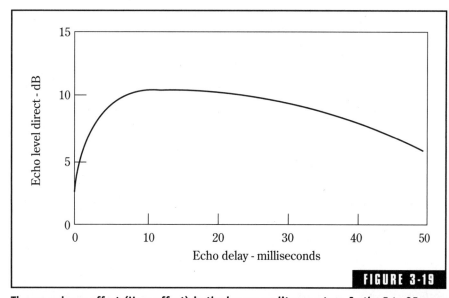

FIGURE 3-19

The precedence effect (Haas effect) in the human auditory system. In the 5 to 35 msec region, the echo level has to be about 10 dB higher than the direct sound to be discernible as an echo. In this region, reflected components arriving from many directions are gathered by the ear. The resulting sound seems louder, because of the reflections and appears to come from the direct source. For delays 50 to 100 msec and longer reflections are perceived as discrete echoes. (After Haas.[21])

Perception of Reflected Sound

In the preceding section, "reflected" sound was considered in a rather limited way. A more general approach is taken in this section. It is interesting that the loudspeaker arrangement Haas used was also used by dozens of other researchers and that this is basically the familiar stereo setup; two separated loudspeakers with the observer (listener) located symmetrically between the two loudspeakers. The sound from one loudspeaker is designated as the *direct* sound, that from the other, the delayed sound (the *reflection*). The delay injected between the two signals and their relative levels is adjustable. Speech is used as the signal.[22]

With the sound of the direct loudspeaker set at a comfortable level, and with a delay of, say 10 ms, the level of the reflected, or delayed, loudspeaker sound is slowly increased from a very low value. The sound level of the reflection at which the observer first detects a difference in the sound is the threshold of reflection detection. For levels less than this, the reflection is inaudible; for levels greater than this, the reflection is clearly audible.

As the reflection level is gradually increased above the threshold value, a sense of spaciousness is imparted to the combined sound. This sense of spaciousness prevails, even though the experiment is conducted in an anechoic space. As the level of the reflection is increased about 10 dB above the threshold value, another change is noticed in the sound; a broadening of the sound image and possibly a shifting of the image toward the direct loudspeaker is now added to the increasing spaciousness. As the reflection level is increased another 10 dB or so above the image broadening threshold, another change is noted; *discrete echoes* are heard.

This is all very interesting, but what practical value does it have? Consider a specific example: a listening room in which recorded music will be played. Figure 3-20 contains answers to the effect of sound reflected from floor, ceiling, and walls being added to the direct sound from the loudspeakers. Reflections below the threshold of perception are unusable; reflections perceived as discrete echoes are also unusable. The usable area is the unshaded area between those two threshold curves, A and C. Simple calculations can give estimates of the level and delay of any specific reflection, knowing the speed of sound, the distance traveled and applying the inverse square law. Figure 3-20 gives the subjective reactions the listener will probably have to the combination of any reflection and the direct sound.

To assist in the "simple" calculations mentioned previously, the following equations can be applied:

$$\text{Reflection delay} = \frac{(\text{reflected path, ft}) - (\text{direct path, ft})}{1,130 \text{ ft/sec}}$$

This assumes 100% reflection at the reflecting surface.

$$\text{Reflection level at listening position} = 20 \log \frac{\text{direct distance, ft}}{\text{reflection distance, ft}}$$

This assumes the inverse square propagation.

Occupational and Recreational Deafness

The hearing of workers in industry is now protected by law. The higher the environmental noise, the less exposure allowed (Table 3-2).

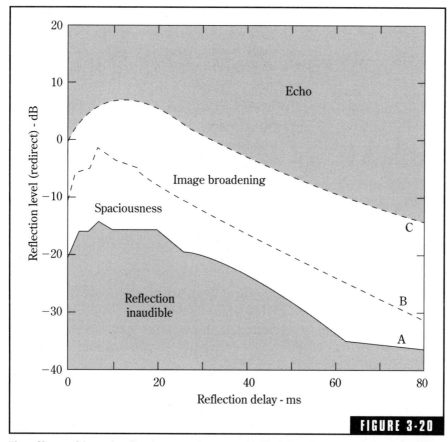

FIGURE 3-20

The effects of lateral reflections on the perception of the direct sound in a simulated stereo arrangement. These measurements were made in anechoic conditions, lateral angles 45–90 degrees, with speech as the signal. (A) Absolute threshold of audibility of the reflection. (B) Image shift/broadening threshold (A & B After Olive and Toole,[22] and Toole.[23]) (C) Lateral reflection perceived as a discrete echo (After Meyer and Schodder,[24] and Lochner and Burger[25]).

Researchers are trying to determine what noise exposure workers are subjected to in different plants. This is not easy as noise levels fluctuate and workers move about, but wearable *dosimeters* are often used to integrate the exposure over the work day. Industries are hard pressed to keep up with changes in regulations, let alone the installation of noise shields around offending equipment and keeping ear plugs in or ear muffs on the workers. Nerve deafness resulting from occupational noise is recognized as a distinct health hazard.

Table 3-2. OSHA permissible noise exposure times.*

Sound pressure level, dB, A-weighting, slow response	Maximum daily exposure hours
85	16
90	8
92	6
95	4
97	3
100	2
102	1.5
105	1
110	0.5
115	0.25 or less

*Reference: OSHA 2206 (1978)

It is especially bad when one works all day in a high-noise environment, then engages in motorcycle or automobile racing, listens to a 400-watt stereo at high level, or spends hours in a discotheque. The professional audio engineer operating with high monitoring levels is risking irreparable injury to the basic tools of the trade—his ears. As high-frequency loss creeps in, the volume control is turned up to compensate, and the rate of deterioration is accelerated.

The key to conservation of hearing is the audiogram. Comparing today's audiogram with earlier ones establishes the trend; if downward, steps can be taken to check it. The audiogram of Fig. 3-21, which looks something like the Big Dipper

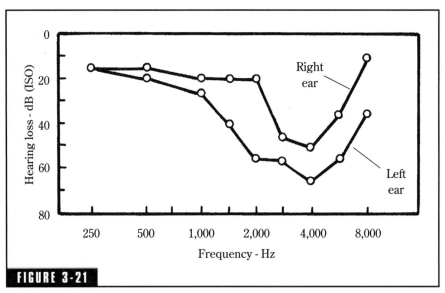

FIGURE 3-21

Audiograms showing serious loss centered on 4 kHz, presumably resulting from years of exposure to high-level sound in the control room of a recording studio.

constellation, is that of a 50-ish sound mixer in a recording studio. The indications are that this loss, centered on 4 kHz, is the accumulation of many years of listening to high-level sounds in the control room.

Summary

■ The ear is sensitive enough to hear the tattoo of air particles on the eardrums in the quietude of an anechoic chamber.

■ The auditory canal, acting as a quarter-wave pipe closed at one end by the eardrum, contributes an acoustical amplification of about 10 dB, and the head diffraction effect produces another 10 dB near 3 kHz. These are vital speech frequencies.

■ The leverage of the ossicle bones of the middle ear and the ratio of areas of the eardrum and oval window successfully match the impedance of air to the fluid of the inner ear.

■ The Eustachian tube and round window provide pressure release and equalization with atmospheric pressure.

■ Waves set up in the inner ear by vibration of the oval window excite the sensory hair cells, which are connected to the brain. There is a "place effect," the peak of hair cell agitation for higher frequencies being nearer the oval window, and low frequencies at the distal end.

■ The area of audibility is bounded by two threshold curves, the threshold of audibility at the lower extreme and the threshold of feeling or pain at the loud extreme. Our entire auditory experience occurs within these two extremes.

■ The loudness of tone bursts decreases as the length of the burst is decreased. Bursts greater than 200 msec have full loudness, indicating a time constant of the ear at about 100 msec.

■ Our ears are capable of accurately locating the direction of a source in the horizontal plane. In a vertical median plane, however, localization ability is less accurate.

- Pitch is a subjective term. Frequency is the associated physical term, and the two have only a general relationship.

- Subjective timbre or quality of sound and the physical spectrum of the sound are related, but not equal.

- The nonlinearity of the ear generates intermodulation products and spurious harmonics.

- The Haas, or precedence, effect describes the ability of the ear to integrate all sound arriving within the first 50 msec, making it sound louder.

- Although the ear is not effective as a measuring instrument yielding absolute values, it is very keen in comparing frequencies, levels, or sound quality.

- Occupational and recreational noises are taking their toll in permanent hearing loss. Definite precautionary steps to minimize this type of environmentally caused deafness are recommended.

Endnotes

[1]Bloom, P.J., *Creating source illusions by special manipulation*, J. Audio Eng. So., 25 (1977) 560-565.

[2]Mehrgardt, S and V. Mellert, *Transformation characteristics of the external human ear*, J. Acous. Soc. Am., 61, 6 (1977) 1567-1576.

[3]Moore, Brian C.J., *An introduction to the psychology of hearing*, New York, Academic Press (1982).

[4]Pickles, James D., *An introduction to the physiology of hearing*, 2nd. ed., San Diego, Academic Press (1988).

[5]Tobias, Jerry V. Ed., *Foundations of modern auditory theory*, Vol. 1, New York, Academic Press (1970).

[6]Tobias, Jerry V., Ed., *Foundations of modern auditory theory*, Vol. 2, New York, Academic Press (1972).

[7]Fletcher, H. and W.A. Munson, *Loudness, its definition, measurement, and calculations*, J. Acous. Soc. Am., 5 (1933) 82-108.

[8]Robinson, D.W. and R.S. Dadson, *A re-determination of the equal-loudness relations for pure tones*, British J. of Appl. Psychology., 7 (1956) 166-181. (Adopted by the International Standards Organization as ISO-226.)

[9]Toole, Floyd E., *Loudness—Applications and implications for audio*, dB Magazine, Part I: 7, 5 (May 1973) 7-30, Part II: 7, 6 (June 1973) 25-28.

[10]Zwicker, C.G., G. Flottorp, and S.S. Stevens, *Critical bandwidths in loudness summation*, J. Acous. Soc. Am., 29, 5 (1957) 548-557.

[11]Moore, Brian C.J. and Brian Glasberg, *Suggested formulae for calculating auditory filter bandwidths and excitation patterns*, J. Acous. Soc. Am., 74, 3 (1983) 750-753.

[12]Meyer, Erwin, *Physical measurements in rooms and their meaning in terms of hearing conditions*, Proc. 2nd Int. Congr. on Acous. (1956) 59-68.

[13]Stevens, S.S. and J. Volkman, *The relation of pitch to frequency: a revised scale*, Am. J. Psychology, 53 (1940) 329-353.

[14]Plomb, R. and H.J.M. Steeneken, *Place dependence of timbre in reverberant sound fields*, Acustica, 28, 1 (1973) 50-59.

[15]Blauert, Jens, *Spatial Hearing*, Cambridge, MIT Press (1983).

[16]Stark, Craig, *The sense of hearing*, Stereo Review, (September 1969) 66, 71-74.

[17]Fletcher, Harvey, *The ear as a measuring instrument*, J. Audio Eng. Soc., 17, 5 (1969) 532-534.

[18]Everest, F. Alton, *The filters in our ears*, Audio, 70, 9 (1986) 50-59.

[19]Schroeder, M.R., D. Gottlob, and K.F. Siebrasse, *Comparative study of European concert halls: correlation of subjective presence with geometric and acoustic parameters*, J. Acous. Soc. Am., 56, 4 (1974) 1195-1201.

[20]Hawkes, R.J. and H. Douglas, *Subjective experience in concert auditoria*, Acustica, 28, 5 (1971) 235-250.

[21]Haas, Helmut, *The influence of a single echo on the audibility of speech*, J. Audio Eng. Soc., 20, 2 (1972) 146-159. (This is an English translation from the German by Dr. Ingr. K.P.R. Ehrenberg of Haas' original paper in Acustica, 1, 2 (1951).

[22]Olive, S.E. and F.E. Toole, *The detection of reflections in typical rooms*, J. Audio Engr. Soc. 37 (1989) 539-553.

[23]Toole, F.E., *Loudspeakers and rooms for stereophonic sound reproduction*, Proc. AES 8th International Conference, Washington, D.C., 3-6 May, 1990, pp 71-91.

[24]Meyer, E. and G.R. Schodder, *On the influence of reflected sounds on directional localization and loudness of speech*, Nachr. Akad. Wiss., Göttingen, Math., Phys., Klasse IIa, 6 (1952) 31-42.

[25]Lochner, J.P.A. and J.F. Burger, *The subjective masking of short time delayed echoes by their primary sounds and their contribution to the intelligibility of speech*, Acustica, 8 (1958) 1-10.

CHAPTER 4

Sound Waves in the Free Field

Practical acoustic problems are invariably associated with people, buildings, rooms, airplanes, automobiles, etc. These can generally be classified either as problems in physics (sound as a stimulus) or problems in psychophysics (sound as a perception), and often as both. Acoustical problems can be very complex in a physical sense, for example, thousands of reflected components might be involved or obscure temperature gradients might bend the sound in such a way as to affect the results. When acoustical problems involve human beings and their reactions, "complexity" takes on a whole new meaning.

Don't be discouraged if you want a practical understanding of acoustics, but your background is in another field, or you have little technical background at all. The inherent complexity of acoustics is pointed out only to justify going back to the inherent simplicity of sound in a free field as a starting point in the study of other types of practical sound fields.

Free Sound Field: Definition

Sound in a free field travels in straight lines, unimpeded and undeflected. Unimpeded sound is sound that is unreflected, unabsorbed, undiffracted, unrefracted, undiffused, and not subjected to resonance effects. These are all hazards that could (and do) face a simple ray of sound leaving a source.

Free space must not be confused with cosmological space. Sound cannot travel in a vacuum; it requires a medium such as air. Here, free space means any air space in which sound acts as though it is in the theoretical free space. Limited free space can even exist in a room under very special conditions.

Sound Divergence

The point source of Fig. 4-1 radiates sound at a fixed power. This sound is of uniform intensity (power per unit area) in all directions. The circles represent spheres having radii in simple multiples. All of the sound power passing through the small square area at radius d also passes through the areas at $2d$, $3d$, $4d$, etc. This increment of the total sound power traveling in this single direction is spread over increasingly greater areas as the radius is increased. Intensity decreases with distance. As the area of a sphere is $4\pi r^2$, the area of a small segment on the surface of the sphere also varies as the square of the radius. Doubling the distance from d to $2d$ reduces the intensity to $1/4$, tripling the distance reduces the intensity to $1/9$, and quadrupling the distance reduces intensity to $1/16$. *Intensity of sound is inversely proportional to the square of the distance in a free field.*

Intensity of sound (power per unit area) is a difficult parameter to measure. Sound pressure is easily measured. As intensity is proportional to the square of sound pressure, the *inverse square law* (for intensity) becomes *the inverse distance law* (for sound pressure). In other words, sound pressure varies inversely as the first power of the distance. In Fig. 4-2, the sound-pressure level in decibels is plotted against distance. This illustrates the basis for the common and very useful expression, *6 dB per doubling of the distance* that, again, applies only for a free field.

Examples: Free-Field Sound Divergence

When the sound-pressure level L_1 at distance d_1 from a point source is known, the sound-pressure level L_2 at another distance d_2 can be calculated from:

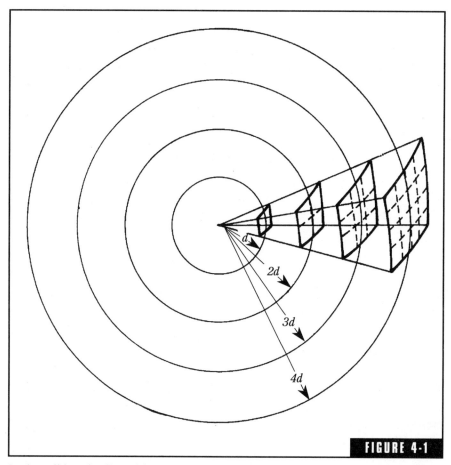

FIGURE 4-1

In the solid angle shown, the same sound energy is distributed over spherical surfaces of increasing area as *d* is increased. The intensity of sound is inversely proportional to the square of the distance from the point source.

$$L_2 = L_1 - 20 \log \frac{d_2}{d_1}, \text{ decibels} \qquad \text{(4-1)}$$

In other words, the difference in sound-pressure level between two points that are d_1 and d_2 distance from the source is:

$$L_2 - L_1 = 20 \log \frac{d_2}{d_1}, \text{ decibels} \qquad \text{(4-2)}$$

For example, if a sound-pressure level of 80 dB is measured at 10 ft, what is the level at 15 ft?

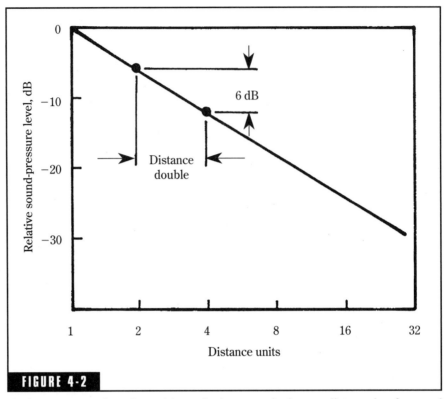

FIGURE 4-2

The inverse square law of sound intensity becomes the inverse distance law for sound pressure. This means that sound-pressure level is reduced 6 dB for each doubling of the distance.

Solution:

20 log 10/15 = 3.5 dB; the level is 80 − 3.5 = 76.5 dB.

What is the sound-pressure level at 7 ft?

Solution:

20 log 10/7 = + 3.1 dB, and level is 80 + 3.1 = 83.1 dB.

All this is for a free field in which sound diverges spherically, but this procedure may be helpful for rough estimates even under other conditions.

If a microphone is 5 feet from an enthusiastic soprano and the VU meter in the control room peaks +6, moving the microphone to 10 feet would bring the reading down *approximately* 6 dB. The word "approximately" is important. The inverse square law holds true only for free field conditions. The effect of sound energy reflected from walls would be to make the change for a doubling of the distance something less than 6 dB.

An awareness of the inverse square law is of distinct help in estimating acoustical situations. For instance, a doubling of the distance from 10 to 20 feet would, for free space, be accompanied by the same sound-pressure level decrease, 6 dB, as for a doubling from 100 to 200 feet. This accounts for the great carrying power of sound outdoors.

Inverse Square in Enclosed Spaces

Free fields exist in enclosed spaces only in very special and limited circumstances. The reflections from the enclosing surfaces affect the way sound level decreases with distance. No longer does the inverse square law or the inverse distance law describe the entire sound field. For example, assume that there is an installed loudspeaker in an enclosed space that is capable of producing a sound-pressure level of 100 dB at a distance of 4 ft. As shown in the graph of Fig. 4-3, free field

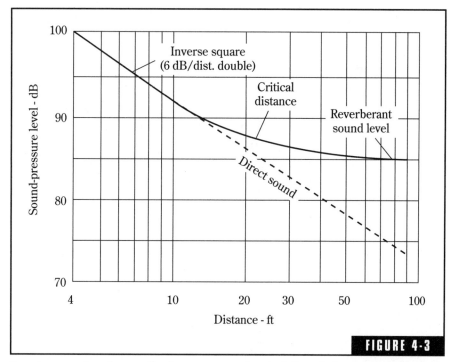

FIGURE 4-3

Even in an enclosed space the inverse square law is followed close to the source. By definition, the *critical distance* is that distance at which the direct sound pressure is equal to the reverberant sound pressure.

conditions exist close to the loudspeaker. This means that spherical divergence prevails in this limited space, and reflections from the surfaces are of negligible comparative level. Moving away from the loudspeaker, the effects of sound reflected from the surfaces of the room begin to be effective. At the *critical distance* the direct and the reflected sound are equal. The critical distance may be taken as a rough single-figure description of the acoustics of the environment.

In the region very close to the loudspeaker, the sound field is in considerable disarray. The loudspeaker, at such close distances, can in no way be considered a point source. This region is called the *near field*. Only after moving several loudspeaker dimensions away from it can significant measurements be made in the *far field*.

Hemispherical Propagation

True spherical divergence implies no reflecting surfaces at all. Tied to this earth's surface as we are, how about hemispherical sound propagation over the surface of this planet? Estimates made by the very convenient "6 dB per distance double" rule are only rough approximations.

Reflections from the surface of the earth outdoors usually tend to make the sound level with distance something less than that indicated by the 6 dB per distance double. The reflective efficiency of the earth's surface varies from place to place. Note the sound level of a sound at 10 ft and again at 20 ft from the source. The difference between the two will probably be closer to 4 dB than 6 dB. For such outdoor measurements the distance law must be taken at "X dB (4?, 5?) per distance double." There is also the effect of general environmental noise that can influence the measurement of specific sound sources.

Speech, Music, and Noise

Speech, music, and noise are common in that they are within the experience of everyone. Noise is also a common thread that runs through speech and music. Speech sounds are but modulated noise. Noise is a close companion to every musical instrument. The highest skill of every musician must be exerted to minimize such incidental noises as thumps, scratches, and wheezes. The close relationship of speech, music, and noise is made evident in this chapter.

The Voice System

One of the many amazing things about the human body is the high degree of efficiency associated with the multiple use of organic systems. The functions of eating, breathing, and speaking all take place in relative simultaneous harmony. We can eat, breathe, talk practically at the same time through the interworking of muscle action and valves, without food going down the wrong hatch. If we sometimes try to do too many things at once, the system is momentarily thwarted, and we agonize as a bit of food is retrieved from the wrong pipe.

Artificial Larynx

Noise that contains energy over a wide range of constantly shifting frequencies, phases, and amplitudes can be shaped even into speech.

Sometimes people lose their voices. Perhaps the vocal cords are paralyzed, or the larynx was removed surgically. For these people, the Western Electric Company offers a prosthetic device, which when held against the throat, produces pulses of sound that simulate the sounds produced by the natural vocal cords as they interrupt the air stream. This battery-operated device even has a pitch control for controlling "voice" pitch. Then the tongue, lips, teeth, nasal passages, and throat perform their normal function of molding the pulsed noise into words. Even if the overall effect has a somewhat duck-like quality, it enables the user to speak by shaping the noise appropriately.

Sound Spectrograph

An understanding of speech sounds is necessary to understand how the sounds are produced. Speech is highly variable and transient in nature, comprising energy chasing up and down the three-dimensional scales of frequency, sound level, and time. It takes the sound spectrograph to show all three on the same flat surface such as the pages of this book. Examples of several commonly experienced sounds revealed by the spectrograph are shown in Fig. 5-1. In these spectrographs, time progresses horizontally to the right, frequency increases from the origin upward, and the sound level is indicated roughly by the density of the trace—the blacker the trace, the more intense the sound at that frequency and at that moment of time. Random noise on such a plot shows up as a gray, slightly mottled rectangle as all frequencies in the audible range and all intensities are represented as time progresses. The snare drum approaches random noise at certain points, but it is intermittent. The "wolf whistle" opens on a rising note followed by a gap, and then a similar rising note that then falls in frequency as time goes on. The police whistle is a tone, slightly frequency modulated. Each common noise has its spectrographic signature that reveals the very stuff that characterizes it.

The human voice mechanism is capable of producing many sounds other than speech. Figure 5-2 shows a number of these as revealed by sound spectrograms. It is interesting to note that harmonic trains appear on a spectrogram as more or less horizontal lines spaced vertically in frequency. These are particularly noticeable in the trained soprano's voice and the baby's cry, but traces are evident

Miscellaneous sounds

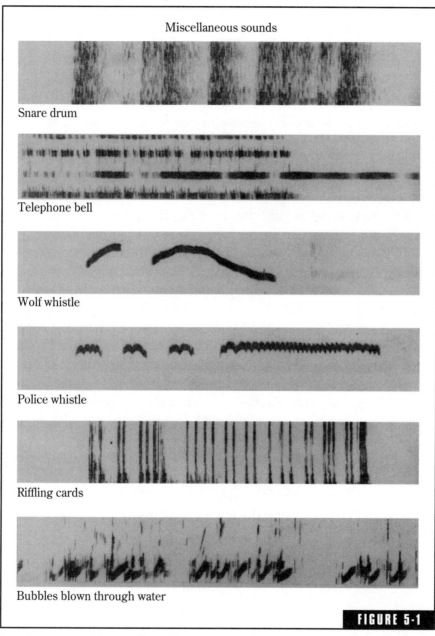

Snare drum

Telephone bell

Wolf whistle

Police whistle

Riffling cards

Bubbles blown through water

FIGURE 5-1

Sound spectrographic recordings of common sounds. Time progresses to the right, the vertical scale is frequency, and the intensity of components is represented by the intensity of the trace. *AT&T Bell Laboratories.*

in other spectrograms. The following discussion owes much to the clear presentation of Flanagan.[1]

Sound Sources for Speech

The artificial larynx is based on the fact that there are really two more or less independent functions in the generation of speech sounds: the sound source and the vocal system. In general, it is a series flow as pictured in Fig. 5-3A, in which the raw sound is produced by a source and subsequently shaped in the vocal tract. To be more exact, there are really three different sources of sound to be shaped by the vocal tract as indicated in Fig. 5-3B. First, there is the one we naturally think of— the sounds emitted by the vocal cords. These are formed into the *voiced sounds.* They are produced by air from the lungs flowing past the slit between the vocal cords (the glottis), which causes the cords to vibrate. The air stream, broken into pulses of air, produces a sound that can almost be called periodic, that is, repetitive in the sense that one cycle follows another.

The second source of sound is that made by forming a constriction at some point in the vocal tract with the teeth, tongue, or lips and forcing air through it under high enough pressure to produce significant turbulence. Turbulent air creates noise. This noise is shaped by the vocal tract to form the *fricative sounds* of speech such as the consonants f, s, v, and z. Try making these sounds, and you will see that high-velocity air is very much involved.

The third source of sound is produced by the complete stoppage of the breath, usually toward the front, a building up of the pressure, and then the sudden release of the breath. Try speaking the consonants k, p, and t, and you will sense the force of such *plosive* sounds. They are usually followed by a burst of fricative or turbulent sound. These three types of sounds—voiced, fricative, and plosive—are the raw sources that are shaped into the words we casually speak without giving a thought to the wonder of their formation.

Vocal Tract Molding of Speech

The vocal tract can be considered as an acoustically resonant system. This tract, from the lips to the vocal cords, is about 6.7 in (17 cm) long. Its cross-sectional area is determined by the placement of the lips, jaw, tongue, and velum (a sort of trapdoor that can open or close off the

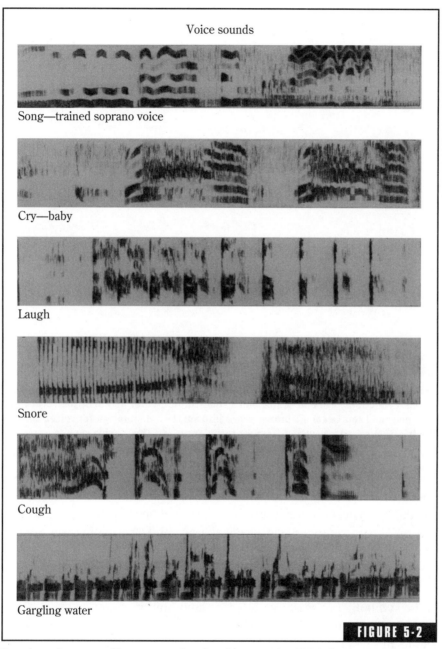

Voice sounds

Song—trained soprano voice

Cry—baby

Laugh

Snore

Cough

Gargling water

FIGURE 5-2

Sound spectrograms of human sounds other than speech. *AT&T Bell Laboratories.*

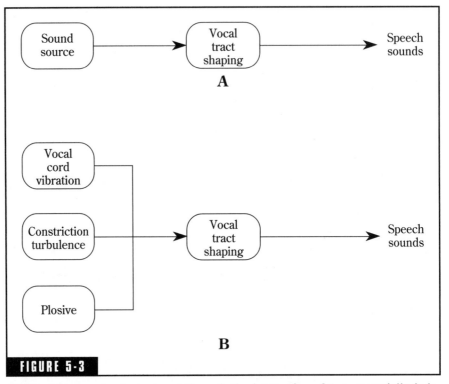

FIGURE 5-3

(A) The human voice is produced through the interaction of two essentially independent functions, a sound source and a time-varying-filter action of the vocal tract. (B) The sound source can be broken down into vocal-cord vibration for voiced sounds, the fricative sounds resulting from air turbulence, and the plosive sounds.

nasal cavity) and varies from zero to about 3 sq in (20 sq cm). The nasal cavity is about 4.7 in (12 cm) long and has a volume of about 3.7 cu in (60 cu cm). These dimensions are mentioned because they have a bearing on the resonances of the vocal tract and their effect on speech sounds.

Formation of Voiced Sounds

If the symbolic boxes of Fig. 5-3 are elaborated into source spectra and modulating functions, we arrive at something everyone in audio is interested in—the spectral distribution of energy in the voice. We also get a better understanding of the aspects of voice sounds that contribute to the intelligibility of speech in reverberation, noise, etc. Figure 5-4 shows the steps in producing voiced sounds. First, there is the sound produced by the vibration of the vocal cords, pulses of sound having a fine spectrum that falls off at about 10 dB per octave as frequency is increased as shown in Fig. 5-4A. The sounds of the vocal cords pass through the vocal tract, which acts as a filter varying with time. The humps of Fig. 5-4B are due to the acoustical resonances, called *formants* of the vocal pipe, which is open at the mouth end and essentially closed at the vocal cord end. Such an acoustical pipe 6.7 inches long has resonances at odd quarter wavelengths, and these peaks occur at approximately 500, 1,500, and 2,500 Hz. The output sound, shaped by the resonances of the vocal tract, is shown in Fig. 5-4C. This applies to the voiced sounds of speech.

Formation of Unvoiced Sounds

Unvoiced sounds are shaped in a similar manner as indicated in Fig. 5-5. Unvoiced sounds start with the distributed, almost random-noise-like spectrum of the turbulent air as fricative sounds are produced. The distributed spectrum of Fig. 5-5A is generated near the mouth end of the vocal tract, rather than the vocal cord end; hence, the resonances of Fig. 5-5B are of a somewhat different shape. Figure 5-5C shows the sound output shaped by the time-varying filter action of Fig. 5-5B.

Putting It All Together

The voiced sounds, originating in vocal cord vibrations, the unvoiced sounds, originating in turbulences, and plosives, which originate near the lips, go together to form all of our speech sounds. As we speak, the

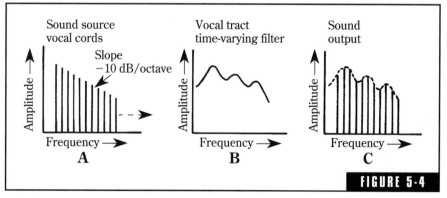

Sound spectrograms of human sounds other than speech. *AT&T Bell Laboratories.*

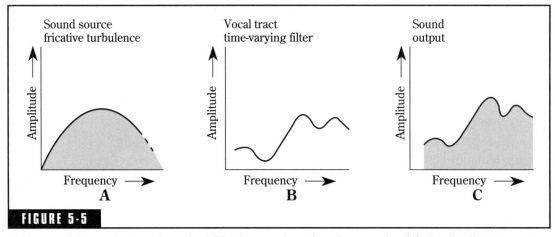

FIGURE 5-5

A diagram of the production of unvoiced fricative sounds such as f, s, v, and z. (A) The distributed spectrum of noise due to air turbulence resulting from constrictions in the vocal tract. (B) The time-varying filter action of the vocal tract. (C) The output sound resulting from the filter action of the distributed sound of (A).

formant resonances shift about in frequency as the lips, jaw, tongue, and velum change position to shape the desired words. The result is the unbelievable complexity of human speech evident in the spectrogram of Fig. 5-6. Information communicated via speech is a pattern of frequency and intensity shifting rapidly with time. Notice that there is little speech energy above 4 kHz in Fig. 5-6, nor (which does not show) below 100 Hz. Now it's understandable why the presence filter peaks in the 2- to 3-kHz region; that is where the pipes resonate!

Synthesized Speech

Mechanical speaking machines date back to 1779, when Kratzenstein of St. Petersburg constructed a set of acoustical resonators to emulate the human mouth. These were activated with reeds such as those of a mouth organ. He was able to produce reasonably recognizable vowel sounds with the contraption. Wolfgang von Kempelen of Vienna did a much better job in 1791, which Wheatstone later improved upon. This machine used a bellows to supply air to a leather tube that was manipulated by hand to simulate mouth action and included an "S" whistle, a "SH" whistle, and a nostril cutoff valve. After experimenting with a copy of Kempelen's machine in boyhood, Alexander Graham Bell patented a procedure for producing speech in 1876. One important precursor of the modern digital devices for synthesizing speech was

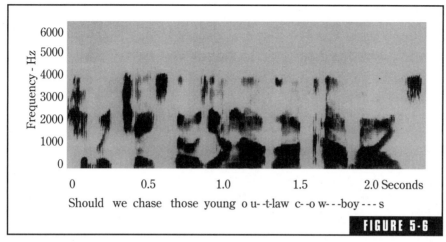

Sound spectrogram of a sentence spoken by a male voice. *AT&T Bell Laboratories.*

the analog Voder from Bell Laboratories that was demonstrated at the World Fairs in New York (1939) and in San Francisco (1940). It took a year to train operators to play the machine to produce simple, but recognizable, speech.

Digital Speech Synthesis

Techniques for storing human speech in computer memory and playing it back under specified, fixed conditions are widely used. Electrical machines of this type now talk to us in the form of language translators, talking calculators, spelling machines, as well as telephone-information services. We will be seeing (rather, hearing) a stream of other answer-back applications of this technique in the days ahead, including both storage and recall, and true speech synthesis.

It is interesting to note that to program a computer to talk, a model of speech production is necessary and that the models of Figs. 5-3, 5-4, and 5-5 have been applied in just this way. Figure 5-7 shows a diagram of a digital synthesis system. A random-number generator produces the digital equivalent of the s-like sounds for the unvoiced components. A counter produces pulses simulating the pulses of sound of the vocal cords for the voiced components. These are shaped by time-varying digital filters simulating the ever-changing resonances of the vocal tract. Special signals control each of these to form digitized speech, which is then changed to analog form in the digital-to-analog converter.

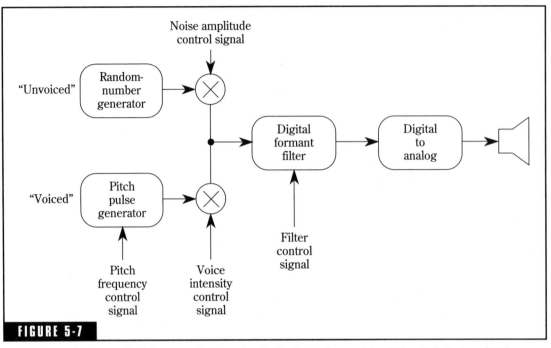

FIGURE 5-7

A digital system for synthesizing speech. Note the similarity to the models of Figs. 5-3, 5-4, and 5-5.

Other applications of digital speech synthesis include voice-recognition systems for "hands-free" typing and computer programs for recording that automatically adjust the intonation of a vocalist who might otherwise be singing sharp or flat.

Directionality of Speech

Speech sounds do not have the same strength in all directions. This is due primarily to the sound shadow cast by the head. A question arises as to just how such directionality can be measured. Should a sound source be placed in the mouth? Well, the mouth itself is a continuous source of speech sounds, so why not use these speech sounds for the measurement of directionality? That is what Kuttruff has done with the results shown in Fig. 5-8.[2] Because speech sounds are highly variable and extremely complex, careful averaging is necessary to give an accurate measure of directional effects.

The horizontal directional effects, shown in Fig. 5-8A, show only a modest directional effect of about 5 dB in the 125- to 250-Hz band.

This is to be expected because the head is small compared to wavelengths of 4.5 to 9 feet associated with this frequency band. There are significant directional effects, however, for the 1,400- to 2,000-Hz band. For this band, which contains important speech frequencies, the front-to-back difference is about 12 dB.

In the vertical plane, Fig. 5-8B, the 125- to 250-Hz band shows about 5 dB front-to-back difference again. For the 1,400- to 2,000-Hz band, the front-to-back difference is also about the same as the horizontal plane, except for the torso effect. The discrimination against high speech frequencies picked up on a lapel microphone becomes obvious in Fig. 5-8B, although the measurements were not carried to angles closer to 270 degrees.

Music

Musical sounds are extremely variable in their complexity and can range from a near sine-wave form of a single instrument or voice to the highly complex mixed sound of a symphony orchestra. Each instrument and each voice has a different tonal texture for each note. Many musical instruments, such as the violin, viola, cello, or bass, produce their tones by vibration of strings. On a stretched string, the overtones are all exact multiples of the fundamental, the lowest tone produced. These overtones may thus be called harmonics. If the string is bowed in the middle, odd harmonics are emphasized because the fundamental and odd harmonics have maximum amplitude

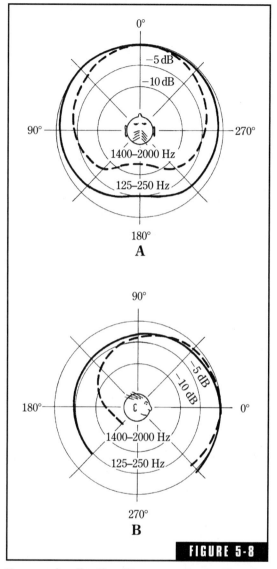

FIGURE 5-8

Human voice directionality measured using the voice as a sound source. (A) Front-to-back directional effects of about 12 dB are found for the important speech frequencies. (B) In the vertical plane, the front-to-back directional effects for the 1,400- to 2,000-Hz band are about the same as for the horizontal plane. *By permission of Heinrich Kuttruff and Applied Science Publishers Ltd., London.*

there. Because the even harmonics have nodes in the center of the string, they will be subdued if bowed there. The usual place for bowing is near one end of the strings, which gives a better blend of even and odd harmonics. There is a problem with the seventh harmonic because it belongs to a different musical family. By bowing $1/7$ of the distance from one end, this harmonic is decreased.

The harmonic content of the E and G notes of a violin are displayed graphically in Fig. 5-9. Harmonic multiples of the higher E tone are spaced wider and hence have a "thinner" timbre. The lower frequency tone, on the other hand, has a closely spaced spectral distribution and a richer timbre. The small size of the violin relative to the low frequency of the G string means that the resonating body cannot produce a fundamental at as high a level as the higher harmonics. The harmonic content and spectral shape depend on the shape and size of the resonating violin body, the type and condition of the wood, and even the varnish. Why there are so few superb violins among the many good ones is a problem that has not yet been solved completely.[3]

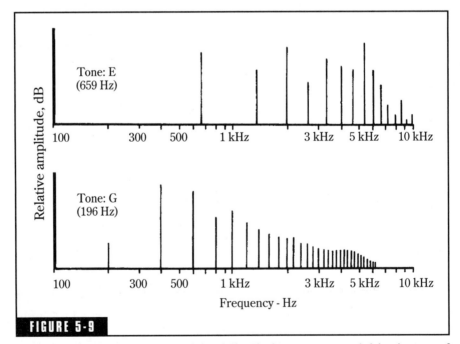

FIGURE 5-9

Harmonic content of open strings of the violin. The lower tones sound richer because of the closely packed harmonics.

Wind Instruments

Resonances in the three-dimensional room are discussed in detail in Chap. 15. In many musical instruments, resonance in pipes or tubes must be considered primarily one dimensional. Standing-wave effects are dominant in pipes. If air is enclosed in a narrow pipe closed at both ends, the fundamental (twice the length of the pipe) and all its harmonics will be formed. Resonances are formed in a pipe open at only one end at the frequency at which the pipe length is four times the wavelength, and results in odd harmonics. Wind instruments form their sounds this way; the length of the column of air is continuously varied, as in the slide trombone, or in jumps as in the trumpet or French horn, or by opening or closing holes along its length as in the saxophone, flute, clarinet, and oboe.

The harmonic content of several wind instruments is compared to that of the violin in the spectrograms of Fig. 5-10. Each instrument has its characteristic timbre as determined by the number and strength of its harmonics and by the formant shaping of the train of harmonics by the structural resonances of the instrument.

Nonharmonic Overtones

Harvey Fletcher tried to synthesize piano sounds.[4] It was emphasized that piano strings are stiff strings and vibrate like a combination of solid rods and stretched strings. This means that the piano overtones are not strictly harmonic. Bells produce a wild mixture of overtones, and the fundamental is not even graced with that name among specialists in the field. The overtones of drums are not harmonically related, although they give a richness to the drum sound. Triangles and cymbals give such a mixture of overtones that they blend reasonably well with other instruments. Nonharmonic overtones produce the difference between organ and piano sounds and give variety to musical sounds in general.

Dynamic Range of Speech and Music

In the concert hall, a full symphony orchestra is capable of producing some very loud sounds when the score says so, but also soft, delicate passages. Seated in the audience, one can fully appreciate this grand sweep of sound due to the great dynamic range of the human ear. The dynamic range between the loudest and the softest passage will be on the order of 60 to 70 dB. To be effective, the soft passages must still be

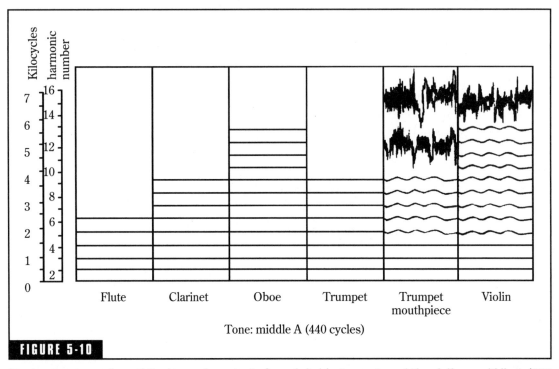

FIGURE 5-10

Spectrogram comparison of the harmonic content of woodwind instruments and the violin as middle A (440 Hz) is played. The differences displayed account for the difference in timbre of the different instruments. *AT&T Bell Laboratories.*

above the ambient background noise in the hall, hence the emphasis on adequate structural isolation to protect against traffic and other outside noises, and precautions to ensure that air-handling equipment noise is low.

For those not present in the music hall, AM or FM radio, television, magnetic recordings, or disc recordings must suffice. These conventional media are unable to handle the full dynamic range of the orchestra. Noise at the lower extreme and distortion at the upper extreme introduce limitations. In the case of broadcast media, there are the added regulatory restrictions prohibiting interference with adjacent channels.

Digital audio has brought some major revisions of our thinking in regard to dynamic range and signal-to-noise ratio. The dynamic range in a digital system is directly related to the range of binary digits (bits).

Number of binary digits	Dynamic range, dB
4	24
8	48
12	72
16	96
24	144

The theoretical 96-dB dynamic range provided by the 16-bit digital system is staggering to someone steeped in the traditional techniques. At last there is a recording system that handles the concert hall dynamic range reasonably well. The commercial compact disc (CD) is capable of carrying 74 minutes of full-fidelity music with a 96 dB signal-to-noise ratio. The digital audio cassette is another worthy addition to the recording/playback arsenal, as well as digital audio tape (DAT) systems. Digital techniques have transferred dynamic range limitations from the medium to the concert hall on the one hand and the playback environment on the other.

However, in recent years the audio community has grown dissatisfied with the quantization noise and "graininess" of 16-bit digital audio. New 24-bit formats are on the rise, such as "Super Audio CD" and "Audio DVD." In addition, professional audio mastering engineers are now working in 24-bit resolution to avoid the audibility of digital artifacts resulting from lower-resolution processing.

Power in Speech and Music

In learning more about the various signals to be handled, one must consider the peak power of various sources. For speech, the average power is only about 10 microwatts, but peaks might reach a milliwatt. Most of the power of speech is in the low frequencies, with 80 percent below 500 Hz, yet there is very little power below 100 Hz. On the other hand, the small amount of power in the high frequencies determines the intelligibility of speech and thus is very important because that is where the consonants are. The peak power of various musical instruments is listed in Table 5-1.

Table 5-1 Power of Musical Sources.[5]

Instrument	Peak, power (watts)
Full orchestra	70
Large bass drum	25
Pipe organ	13
Snare drum	12
Cymbals	10
Trombone	6
Piano	0.4
Trumpet	0.3
Bass saxophone	0.3
Bass tuba	0.2
Double bass	0.16
Piccolo	0.08
Flute	0.06
Clarinet	0.05
French horn	0.05
Triangle	0.05

Frequency Range of Speech and Music

It is instructive to compare the frequency range of the various musical instruments with that of speech. This is best done graphically. Figure 5-11 includes the ranges only of the fundamental tones, and not of the harmonic tones of the instruments. The very low piano and organ notes, which are below the range of audibility of the ear, are perceived by their harmonics. Certain high-frequency noise accompanying musical instruments is not included, such as reed noise in woodwinds, bowing noise of strings, and key clicks and thumps of piano and percussion instruments.

Future Dynamic-Range Requirements

If the peak instantaneous sound levels and noise thresholds are regarded as determining dynamic range requirements, much greater ranges are required. Fiedler's study[6] has shown that a dynamic range of up to 118 dB is necessary for subjectively noise-free reproduction of music (see Fig. 5-12). He considered the peak instantaneous sound level of various sources, as shown at the top of the figure, and the just-audible threshold for white noise added to the program source when the listener is in a normal listening situation, as shown at the bottom of the figure. He used musical performances of high peak levels in a quiet environment and a very simple recording setup. The results are summarized in Fig. 5-12. The signal-to-noise ratio offered by a 16-bit PCM (pulse code modulation) system is shown to be inadequate for all but the piano solo. Future developments will undoubtedly require greater dynamic range than that offered by 16-bit digital systems.

Auditory Area

The frequency range and the dynamic range of speech, music, and all others sounds places varying demands on the human ear. The auditory area back in Fig. 3-8 describes the capability of the ear. Both speech

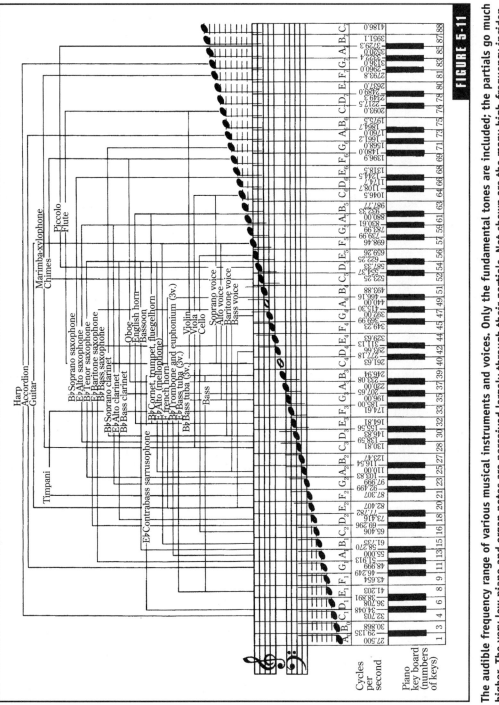

The audible frequency range of various musical instruments and voices. Only the fundamental tones are included; the partials go much higher. The very low piano and organ notes are perceived largely through their partials. Not shown are the many high-frequency incidental noises produced by the instruments. *C. G. Conn, Ltd., Oak Brook, Illinois.*

FIGURE 5-11

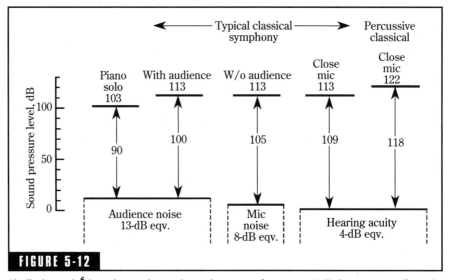

FIGURE 5-12

Fiedler's study[6] has shown that a dynamic range of up to 118 dB is necessary for subjectively noise-free reproduction of music.

and music use only a portion of this area. The portion of the auditory area used in speech is shown by the shaded area of Fig. 5-13. This kidney-bean shape is located centrally in the auditory area, which makes sense because neither the extremely soft or extremely loud sounds, nor sounds of very low or very high frequency, are used in common speech sounds. The speech area of Fig. 5-13 is derived from long-time averages, and its boundaries should be fuzzy to represent the transient excursions in level and frequency. The speech area, as represented, shows an average dynamic range of about 42 dB. The 170- to 4,000-Hz frequency range covers about 4.5 octaves.

The music area of Fig. 5-14 is much greater than the speech area of Fig. 5-13. Music uses a much greater proportion of the full auditory area of the ear. Its excursions in both level and frequency are correspondingly greater than speech, as would be expected. Here again, long-time averages are used in establishing the boundaries of the music area, and the boundaries really should be fuzzy to show extremes. The music area shown has a dynamic range of about 75 dB and a frequency range of about 50 to 8,500 Hz. This frequency span is about 7.5 octaves, compared to the 10-octave range of the human ear. High-fidelity standards demand a much wider frequency range than this, and rightly so. Without the averaging process involved in establishing the speech and music areas, both the dynamic range and the frequency range would be greater to accommodate the short-term transients that contribute little to the overall average, but are still of great importance.

Noise

The word "signal" implies that information is being conveyed. How can noise be considered an information carrier? An enthusiastic "Bronx cheer" raspberry conveys considerable information about the perpetrator's attitude toward someone or something. Noise is the basic part of such a communication, modulated in just the right way. Interrupting noise to form dots and dashes is another way to shape noise into communication. We will also see how a decaying band of noise can give information on the acoustical quality of a room. There are types of noise that are undesirable. Sometimes it is difficult to tell whether it is the

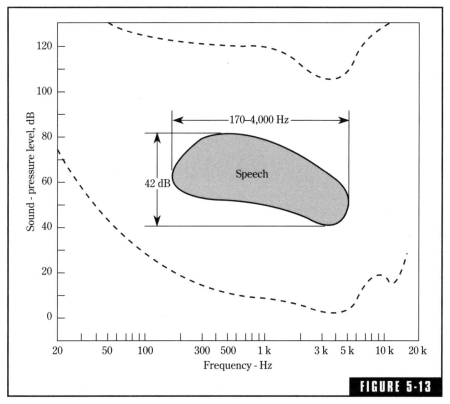

FIGURE 5-13

The portion of the auditory region utilized for typical speech sounds.

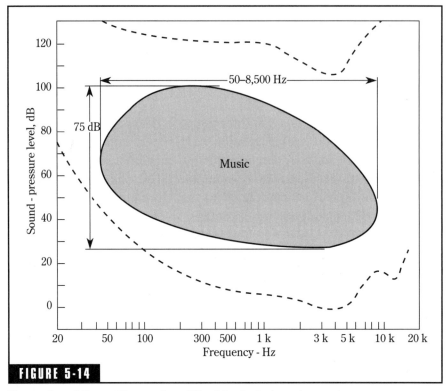

FIGURE 5-14

The portion of the auditory region utilized for typical music sounds.

unpleasant thing we call noise or a carrier of information. The noise of an automobile conveys considerable information on how well it is running. One person's noise might be someone else's communication. A high-fidelity system can produce some beautiful sounds deemed very desirable by the owner, but to a neighbor they might not be considered beautiful at all. Sometimes it isn't easy to distinguish between information and noise. The same sound can be both. Society establishes limits to keep objectionable noise to a minimum while ensuring that information-carrying sounds can be heard by those who need to hear them.

Noise—The Good Kind

A good kind of noise? Defining noise as unwanted sound fits the system noise considered previously, but noise is becoming an increasingly important tool for measurements in acoustics as discussed in Chaps. 5 and 7. This good noise is not necessarily different from the bad noise interfering with our listening to a favorite recording, it is just that the noise is put to a specific use.

In acoustical measurements, the use of pure tones is often very difficult to handle while a narrow band of noise centered on the same frequency would make satisfactory measurements possible. For example, if a studio is filled with a pure tone signal of 1,000 Hz from a loudspeaker, a microphone picking up this sound will have an output that varies greatly from position to position due to room resonances. If, however, a band of noise one octave wide centered at 1,000 Hz were radiated from the same loudspeaker, the level from position to position would tend to be more uniform, yet the measurement would contain information on what is happening in the region of 1,000 Hz. Such measuring techniques make sense because we are usually interested in how a studio or listening room reacts to the very complex sounds being recorded or reproduced, rather than to steady, pure tones.

Random Noise

Random noise is generated in any electrical circuit and minimizing its effect often becomes a very difficult problem. Heavy ions falling back on the cathode of a thermionic vacuum tube produce noise of a relatively high amplitude and wide spectrum. Furthermore, the introduction of some gas molecules into the evacuated space will produce even more noise. Today a random noise generator is made with a silicon diode or other solid-state device followed by an amplifier, voltmeter, and attenuator.

In Fig. 5-15 a pure sine wave and a random noise signal are compared as viewed on a cathode ray oscilloscope. The regularity of the one is in stark contrast to the randomness of the other. If the horizontal sweep of the oscilloscope is expanded sufficiently and a snapshot is taken of the random noise signal, it would appear as in Fig. 5-16.

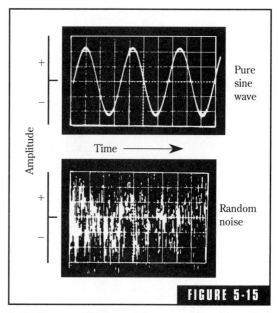

Cathode ray oscillograms of a pure sine wave and of random noise. Random noise may be considered made up of sine waves, which are continually shifting in amplitude, phase, and frequency.

Noise is said to be purely random in character if it has a "normal" or "Gaussian" distribution of amplitudes.[7] This simply means that if we sampled the instantaneous voltage at a thousand equally spaced times, some readings would be positive, some negative, some greater, some smaller, and a plot of these samples would approach the familiar Gaussian distribution curve of Fig. 5-17.

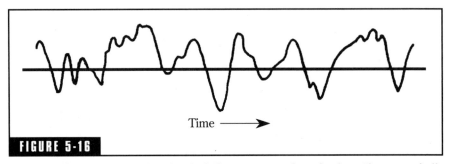

Time ⟶

FIGURE 5-16

A section of the random noise signal of Fig. 5-15 spread out in time. The nonperiodic nature of a noise signal is evident, the fluctuations are random.

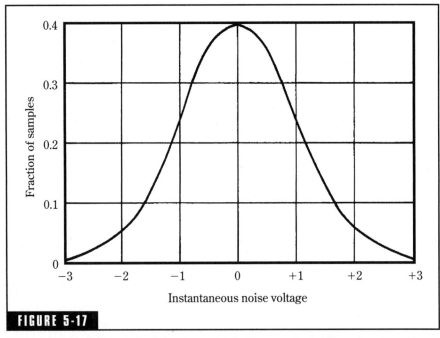

FIGURE 5-17

The proof of randomness of a noise signal lies in the sampling of instantaneous voltage, say, at 1,000 points equally spaced in time and plotting the results. The familiar bell-shaped Gaussian distribution curve results if the noise is truly random.

White and Pink Noise

References to *white noise* and *pink noise* are common and sometimes confusing. What is the difference? White noise is analogous to white light in that the energy of both is distributed uniformly throughout the spectrum. In other words, white noise energy exhibits a *flat distribution* of energy with frequency (Fig. 5-18A).

White light sent through a prism is broken down into a range of colors. The red color is associated with the longer wavelengths of light, that is, light in the lower frequency region. *Pink noise* is noise having higher energy in the low frequencies. In fact, pink noise has come to be identified specifically as noise exhibiting high energy in the low-frequency region with a specific downward slope of 3 dB per octave (Fig. 5-18C). There is a practical reason for this specific slope.

These two colorful terms arose because there are two types of spectrum analyzers in common use. One is the *constant bandwidth*

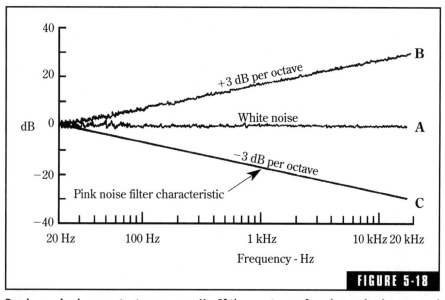

FIGURE 5-18

Random noise has constant energy per Hz. If the spectrum of random noise is measured (white) with a wave analyzer of fixed bandwidth, the resulting spectrum will be flat with frequency as in A. If measured with an analyzer whose passband width is a given percentage of the frequency to which it is tuned, the spectrum will slope upward at 3 dB per octave, as in B. By processing the white noise spectrum of A with a filter that slopes downward at 3 dB per octave, such as in C, a flat response results when constant percentage bandwidth filters are used such as octave or one-third octave filters. In measuring a system, pink noise is applied to the input and, if the system is flat, the read out response will be flat if one-third octave filters, for example, are used.

analyzer, which has a passband of fixed width as it is tuned throughout the spectrum. One well-known analyzer of this type has a bandwidth of 5 Hz. If white noise with its flat spectrum were measured with a constant-bandwidth analyzer, another flat spectrum would result because the fixed bandwidth would measure a constant energy throughout the band shown in Fig. 5-18A.

Another very popular and convenient spectrum analyzer is the *constant percentage bandwidth* analyzer. In this instrument the bandwidth changes with frequency. An example of this is the one-third-octave analyzer, commonly used because its bandwidth follows reasonably well with the critical bandwidth of the human ear throughout the audible frequency range. At 100 Hz the bandwidth of the one-third-octave analyzer is only 23 Hz but at 10 kHz the bandwidth is 2,300 Hz. Obviously, it intercepts much greater noise energy in a one-third octave band centered at 10 kHz than one centered at 100 Hz. Measuring white noise with a constant-percentage analyzer would give an upward-sloping result with a slope of 3 dB/octave, as shown in Fig. 5-18B.

In audio-frequency measurements, the desired characteristic of many instruments, rooms, etc. is a flat response throughout the frequency range. Assume that the system to be measured has a characteristic almost flat with frequency. If this system is excited with white noise and measured with the very convenient constant-percentage analyzer, the result would have an upward slope of 3 dB/octave. It would be far more desirable if the measured result would be close to flat so that deviations from flatness would be very apparent. This can be accomplished by using a noise with a downward slope of 3 dB/octave. By passing white noise through a filter, such as that of Fig. 5-19, such a downward sloping excitation noise can be obtained. Such a noise, sloping downward at 3 dB/octave, is called *pink noise*. A close-to-flat system (amplifier, room) excited with this pink noise would yield a close-to-flat response, which would make deviations from flatness very obvious. For such reasons pink noise is here to stay.

Signal Distortion

Our discussion of the various signals encountered in audio is incomplete without at least an acknowledgment of what can happen

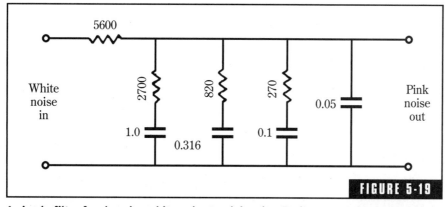

FIGURE 5-19

A simple filter for changing white noise to pink noise. It changes random noise of constant energy per Hz to pink noise of constant energy per octave. Pink noise is useful in acoustical measurements utilizing analyzers having passbands with bandwidth of a constant percentage of the center frequency. *General Radio Company*[1]

to the precious signal in passing through transducers, amplifiers, and various forms of signal processing gear. Here is an incomplete list:[8,9,10]

Bandwidth limitation   If the passband of an amplifier cuts lows or highs, the signal output is different from the input. If the scratch filter reduces record surface noise, the overall effect can be improved, but basically the signal itself is the poorer for it.

Nonuniform response   Peaks and valleys within the passband also alter the signal waveshape.

Distortions in time   If tape travels across the head at any other than the recording speed, the frequency components are shifted up or down in frequency. If there are slow or fast fluctuations in that speed, wow and flutter are introduced and the signal is degraded.

Phase distortion   Any phase shifts introduced upset the time relationship between signal components.

Dynamic distortion   A compressor or expander changing the original dynamic range of a signal is a form of distortion.

Crossover distortion   In class-B amplifiers, in which the output devices conduct for only half of the cycle, any discontinuities near zero output result in what is called *crossover* distortion.

Nonlinear distortion   If an amplifier is truly linear, there is a one-to-one relationship between input and output. Feedback helps to control nonlinear tendencies. The human ear is not linear. When a pure tone is impressed on the ear, harmonics can be heard. If two loud tones are presented simultaneously, sum and difference tones are generated in the ear itself; and these tones can be heard as can their harmonics. A cross-modulation test on an amplifier does essentially the same thing. If the amplifier (or the ear) were perfectly linear, no sum or difference tones or harmonics would be generated. The production within the component of frequency elements that were not present in the input signal is the result of nonlinear distortion.

Transient distortion   Strike a bell and it rings. Apply a steep wavefront signal to an amplifier and it might ring a bit too. For this reason, signals such as piano notes are difficult to reproduce. Tone burst test signals are an attempt to explore the transient response characteristics of equipment, as are square waves. Transient intermodulation (TIM) distortion, slew induced distortion, and other sophisticated measuring techniques have been devised to evaluate transient forms of distortion in systems.

Harmonic Distortion

The harmonic distortion method of evaluating the effects of circuit nonlinearities is probably the oldest and the most universally accepted method. It certainly is the easiest to understand. In this method the device under test is driven with a sine wave of high purity. If the signal encounters any nonlinearity, the output waveshape is changed, i.e., harmonic components appear that were not in the pure sine wave. A spectral analysis of the output signal is made to measure these harmonic distortion products. The most revealing method is to use a wave analyzer having a constant passband width of, say, 5 Hz, which can be swept through the audio spectrum. Figure 5-20 shows illustrated results of such a measurement. The wave analyzer is first tuned to the fundamental, f_0 = 1 kHz, and the level is set for a convenient 1.00 volt. The wave analyzer is then tuned to the 2 kHz region until the $2f_0$ second harmonic is found. The voltmeter, which is a part of the analyzer, reads 0.10 volt. The third harmonic at 3 kHz gives a reading of 0.30 volt, the fourth a reading of 0.05 volt and so on up the frequency scale. Beyond $6f_0$ = 6 kHz no measurable components were found after diligent search. The data are then assembled in Table 5-2.

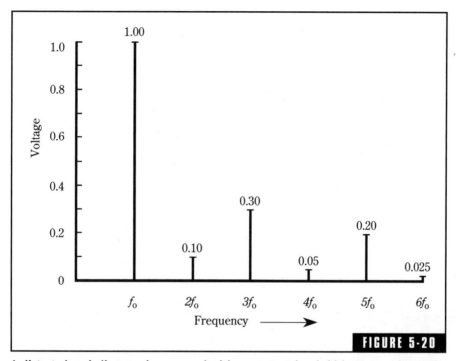

A distorted periodic wave is measured with a constant bandwidth wave analyzer. The fundamental, f_0, is set for some reference voltage, taken here as 1.00 volt. Tuning the wave analyzer to $2f_0$ the second harmonic amplitude is measured as 0.10 volt. The wave analyzer is tuned successively to $3f_0$, $4f_0$ and other harmonics yielding amplitudes of each harmonic as shown. The root-mean-square of the harmonic voltages is then compared to the 1.00 volt fundamental to find the total harmonic distortion expressed in percentage.

Table 5-2 Harmonic Distortion Products.

Harmonic	Volts	(Volts)^2
2nd Harmonic $2f_0$	0.10	0.01
3rd Harmonic $3f_0$	0.30	0.09
4th Harmonic $4f_0$	0.05	0.0025
5th Harmonic $5f_0$	0.20	0.04
6th Harmonic $6f_0$	0.025	0.000625
7th and higher	(negligible)	
		Sum 0.143125

Fundamental f_0 =1 kHz, 1.00 volt amplitude

The total harmonic distortion (THD) may then be found from the expression:

$$\text{THD} = \frac{\sqrt{(e_2)^2 + (e_3)^2 + (e_4)^2 \dots (e_n)^2}}{e_o} \times 100 \qquad (5\text{-}1)$$

where e_2, e_3, $e_4 \dots e_n$ = voltages of 2nd, 3rd, 4th, etc. harmonics

e_o = voltage of fundamental

In Table 5-2 the harmonic voltages have been squared and added together reducing Eq. 5-1 to:

$$\text{THD} = \frac{\sqrt{0.143125}}{1.00} \times 100$$

$$= 37.8\%$$

A total harmonic distortion of 37.8% is a very high distortion that would make any amplifier sound horrible on any type of signal, but the example has served our purpose in illustrating just what THD is and one method for obtaining it.

Wave analyzers are expensive, high-precision instruments that are rarely found in equipment service shops. A very simple adaptation of the THD method is, however, widely used. Consider Fig. 5-20 again. If the f_o fundamental were adjusted to some known value and then a notch filter were adjusted to f_o essentially eliminating it, only the harmonics would be left. Measuring these harmonics all lumped together with an RMS (root mean square) meter, accomplishes what was done in the square root portion of Eq. 5-1. Comparison of this RMS measured value of the harmonic components with that of the fundamental and expressing it as a percentage gives the total harmonic distortion.

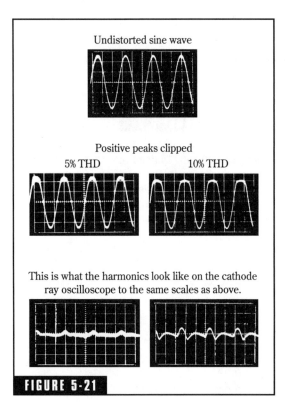

Undistorted sine wave

Positive peaks clipped

5% THD   10% THD

This is what the harmonics look like on the cathode ray oscilloscope to the same scales as above.

FIGURE 5-21

Cathode ray oscillograms show pure, undistorted sine wave, which is applied to the input of an amplifer that clips the positive peaks of the signal. The appearance of the clipped sine wave for 5% and 10% total harmonic distortion is shown. If the fundamental is rejected by a notch filter, the summed harmonics appear as shown.

In Fig. 5-21 an undistorted sine wave is sent through an amplifier, which clips positive peaks. On the left, the flattening of the positive peaks with 5% THD is evident, and shown below is what the combined total of all the harmonic products look like with the fundamental rejected. On the right is shown the effect of greater clipping to yield 10% THD. Figure 5-22 shows what happens when the sine wave passing through the amplifier is symmetrically clipped on both positive and negative peaks. The combined distortion products for symmetrical clipping have a somewhat different appearance, but they measure the same 5% and 10% THD.

In all of this exercise keep in mind that consumer-type power amplifiers commonly have specifications listing total harmonic distortion nearer 0.05% rather than 5% or 10%. In a series of double-blind subjective tests Clark found that 3% distortion was audible on different types of sounds.[11] With carefully selected material (such as a flute solo) detecting distortions down to 2% or 1% might be possible. A distortion of 1% with sine waves is readily audible.

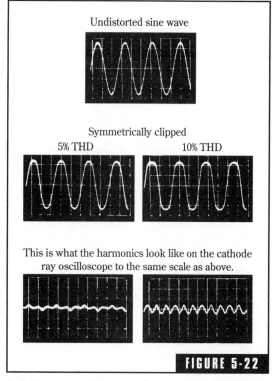

Undistorted sine wave

Symmetrically clipped

5% THD   10% THD

This is what the harmonics look like on the cathode ray oscilloscope to the same scale as above.

FIGURE 5-22

Cathode ray oscillograms show a pure, undistorted sine wave, which is applied to the input of an amplifer which clips both positive and negative peaks in a symmetrical fashion. The appearance of the clipped sine wave is shown for 5% and 10% total harmonic distortion. The appearance of the harmonics alone, with the fundamental filtered out, is also shown.

Endnotes

[1]Flanagan, J. L, *Voices of Men and Machines,* J. Acous. Soc. Am., 51, 5, Part 1, May 1972, 1375–1387.

[2]Kuttruff, Heinrich, *Room Acoustics,* London, Applied Science Publishers, Ltd., 1979, 16, 17.

[3]Hutchins, Carleen M., and Francis L. Fielding, *Acoustical Measurement of Violins,* Physics Today, July 1968, 35–41. Contains extensive bibliography.

[4]Fletcher, Harvey, *The Ear As a Measuring Instrument,* J. Audio Eng. Soc., 17, 5, October 1969, 532–534.

[5]Sivian, L.J., H.K. Dunn, and S.D. White, *Absolute Amplitudes and Spectra of Certain Musical Instruments and Orchestras,* J. Acous. Soc. Am., 2, 3, January 1931, 330–371.

[6]Fiedler, Louis D., *Dynamic-Range Requirements for Subjectively Noise-Free Reproduction of Music,* J. Aud. Eng. Soc., 30, 7/8, 1982, 504–511.

[7]Peterson, Arnold P.G. and Ervin E. Gross, Jr., *Handbook of Noise Measurements,* Concord, Mass., GenRad. 7th Ed., 1974.

[8]Cabot, Richard C., *A Comparison of Nonlinear Distortion Measurement Methods,* Presented at the 66th convention of the Audio Eng. Soc., Los Angeles, 1980, Preprint 1638.

[9]Buff, Paul C., *Perceiving Audio Noise and Distortion,* Recording Eng./Prod., 10, 3, June 1979, 84.

[10]Jung, Walter G., Mark L. Stephens, and Craig C. Todd, *An Overview of SID and TIM,* Audio,
Part I—63, 6 (June 1979), 59–72
Part II—63, 7 (July 1979), 38–47

[11]Clark, David, *High-Resolution Subjective Testing Using a Double-Blind Comparator,* J. Audio Eng. Soc., 30, 5 (May 1982), 330–338.

Analog and Digital Signal Processing

In the early days of sound recording, signal storage was a major problem. The final recording was laid down directly, without benefit of stop-and-go recording of portions that could be patched together later. This had the advantage of minimizing the number of recording generations, but the format was a stringent one with little latitude for artistic enhancement.

With the introduction of high-quality signal storage (magnetic tape or digital memory) many creative decisions, normally reserved for the recording session, were moved to the mix-down session. This opens up the opportunity of making quality enhancements in the mix-down session, long after the recording session is finished and forgotten. These quality enhancements are often made with filters of one kind or another. Is there a short traffic rumble on take 6 of the recording? A high-pass filter might cure it. Did that narrator's dentures contribute an occasional high-frequency hiss? Run it through the "de-esser." How about that congenital 3 dB sag in the response at 4 kHz? Easy, add a 3 dB peak of appropriate width with the parametric equalizer, and so on.

The introduction of integrated circuits in the 1960s made signal processing equipment lighter, more compact, and less expensive. The coming of the digital revolution made it possible to routinely accomplish sophisticated signal processing tasks, heretofore impractical or impossible. This chapter is a very brief overview of both analog and digital sound processing principles and practice, especially the principles.

Resonance

The man on the stage is doing the old trick of breaking the wine glass with sound. Instead of doing it with the diva's voice, he holds the goblet in front of the loudspeaker, which shatters it as a high-intensity tone is emitted. The secret lies in the brief preparation made before the audience was assembled. At that time he placed a small coin in the goblet and held it in front of the loudspeaker as the frequency of the sine generator was varied at a low level. He carefully adjusted the generator until the frequency was found at which the coin danced wildly in the glass. During the demonstration no tuning was necessary, a blast of sound at this predetermined frequency easily shattered the glass.

The wild dancing of the coin in the glass in the preliminary adjustment indicated that the excitation frequency from the loudspeaker was adjusted to the *natural frequency* or *resonance* of the goblet. At that frequency of resonance a modest excitation resulted in very great vibration of the glass, exceeding its breaking point. As shown in Fig. 6-1, the amplitude of the vibration of the glass changes as the frequency of excitation is varied, going through a peak response at the frequency of resonance, f_o.

Such resonance effects appear in a wide variety of systems: the interaction of mass and stiffness of a mechanical system, such as a tuning fork, or the acoustical resonance of the air in a bottle, as the mass of the air in the neck of the bottle reacts with the springiness of the air entrapped in the body of the bottle. See Helmholtz resonators, Chap. 9.

Resonance effects are also dominant in electronic circuits as the inertia effect of an inductance reacts with the storage

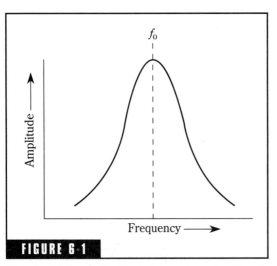

FIGURE 6-1

The amplitude of vibration of any resonant system is maximum at the *natural frequency* or *resonant frequency* (*f*) and is less at frequencies below and above that frequency.

effect of a capacitance. Figure 6-2 shows the symbols for *inductance* (*L*), commonly a coil of wire, and *capacitance* (*C*), commonly made of sheets of conducting material separated by nonconducting sheets. Energy can be stored in the magnetic field of an inductance as well as in the electrical charges on the plates of a capacitance. The interchange of energy between two such storage systems can result in a resonance effect. Perhaps, the simplest example of this is a weight on a spring.

Figure 6-3 shows two forms in which an inductance and a capacitance can exhibit resonance. Let us assume that an alternating current of constant amplitude, but varying frequency is flowing in the *parallel resonant* circuit of Fig. 6-3A. As

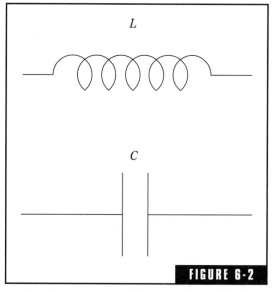

FIGURE 6-2

Symbols for inductance (*L*) and capitance (*C*).

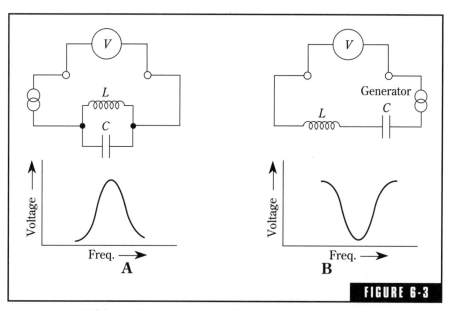

FIGURE 6-3

A comparison of (A) parallel resonance and (B) series resonance. For a constant alternating current flowing, the voltage across the parallel resonant circuit peaks at the resonance frequency while that of the series resonant circuit is a minimum.

the frequency is varied, the voltage at the terminals reaches a maximum at the natural frequency of the LC system, falling off at lower and higher frequencies. In this way the typical resonance curve shape is developed. Another way of saying this is that the parallel resonant circuit exhibits maximum impedance (opposition to the flow of current) at resonance.

Figure 6-3B illustrates the *series resonant* arrangement of an inductance *L* and a capacitance *C*. As the alternating current of constant magnitude and varying frequency flows in the circuit, the voltage at the terminals describes an inverted resonance curve in which the voltage is minimum at the natural frequency and rising at both lower and higher frequencies. It can also be said that the series resonant circuit presents minimum impedance at the frequency of resonance.

Filters

The common forms of filters are the *low-pass filter*, the *high-pass filter*, the *band-pass filter*, and the *band-reject filter* as illustrated in Fig. 6-4. Figure 6-5 shows how inductors and capacitors may be arranged in numerous ways to form very simple high- and low-pass filters. Filters of Figure 6-5C will have much sharper cut-offs than the simpler ones in (A) and (B).

There are many other highly specialized filters with specific and unusual features. With such filters, a wideband signal such as speech or music can be altered at will.

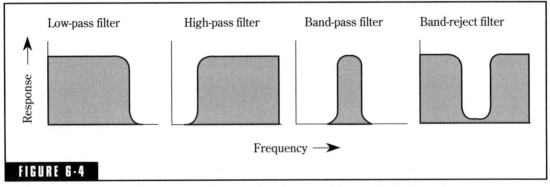

FIGURE 6-4

Basic response shapes for the low-pass, high-pass, band-pass, and band-reject filters.

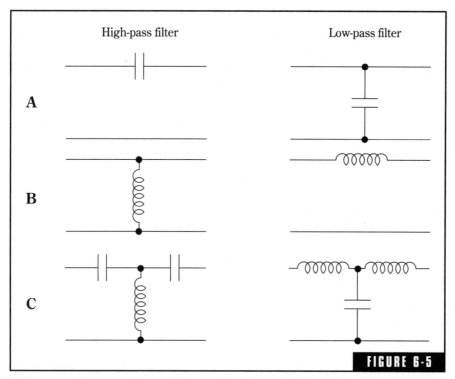

High-pass filter   Low-pass filter

A

B

C

FIGURE 6-5

High-pass and low-pass filters of the simplest form. The filters in (C) will have sharper cut-off than the others.

Adjustable filters can be readily shifted to any frequency within their design band. One type is the constant bandwidth filter which offers the same bandwidth at any frequency. For example, a spectrum analyzer may have a 5-Hz bandwidth whether it is tuned to 100 Hz or 10,000 Hz, or any other frequency within its operating band. An even more widely used adjustable filter offers a pass band-width that is a constant percentage of the frequency to which it is tuned. The ⅓-octave filter is such a device. If it is tuned to 125 Hz the ⅓-octave bandwidth is 112 to 141 Hz. If it is tuned to 8,000 Hz the ⅓-octave bandwidth is 7,079 Hz to 8,913 Hz. The bandwidth is about 23% of the frequency to which it is tuned in either case.

Active Filters

Active filters depend on integrated circuits for their operation. An integrated circuit can have many hundreds of components in a small enclosure. Their fabrication depends on growing transistors and resistors on a semiconductor wafer and interconnecting the components by an evaporated metal pattern. Great circuit complexity can be compressed into unbelievably small space in this way.

A low-pass filter assembled from inductors and capacitors in the old-fashioned way is shown in Fig. 6-6A. Another low-pass filter based in an integrated circuit is shown in Fig. 6-6B. The four small resistors and two small capacitors plus the integrated circuit illustrate the space-saving advantage of the active filter.

It is interesting that the feedback capacitor C_1 in Fig. 6-6B has the electrical effect of an inductance.

Analog vs. Digital Filters

Filters can be constructed in analog or digital form. All the filters discussed to this point have been of the analog type and applied widely in equalizers. By adjusting the values of the resistors, inductors, and capacitors any type of analog filter can be constructed to achieve almost any frequency and impedance matching characteristic desired.

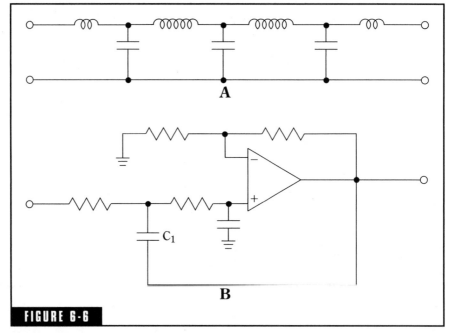

FIGURE 6-6

Two low-pass filters, (A) traditional analog type, (B) an active filter utilizing an integrated circuit.

Digitization

A digital filter contains no such physical components as inductors and capacitors. It is basically a computer program that operates on a sample of the signal. This process is described in Fig. 6-7. The incoming analog signal is represented in Fig. 6-7A. Through a multiplication or modulation process, the analog signal of Fig. 6-7A is combined with the sam-

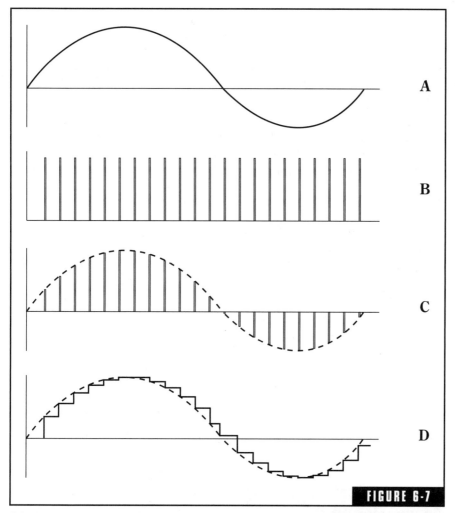

FIGURE 6-7

(A) the analog signal, (B) digitizing pulses, (C) the digitized analog signal resulting from the modulation of (A) with (B). Application of a sample-and-hold circuit to (D) which completes the quantization process reducing the analog samples to discrete values suitable for storage in memory.

pling pulses of Fig. 6-7B. These pulses, in effect, break down the analog signal into a series of very brief samples having amplitudes equal to the instantaneous value of the signal amplitudes, as shown in Fig. 6-7C. This process is called *digitization*. With no loss of information, energy between the sampling points is discarded. The sampling rate must be at least twice the highest frequency of interest. If a sampling rate less than this amount is used, spurious signals are generated.

Quantization

It is now necessary to convert the samples of Fig. 6-7C to discrete values that can be stored in a computer. This is done by a "sample-and-hold" circuit so that the amplitude of each digitized pulse is converted into discrete values suitable for computer storage. The sample-and-hold circuit forces the amplitude of the sample to have a constant value throughout the sample period. This process is essentially a stepping backward through a digitized signal, sample by sample, subtracting from each sample some large proportion of the sample before it. The resulting samples are thus mainly changes in the signal sample.

The closer the spacing of the digitized samples, the more accurately the analog signal is represented. However, a restraining influence on increasing the number of samples is that more computer memory is required to store the data. The calculations required are intensive in multiplication and accumulation operations.

Digital Filters

Digital filters can be made without benefit of inductors or capacitors. A typical digital filter of the so-called FIR type is shown in Fig. 6-8. The analog signal is applied to the input on the left. The analog-to-digital (A/D) converter digitizes and quantisizes the analog signal. An oscillator (clock) determines the number of digitizing pulses per second and controls all timing of the device. The type of filter is determined by the program in the read-only memory.

Application of Digital Signal Processing (DSP)

Digital signal processing has been successfully applied in various ways, including[1]:

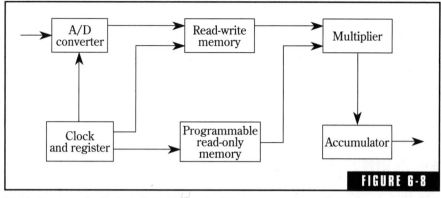

FIGURE 6-8

Block diagram of a typical digital filter, in this case one of the finite impulse response (FIR) type. The type of filter is determined by the read-only program.

- Mixing two signals (convolution)
- Comparing two signals (correlation)
- Changing ac signals to dc (rectification)
- Amplifying signals
- Acting as a transformer
- Spectral analysis
- Speech processing, recognition, etc.
- Noise cancellation
- Music synthesis and processing.

Specific tasks ideally suited to digital signal processing include:

- Subsonic filters
- Ultrasonic filters
- Limiters
- Compressors
- Expanders
- Companders
- Noise gates
- Bass correction

- Noise reduction

- Image enhancement

- Stereo synthesizer

This very brief treatment of digital signal processing has skipped over such vital aspects as aliasing, the multiplier/accumulator, sampling rates, quantization levels, etc.

Application of DSP to Room Equalization

The application of digital signal processing to the loudspeaker-room-listener problem is currently being pursued vigorously. Huge errors are common in both the room and the loudspeakers. The problems of the listener are primarily the great differences in sensitivity between listeners.

The basic idea is to measure the frequency and phase response of the speaker/room combination and apply an equalizer that perfectly compensates for the defects. This is a very complicated operation, but theoretically possible through the great potential of digital signal processing. One problem to be solved is how to generalize the solution based on data picked up at one point in the room. There are many other problems but the next decade should bring major advances in this field.

Endnote

[1]Burkhard, Mahlon, *Filters and Equalizers*, Chapter 20, Handbook For Sound Engineers, Glen Ballou, Ed., Carmel, IN, Howard W. Sams Co., 2nd Edition (1987).

Reverberation

Pressing the gas pedal of an automobile results in acceleration of the vehicle to a certain speed. If the road is smooth and level, this speed will remain constant. With this accelerator setting the engine produces just enough torque to overcome all the frictional losses and a balanced (steady-state) condition results.

So it is with sound in a room. When the switch is closed, a loudspeaker arranged to emit random noise into a room will produce a sound that quickly builds up to a certain level. This is the steady-state or equilibrium point at which the sound energy radiated from the loudspeaker is just enough to supply all the losses in the air and at the boundaries of the room. A greater sound energy radiated from the loudspeaker will result in a higher equilibrium level, less power to the loudspeaker will result in a lower equilibrium level.

When the loudspeaker switch is opened, it takes a finite length of time for the sound level in the room to decay to inaudibility. This "hanging-on" of the sound in a room after the exciting signal has been removed is called *reverberation* and it has a very important bearing on the acoustic quality of the room.

In England, a symphony orchestra was recorded as it played in a large anechoic (echo-free) chamber. This music, recorded with almost no reverberation for research purposes, is of very poor quality for normal listening. This music is even thinner, weaker, and less resonant than outdoor recordings of symphonic music, which are noted for their flatness. Clearly, symphonic and other music requires reverberation and the amount of it is a lively technical topic to be covered.

At one time reverberation was considered the single most important characteristic of an enclosed space for speech or music. It has shriveled in importance under the influence of important research on the acoustics of enclosures. At the present time it is only one of several measurable parameters that define the quality of a space.

Reverberation and Normal Modes

The natural resonances of a small room are revealed in the various vibrational modes as described in chapter 15. It is necessary to anticipate this chapter a bit to understand the relationship of these natural room resonance frequencies and the reverberation of the room.

Our introduction to the measurement of reverberation will be through a method wholly unsuited to practical use, but as an investigative tool, it reveals some very important factors that focus attention on the normal modes of rooms. Historically, in the Broadcasting House in New Delhi, India, is (or was) a Studio 10 used for news broadcasts. Measurements of reverberation time in this studio were reported by Beranek[1] and later analyzed by Schultz. The first set of measurements were made in the completely bare, untreated room. Knowing construction in India, it can be safely assumed that concrete and ceramic tile dominated the room surfaces. The measurements were made with sine-wave signals, and great patience and care were exercised to obtain the detailed results.

Starting with the oscillator set to about 20 Hz below the first axial mode, the acoustics of the room do not load the loudspeaker and a relatively weak sound is produced with the amplifier gain turned up full (even assuming the use of a good subwoofer). As the oscillator frequency is adjusted upward, however, the sound becomes very loud as the 1, 0, 0 mode (24.18 Hz) is energized (Fig. 7-1). Slowly adjusting the oscillator upward we go through a weak valley but at the frequency of the 0, 1, 0 mode (35.27 Hz) there is high level sound once more. Similar peaks are found at the 1, 1, 0 tangential mode (42.76 Hz), the 2, 0, 0 axial mode (48.37 Hz), and the 0, 0, 1 axial mode (56.43 Hz).

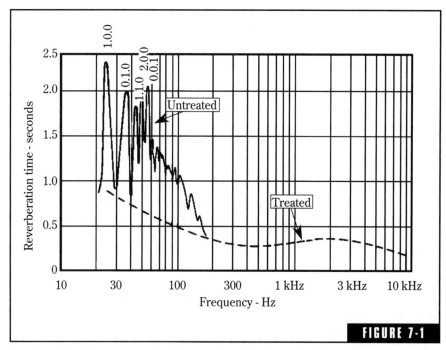

Reverberation time measured with pure sine signals at low frequencies reveals slow sound decay (long reverberation time) at the modal frequencies. These peaks apply only to specific modes and are not representative of the room as a whole. High modal density, resulting in uniformity of distribution of sound energy and randomizing of directions of propagation, is necessary for reverberation equations to apply. (Beranek,[1] and Schultz.[2])

Now that the loudness of peaks and valleys have been explored, let's examine the decay of sound. Exciting the 1, 0, 0 mode at 24.18 Hz, the decay is measured as the source is interrupted and we get a long reverberation time of 2.3 seconds. Similar slow decays are observed at 35.27 Hz, 42.76, 48.37, and 56.43 Hz with faster decays (shorter reverberation times) in between. The decays at the modal frequencies are decay rates characteristic of individual modes, not of the room as a whole.

Long reverberation time implies low absorbance, and short reverberation time implies high absorbance. It is difficult to believe that the sound absorbing qualities of the walls, floor, and ceiling vary this much within a frequency range of a few hertz. For the 1, 0, 0 mode, only the absorbance of the two ends of the room comes into play; the four other surfaces are not involved at all. For the 0, 0, 1 mode, only the floor and ceiling are involved. All we have done in this low-frequency range is to measure the decay rate of individual modes, definitely not the average condition of the room.

We see now why there is that big question mark over applying the concept of reverberation time to small rooms having dimensions comparable to the wavelength of sound. Schultz states that reverberation time is a statistical concept "in which much of the mathematically awkward details are averaged out."[2] In small rooms these details are not averaged out.

The reverberation time formulas of Sabine, Eyring, and others are based on the assumption of an enclosed space in which there is highly uniform distribution of sound energy and random direction of propagation of the sound. At the low-frequency points of Fig. 7-1, energy is distributed very unevenly and direction of propagation is far from random. After the room was treated, reverberation time measurements followed the broken line, but statistical randomness still does not prevail below 200 Hz even though modal frequencies are brought under some measure of control.

Growth of Sound in a Room

Referring to Fig. 7-2A, let us consider a source S and a human receiver H in a room. As source S is suddenly energized, sound travels outward from S in all directions. Sound travels a direct path to H and we shall consider zero time (see Fig. 7-2B) as that time at which the direct sound reaches the ears of listener H. The sound pressure at H instantly jumps to a value less than that which left S due to spherical divergence and small losses in the air. The sound pressure at H stays at this value until reflection R_1 arrives and then suddenly jumps to the $D + R_1$ value. Shortly thereafter R_2 arrives, causing the sound pressure to increase a bit more. The arrival of each successive reflected component causes the level of sound to increase stepwise. These additions are, in reality, vector additions involving both magnitude and phase, but we are keeping things simple for the purposes of illustration.

Sound pressure at receiver H grows step by step as one reflected component after another adds to the direct component. The reason the sound pressure at H does not instantly go to its final value is that sound travels by paths of varying length. Although 1,130 ft/sec, the

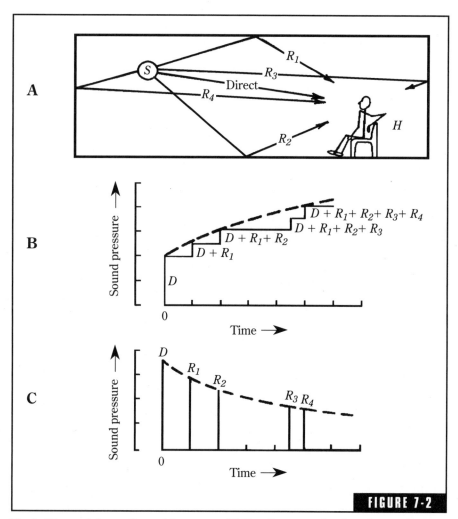

The buildup and decay of sound in a room. (A) The direct sound arrives first at time = 0, reflected components arriving later. (B) The sound pressure at *H* builds up stepwise. (C) The sound decays exponentially after the source ceases.

speed of sound, is about the muzzle velocity of a .22 caliber rifle, reflected components are delayed an amount proportional to the difference in distance between the reflected path and the direct path. The buildup of sound in a room is thus relatively slow due to finite transit time.

The ultimate level of sound in the room is determined by the energy going into the source *S*. The energy it radiates (less than the input by the amount of loss in S) is dissipated as heat in wall reflections and other boundary losses, along with a small amount in the air itself. With a constant input to S, the sound-pressure level builds up as in Fig. 7-2B to a steady-state equilibrium, even as an automobile traveling steadily at 50 miles per hour with the accelerator in a given position. Pushing down on the accelerator pedal increases the energy to the engine, and the automobile stabilizes at a new equilibrium point at which the many frictional losses are just supplied. Increasing the input to the source S means a new equilibrium of room-sound-pressure level as room losses are just supplied.

Decay of Sound in a Room

After opening the switch feeding source S, the room is momentarily still filled with sound, but stability is destroyed because the losses are no longer supplied with energy from S. Rays of sound, however, are caught in the act of darting about the room with their support cut off.

What is the fate of the ceiling reflected component R_1? As S is cut off, R_1 is on its way to the ceiling. It loses energy at the celling bounce and heads toward H. After passing H it hits the rear wall, then the floor, the ceiling, the front wall, the floor again, and so on . . . losing energy at each reflection and spreading out all the time. Soon it is so weak it can be considered dead. The same thing happens to R_2, R_3, R_4, and a multitude of others not shown. Figure 7-2C shows the exponential decrease of the first bounce components, which would also apply to the wall reflections not shown and to the many multiple bounce components. The sound in the room thus dies away, but it takes a finite time to do so because of the speed of sound, losses at reflections, the damping effect of the air, and divergence.

Idealized Growth and Decay of Sound

From the view of geometrical (ray) acoustics, the decay of sound in a room, as well as its growth, is a stepwise phenomenon. However, in the practical world, the great number of small steps involved result in smooth growth and decay of sound. In Fig. 7-3A, the idealized forms of growth and decay of sound in a room are shown. Here the sound pressure is on a linear scale and is plotted against time. Figure

7-3B is the same thing except that the vertical scale is plotted in decibels, i.e., to a logarithmic scale.

During the growth of sound in the room, power is being fed to the sound source. During decay, the power to the source is cut off, hence the difference in the shapes of the growth and decay curves. The decay of Fig. 7-3B is a straight line in this idealized form, and this becomes the basis for measuring the reverberation time of an enclosure.

Reverberation Time

Reverberation time is defined as that time required for the sound in a room to decay 60 dB. This represents a change in sound intensity or sound power of 1 million ($10 \log 1{,}000{,}000 = 60$ dB), or a change of sound pressure or sound-pressure level of 1,000 ($20 \log 1{,}000 = 60$ dB). In very rough human terms, it is the time required for a sound that is very loud to decay to

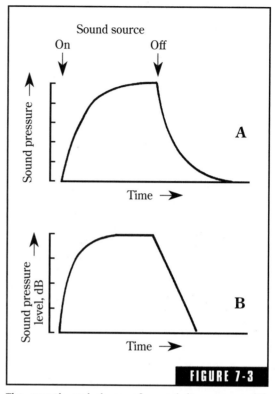

FIGURE 7-3

The growth and decay of sound in a room. (A) Vertical scale in linear sound pressure units. (B) The vertical scale in logarithmic units (decibels).

inaudibility. W. C. Sabine, the Harvard pioneer in acoustics who introduced this concept, used a portable wind chest and organ pipes as a sound source, a stopwatch, and a pair of keen ears to measure the time from the interruption of the source to inaudibility. Today we have better technical measuring facilities, but we can only refine our understanding of the basic concept Sabine gave us.

This approach to measuring reverberation time is illustrated in Fig. 7-4A. Using a recording device that gives us a hardcopy trace of the decay, it is a simple step to measuring the time required for the 60-dB decay. At least it is simple in theory. Many problems are encountered in practice. For example, obtaining a nice, straight decay spanning 60 dB or more as in Fig. 7-4A is a very difficult practical problem. Background noise, an inescapable fact of life, suggests that a higher source level is needed. This may occur if the background noise

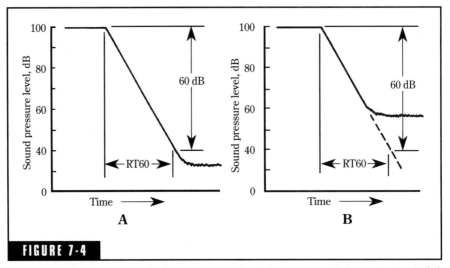

FIGURE 7-4

The length of the decay dependent on strength of the source and the noise level. (A) Rarely do practical circumstances allow a full 60-dB decay. (B) The slope of the limited decay is extrapolated to determine the reverberation time.

level is 30 dB (as in Fig. 7-4A), because source levels of 100 dB are quite attainable. If, however, the noise level is near 60 dB as shown in Fig. 7-4B, a source level greater than 120 dB is required. If a 100-watt amplifier driving a certain loudspeaker gives a sound-pressure level of 100 dB at the required distance, doubling the power of the source increases the sound-pressure level only 3 dB, hence 200 watts gives 103 dB, 400 watts gives 106 dB, 800 watts gives 109 dB, etc. The limitations of size and cost can set a ceiling on the maximum levels in a practical case.

The situation of Fig. 7-4B is the one commonly encountered, a usable trace less than the desired 60 dB. The solution is simply to extrapolate the straight portion of the decay.

Actually, it is important to strive for the greatest decay range possible because we are vitally interested in both ends of the decay. It has been demonstrated that in evaluating the quality of speech or music, the first 20 or 30 dB of decay is the most important to the human ear. On the other hand, the significance of double-slope phenomena is revealed only near the end of the decay. In practice, the highest level of sound source reasonably attainable is used, and filters are often incorporated to improve the signal-to-noise ratio.

Measuring Reverberation Time

There are many approaches to measuring the reverberation time of a room, and many instant-readout devices are on the market to serve those who have only a casual interest in reverberation effects. For example, sound contractors need to know the approximate reverberation time of the spaces in which they are to install a sound-reinforcement system, and measuring it avoids the tedious process of calculating it. The measurements can also be more accurate because of uncertainty in absorption coefficients. Acoustical consultants (at least the old-fashioned ones) called upon to correct a problem space or verify a carefully designed and newly constructed space, generally lean toward the method of recording many sound decays. These sound decays give a wealth of detail meaningful to the practiced eye.

Impulse Sound Sources

The sound sources used to excite the enclosure must have enough energy throughout the spectrum to ensure decays sufficiently above the noise to give the required accuracy. Both impulse sources and those giving a steady-state output are used. For large spaces, even small cannons have been used as impulse sources to provide adequate energy, especially in the lower frequencies. More common impulse sources are powerful electrical spark discharges and pistols firing blanks. Even pricked balloons have been used.

The impulse decays of Fig. 7-5 for a small studio have been included to show their appearance. The sound source was a Japanese air pistol that ruptures paper discs. This pistol was originally intended as an athletic starter pistol but failed to find acceptance in that area. As reported by Sony engineers,[3] the peak sound-pressure level at 1 meter distance is 144 dB, and the duration of the major pulse is less than 1 millisecond. It is ideal for recording echograms, in fact, the decays of Fig. 7-5 were made from impulses recorded for that very purpose.

In Fig. 7-5, the straight, upward, traveling part on the left is the same slope for all decays because it is a result of machine limitation (writing speed 500 mm/sec). The useful measure of reverberation is the downward traveling, more irregular slope on the right side. This slope yields a reverberation time after the manner of Fig. 7-4. Notice

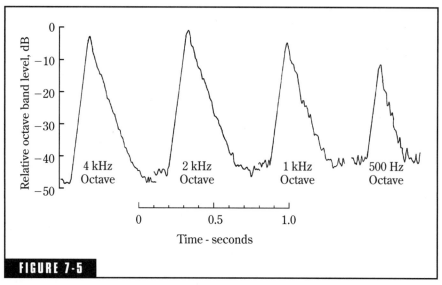

FIGURE 7-5

Reverberatory decays produced by impulse excitation of a small studio. The upgoing left side of each trace is recording-machine limited; the downgoing right side is the reverberatory decay.

that the octave-band noise level is higher for the lower frequency bands. The impulse barely poked its head above noise for the 250 Hz and lower octaves. This is a major limitation of the method unless the heavy artillery is rolled out.

Steady-State Sources

As stated, Sabine used a wind chest and organ pipes. Sine-wave sources providing energy at a single frequency give highly irregular decays that are difficult to analyze. Warbling a tone, which spreads its energy over a narrow band, is an improvement over the fixed tone, but random noise sources have essentially taken over. Bands of random noise give a steady and dependable indication of the average acoustical effects taking place within that particular slice of the spectrum. Octave and $1/3$-octave bands of random noise (white or pink) are most commonly used.

Equipment

The equipment layout of Fig. 7-6 to obtain the following reverberation decays is quite typical. A wideband pink-noise signal is amplified

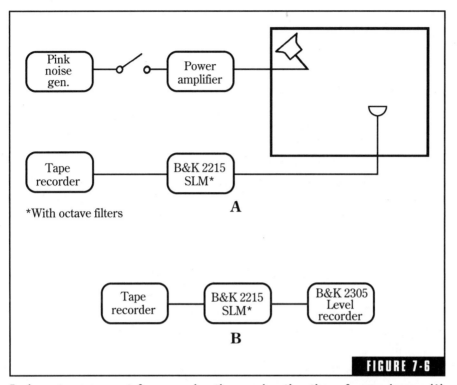

Equipment arrangement for measuring the reverberation time of an enclosure. (A) Recording decays on tape on location. (B) Later recording decays for analysis.

and used to drive a rugged loudspeaker. A switch for interrupting the noise excitation is provided. By aiming the loudspeaker into a corner of the room (especially in smaller rooms), all resonant modes are excited, because all modes terminate in the corners (Chap. 15).

A nondirectional microphone is positioned on a tripod, usually at ear height for a listening room, or microphone height for a room used for recording. The smaller the microphone, the less its directional effects. Some of the larger microphones (e.g., 1-inch-diameter diaphragms) can be fitted with random incidence correctors, but using a smaller microphone (e.g., ½-inch-diameter diaphragm) is considered best for essentially uniform sensitivity to sound arriving from all angles. In Fig. 7-6, the microphone is a high-quality condenser microphone, part of the Brüel & Kjaer 2215 sound-level meter, but separated from it by an extension cable. This provides an excellent preamplifier, built-in octave filters, a calibrated system, and a line-level output signal for the tape recorder.

Measurement Procedure

Every time the switch is closed the room is filled with a very loud wideband pink noise "sh-h-h-h" sound. This is usually loud enough to require the use of ear protectors for everybody in the room. Opening the switch, the sound in the room decays. The microphone, at its selected position, picks up this decay, which is recorded on magnetic tape for later analysis and study.

Signal-to-noise ratio determines the length of the reverberatory decay available for study. As mentioned previously, it is rarely possible to realize the entire 60-dB decay involved in the definition of RT60, nor is it necessary. It is quite possible, however, to get 45- to 50-dB decays with the equipment shown in Fig. 7-6 by the simple expedient of double filtering. For example, the octave filter centered on 500 Hz in the sound level meter is used both in recording and in later playback for analysis.

The analysis procedure outlined in the lower part of Fig. 7-6 uses the same magnetic recorder and B&K 2215 sound-level meter, with the addition of a B&K 2305 graphic-level recorder. The line output of the tape recorder is connected to the front end of the sound-level meter circuit through a 40-dB attenuating pad. To do this, the microphone of the sound-level meter is removed and a special fitting is screwed in its place. The output of the sound-level meter is connected directly to the graphic-level recorder input, completing the equipment interconnection. The appropriate octave filters are switched in as the played-back decay is recorded on the level recorder. The paper drive provides for spreading out the time dimension at adjustable rates. The graphic-level recorder offers a 50-dB recording range for the tracing pen on the paper.

Analysis of Decay Traces

An octave slice of pink noise viewed on a cathode-ray oscilloscope shows a trace that looks very much like a sine wave except that it is constantly shifting in amplitude and phase, which was the definition of random noise in Chap. 5. This characteristic of random noise has its effect on the shape of the reverberatory decay trace. Consider what this constantly shifting random noise signal does to the normal modes of a room (Chap. 15). When the axial, tangential, and oblique resonant modes are considered, they are quite close together on the frequency scale. The number of modes included within an octave band centered on 63 Hz in the specific case to be elaborated later is as follows: 4 axial, 6 tangential, and 2 oblique modes between the −3-dB points. These are graphically shown in Fig. 7-7 in which the taller lines represent the more potent axial modes, the intermediate height the tangential modes, and the shorter lines the oblique modes.

As the switch of Fig. 7-6 is closed, the high-level random noise from the loudspeaker energizes the various modes of the room, exciting mode A, and an instant later exciting mode B. While the shift is being made in the direction of mode B, mode A begins to decay. Before it decays very far, however, the random-noise instantaneous frequency is once more back on A, giving it another boost. All the modes of the room are in constant agitation, alternating between high and somewhat lower levels, as they start to decay in between kicks from the loudspeaker.

At what point will this erratic dance of the modes be as the switch is opened to begin the decay? It is strictly a random situation, but it can be said with confidence that each time the switch is opened for five successive decays, the modal excitation pattern will be somewhat different. The 12 modes in the 63-Hz octave will all be highly energized, but each to a somewhat different level the instant the switch is opened.

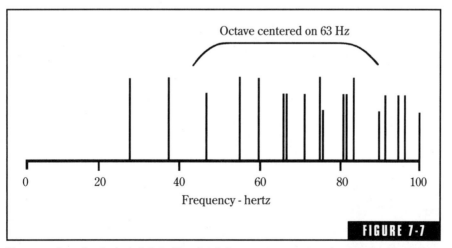

FIGURE 7-7

The normal modes included (−3 dB points) in an octave centered on 63 Hz. The tallest lines axial modes, the intermediate length tangential modes, and the shortest ones oblique modes.

Mode Decay Variations

To make this discussion more meaningful, real-life measurements in a real room are discussed. The room is a rectangular studio for voice recording having the dimensions 20'6" × 15'0" × 9'6", with a volume of 2,921 cubic feet. The measuring equipment is exactly that outlined in Fig. 7-6, and the technique is that described above. Four successive 63-Hz octave decays traced directly from the graphic-level recorder paper are shown in Fig. 7-8A. These traces are not identical, and any differences must be attributed to the random nature of the noise signal because everything else was held constant. The fluctuations in the decays result from beats between closely spaced modes. Because the excitation level of the modes is constantly shifting, the form and degree of the beat pattern shifts from one decay to another depending on

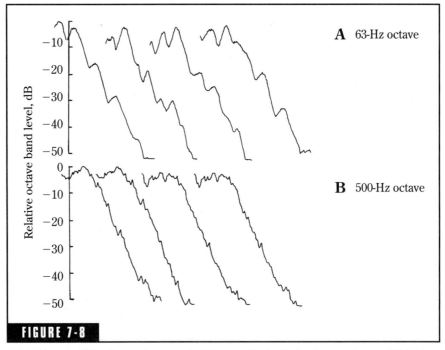

FIGURE 7-8

Actual decays of random noise recorded in a small studio having a volume of 2,921 cu ft; (A) Four successive 63-Hz octave decays recorded under identical conditions. (B) Four successive 500-Hz octave decays also recorded under identical conditions. The differences noted result from the differences in random-noise excitation the instant the switch is opened to start the decay.

where the random excitation happens to be the instant the switch is opened. Even though there is a family resemblance between the four decays, fitting a straight line to evaluate the reverberation time of each can be affected by the beat pattern. For this reason, it is good practice to record five decays for each octave for each microphone position of a room. With eight octaves (63 Hz–8 kHz), five decays per octave, and three microphone positions, this means 120 separate decays to fit and figure for each room, which is laborious. This approach is one way to get a good, statistically significant view of the variation with frequency. A hand-held reverberation time measuring device could accomplish this with less work, but it would not give hard-copy detail of the shape of each decay. There is much information in each decay, and acoustical flaws can often be identified from aberrant decay shapes.

Four decays at 500 Hz are also shown in Fig. 7-8B for the same room and the same microphone position. The 500-Hz octave (354–707 Hz) embraces about 2,500 room modes. With such a high mode density, the 500-Hz octave decay is much smoother than the 63-Hz octave with only a dozen. Even so, the irregularities for the 500-Hz decay of Fig. 7-8B result from the same cause. Remembering that some modes die away faster than others, the decays in Fig. 7-8 for both octaves are composites of all modal decays included.

Writing Speed

The B&K 2305 graphic-level recorder has a widely adjustable writing speed. A sluggish pen response is useful when fast fluctuations need to be ironed out. When detail is desired, faster writing speeds are required. A too slow writing speed can affect the rate of decay as it smooths out the trace, as will be examined.

In Fig. 7-9, the same 63-Hz decay is recorded with five different pen response speeds ranging from 200 to 1,000 mm/sec. The instrument-limited decay for each is indicated by the solid straight lines. A writing speed of 200 mm/sec smooths the fluctuations very well. The decay detail increases as the writing speed is increased, suggesting that a cathode-ray oscilloscope tracing of the decay would show even more modal interference effects during the decay.

The big question is: Does writing speed affect the decay slope from which we read the reverberation time values? Obviously, an extremely slow pen response would record the machine's decay characteristic

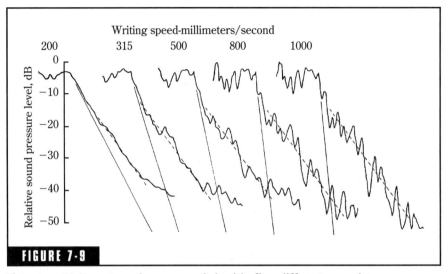

FIGURE 7-9

The same 63-Hz octave decays recorded with five different recorder-pen-response speeds. The solid straight lines indicate instrument-limited decay. The broken lines all have the same slope.

rather than that of the room. For every writing speed and paper speed setting, there is a minimum reverberation time that can be measured. The broken lines drawn through the decays all have the same slope. In Fig. 7-9, it would appear that this particular decay is measured equally well by any of the five traces, although the more detail, the more uncertainty in fitting a straight line. Writing speed is just one of the several adjustments that must be carefully monitored to ensure that important information is not obscured or that errors are not introduced.

Frequency Effect

Typical decays for octave bands of noise from 63 Hz to 8 kHz are included in Fig. 7-10. The greatest fluctuations are in the two lowest bands, the least in the two highest. This is what we would expect from the knowledge that the higher the octave band, the greater the number of normal modes included, and the greater the statistical smoothing. We should not necessarily expect the same decay rate because reverberation time is different for different frequencies. In the particular voice studio case of Fig. 7-10, a uniform reverberation time with frequency was the design goal, which was approximated in practice.

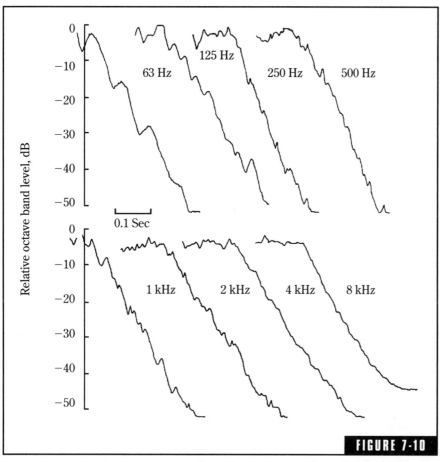

Relative octave band level, dB

63 Hz 125 Hz 250 Hz 500 Hz

0.1 Sec

1 kHz 2 kHz 4 kHz 8 kHz

FIGURE 7-10

Decay of octave bands of noise in a small studio of volume 2,921 cu ft. Fluctuations due to modal interference are greatest for low-frequency octaves containing fewer modes.

Reverberation Time Variation with Position

There is enough variation of reverberation time from one position to another in most rooms to justify taking measurements at several positions. The average then gives a better statistical picture of the behavior of the sound field in the room. If the room is symmetrical, it might be wise to spot all measuring points on one side of the room to increase the effective coverage with a given effort.

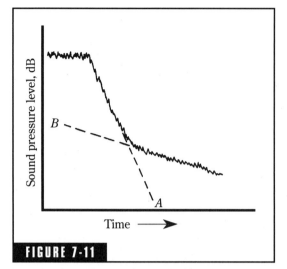

FIGURE 7-11

Reverberatory decay with a double slope due to acoustically coupled spaces. The shorter reverberation time represented by slope A is that of the main room. A second, highly reflective space is coupled through an open doorway. Those seated near the doorway are subjected first to the main-room response and then to the decay of the coupled space.

Acoustically Coupled Spaces

The shape of the reverberation decay can point to acoustical problems in the space. One common effect that alters the shape of the decay is due to acoustically coupled spaces. This is quite common in large public gathering spaces, but is also found in offices, homes, and other small spaces. The principle involved is illustrated in Fig. 7-11. The main space, perhaps an auditorium, is acoustically quite dead and has a reverberation time corresponding to the slope *A*. An adjoining hall with very hard surfaces and a reverberation time corresponding to slope *B* opens into the main room. A person seated in the main hall near the hall opening could very well experience a double-slope reverberation decay. Not until the sound level in the main room falls to a fairly low level would the main room reverberation be dominated by sound fed into it from the slowly decaying sound in the hall. Assuming slope *A* is correct for the main room, persons subjected to slope *B* would hear inferior sound.

Electroacoustically Coupled Spaces

What is the overall reverberant effect when sound picked up from a studio having one reverberation time is reproduced in a listening room having a different reverberation time? Does the listening room reverberation affect what is heard? The answer is definitely yes. This problem has been analyzed mathematically by Mankovsky.[4] In brief, the sound in the listening room is affected by the reverberation of both studio and listening room as follows:

- The combined reverberation time is greater than either alone.

- The combined reverberation time is nearer the longer reverberation time of the two rooms.

■ The combined decay departs somewhat from a straight line.

■ If one room has a very short reverberation time, the combined reverberation time will be very close to the longer one.

■ If the reverberation time of each of the two rooms alone is the same, the combined reverberation time is 20.8% longer than one of them.

■ The character and quality of the sound field transmitted by a stereo system conforms more closely to the mathematical assumptions of the above than does a monaural system.

■ The first five items can be applied to the case of a studio linked to an echo chamber as well as a studio linked to a listening room.

Decay Rate

The definition of reverberation time is based on uniform distribution of energy and random directions of propagation. Because these conditions do not exist in small rooms, there is some question as to whether what we measure should be called reverberation time. It is more properly termed *decay rate*. A reverberation time of 0.3 of a second is equivalent to a decay rate of 60 dB/0.3 sec = 200 dB per second. The use of decay rate instead of reverberation time would tell the experts that we are aware of the basic problems. Speech and music sounds in small rooms do decay even though the modal density is too low to hang the official "reverberation time" tag on the process.

Eliminating Decay Fluctuations

The measurement of reverberation time by the classical method that has been described involves the recording of many decays for each condition and much work in analyzing them. Schroeder has published a new method by which the equivalent of the average of a great number of decays can be obtained in a single decay.[5] One practical, but clumsy, method of accomplishing the mathematical steps required is to:

1. Record the decay of an impulse (noise burst or pistol shot) by the normal method.
2. Play back that decay reversed.
3. Square the voltage of the reversed decay as it builds up.

4. Integrate the squared signal with a resistance-capacitance circuit.

5. Record this integrated signal as it builds up during the reversed decay. Turn it around and this trace will be mathematically identical to averaging an infinite number of traditional decays. Programming this operation into a computer would be easier and more satisfactory.

Influence of Reverberation on Speech

Let us consider what happens to just one tiny word in a reverberant space. The word is *back*. It starts abruptly with a "ba..." sound and ends with the consonant "...ck", which is much lower in level. As measured on the graphic-level recorder, the "ck" sound is about 25 dB below the peak level of the "ba" sound and reaches a peak about 320 milliseconds after the "ba" peak.

Both the "ba" and "ck" sounds are transients that build up and decay after the manner of Fig. 7-3. Sketching these various factors to scale yields something like Fig. 7-12. The "ba" sound builds to a peak at an

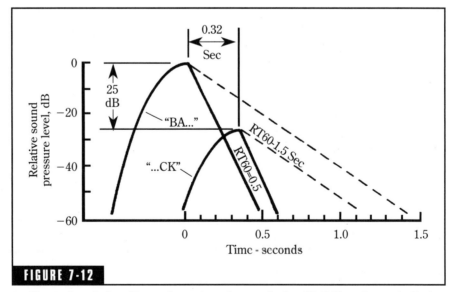

FIGURE 7-12

An illustration of the effects of reverberation on the intelligibility of speech. Understanding the word "back" depends on apprehending the later, lower level consonant "....ck," which is masked by reverberation if the reverberation time is too long.

arbitrary level of 0 dB at time = 0, after which it decays according to the reverberation time of the room, which is assumed to be 0.5 seconds (60 dB decay in 0.5 seconds). The "ck" consonant sound, peaking 0.32 seconds later, is 25 dB below the "ba" sound peak. It too decays at the same rate as the "ba" sound according to the assumed 0.5-second RT60. Under the influence of the 0.5-second reverberation time, the "ck" consonant sound is not masked by the reverberation time of "ba." If the reverberation time is increased to 1.5 seconds, as shown by the broken lines, the consonant "ck" is covered completely by reverberation.

The primary effect of excessive reverberation is to impair the intelligibility of speech by masking the lower level consonants. In the word "back," the word is unintelligible without a clear grasping of the "ck" part. Understanding the "ck" ending is the only way to distinguish "back" from bat, bad, bass, ban, or bath. In this oversimplified way, we can grasp the effect of reverberation on the understandability of speech and the reason why speech is more intelligible in rooms having lower reverberation times.

Sound-reinforcement engineers have been assisted greatly by the work of the Dutch investigators, Peutz[6] and Klein.[7] Because of their work, it is a straightforward procedure to predict with reasonable accuracy the intelligibility of speech in a space from geometrical factors and a knowledge of reverberation time.

Influence of Reverberation on Music

The effect of hall "resonance" or reverberation on music is intuitively grasped but is not generally well understood. This subject has received much attention from scientists as well as musicians, and the final word has yet to appear. Beranek has made a valiant attempt to summarize present knowledge and to pinpoint essential features of concert and opera halls around the world,[8,9] but our understanding of the problem is still quite incomplete. Suffice it to say that the reverberation decay of a music hall is only one important factor among many, another being the echo pattern, especially the so-called "early sound." It is beyond the scope of this book to treat this subject in any detail, but an interesting point or two commonly overlooked are discussed briefly.

Normal modes have been considered in some detail because of their basic importance (see Chap. 15). They are also active in music halls and listening rooms. An interesting phenomenon is pitch change during reverberant decay. In reverberant churches, organ tones have been observed to change pitch as much as a semitone during decay. In searching for an explanation for this phenomenon, two things have been mentioned: shift of energy between normal modes, and the perceptual dependence of pitch on sound intensity. There are problems in both. Balachandran has demonstrated the physical (as opposed to psychophysical) reality of the effect[10] using the Fast Fourier Transform (FFT) technique on the reverberant field created by 2-kHz pulses. He revealed the existence of a primary 1,992-Hz spectral peak, and curiously, another peak at 3,945 Hz. Because a 6-Hz change would be just perceptible at 2 kHz, and a 12-Hz change at 4 kHz, we see that the 39-Hz shift from the octave of 1,992 Hz would give a definite impression of pitch change. The reasons for this are still under study. The reverberation time of the hall in which this effect was recorded was about 2 seconds.

Optimum Reverberation Time

Considering the full range of possible reverberation times, there must be an optimum time between the "too dry" condition of the outdoors and anechoic chambers, and the obvious problems associated with excessively long reverberation times in a stone cathedral. Such an optimum does exist, but there is usually great disagreement as to just what it is because it is a subjective problem and some differences in opinions must be expected. The optimum value depends not only on the one making the judgment, but also on the type of sounds being considered.

Reverberation rooms, which are used for measuring absorption coefficients, are carefully designed for the longest practical RT60 to achieve the maximum accuracy. The optimum here is the maximum attainable.

The best reverberation time for a space in which music is played depends on the size of the space and the type of music. Slow, solemn, melodic music, such as some organ music, is best served by long reverberation time. Quick rhythmic music requires a different reverberation time from chamber music. No single optimum universally fits all types of music, the best that can be done is to establish a range based on subjective judgments of specialists.

Recording studios present still other problems that do not conform to simple rules. Separation recording in which musical instruments are recorded on separate tracks for later mixdown in general require quite dead spaces to realize adequate acoustical separation between tracks. Music directors and band leaders often require different reverberations for different instruments, hence hard areas and absorptive areas may be found in the same studio. The range of reverberation realized in this manner is limited, but proximity to reflective surfaces does affect local conditions.

Spaces for speech require shorter reverberation times than for music because of the general interest in direct sound. In general, long reverberation time tends toward lack of definition and clarity in music and loss of intelligibility in speech. In dead spaces in which reverberation time is very short, loudness and tonal balance may suffer. It is not possible to specify precisely optimum reverberation times for different services, but Figs. 7-13 through 7-15 show at least a rough indication of recommendations given by a host of experts in the field who do not always agree with each other.

The reverberation times for churches in Fig. 7-13 range from highly reverberant liturgical churches and cathedrals to the shorter ranges of

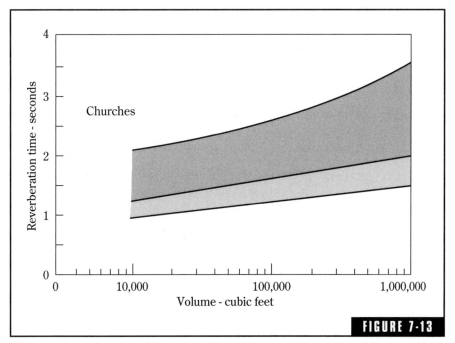

FIGURE 7-13

"Optimum" reverberation time for churches. The upper area applies to the more reverberant liturgical churches and cathedrals, the lower to churches having services more oriented to speech. A compromise between music and speech is required in most churches.

the lower shaded area characteristic of the more talk-oriented churches. Churches generally represent a compromise between music and speech.

Figure 7-14 represents the range of recommended reverberation times for different concert halls. Symphony orchestras are near the top, lighter music somewhat lower. The lower shaded area applies for opera and chamber music.

Those spaces used primarily for speech and recording require close to the same reverberation times as shown in Fig. 7-15. Television studios have even shorter reverberation times to deaden the sounds associated with rolling cameras, dragging cables, and other production noises. It should also be remembered that acoustics in television are dominated by the setting and local furnishings. In many of the spaces represented in Fig. 7-15, speech reinforcement is employed.

Bass Rise of Reverberation Time

The goal in voice studios is to achieve a reverberation time that is the same throughout the audible spectrum. This can be difficult to realize,

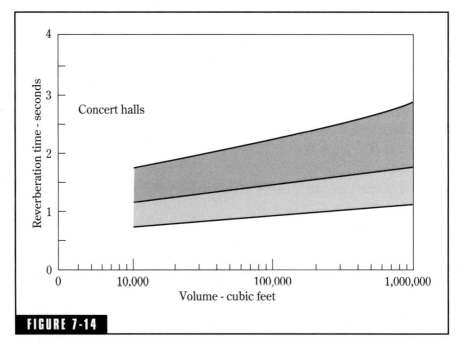

FIGURE 7-14

"Optimum" reverberation time for concert halls. Symphony orchestras are near the top of the shaded areas; lighter music is lower. The lower shaded area applies to opera and chamber music.

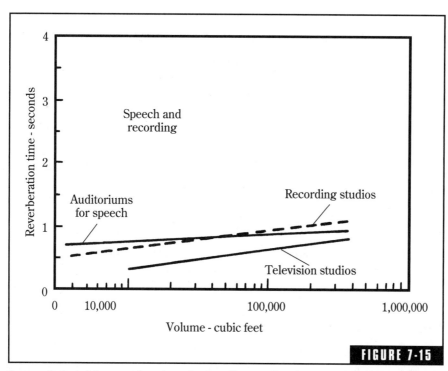

FIGURE 7-15

Spaces designed for speech and music recording require shorter reverberation times.

especially at low frequencies. Adjustment of reverberation time at high frequencies is easily accomplished by adding or removing relatively inexpensive absorbers. At low frequencies, the situation is quite different as absorbers are bulky, difficult to install, and sometimes unpredictable.

Researchers at the British Broadcasting Corporation observed that subjective judgments seemed to indicate a tolerance for a certain amount of bass rise of reverberation time. Investigating this in controlled tests, Spring and Randall[11] found that bass rise to the extent indicated in Fig. 7-16 was tolerated by the test subjects for voice signals. Taking the 1 kHz value as reference, rises of 80% at 63 Hz and 20% at 125 Hz were found to be acceptable. These tests were made in a studio 22 × 16 × 11 feet (volume about 3,900 cu ft) for which the midband reverberation time was 0.4 second (which agrees fairly well with Fig. 7-15).

Bass rise in reverberation time for music has traditionally been accepted to give "sonority" to the music in music halls. Presumably, somewhat greater bass rise than that for speech would be desirable in listening rooms designed for classical music.

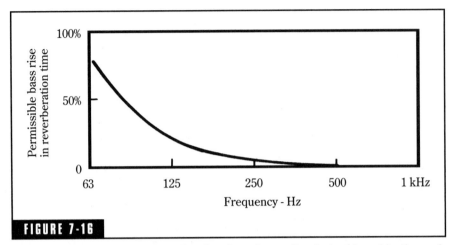

FIGURE 7-16

Permissible bass rise of reverberation time for voice studios derived by subjective evaluation in controlled tests by BBC researchers. (After Spring and Randall.[11])

Living Room Reverberation Time

The reverberation characteristic of the average living room is of interest to the high-fidelity enthusiast, the broadcaster, and the recording specialist. This living room is where the high-fidelity recordings are to be played. Further, the quality control monitoring room of the broadcast and recording studio must have a reverberation time not too far from that of the living room in which the final product will be heard. Generally, such rooms should be "deader" than the living room, which will add its own reverberation to that of recording or broadcast studio.

Figure 7-17 shows the average reverberation time of 50 British living rooms measured by Jackson and Leventhall[12] using octave bands of noise. The average reverberation time decreases from 0.69 second at 125 Hz to 0.4 second at 8 kHz. This is considerably higher than earlier measurements of 16 living rooms made by BBC engineers in which reverberation times between 0.35 and 0.45 were found on the average. Apparently, the living rooms measured by the BBC engineers were better furnished than those measured by Jackson and Leventhall and, presumably, would agree better with living rooms in the United States.

The 50 living rooms of the Jackson-Leventhall study were of varying sizes, shapes, and degree of furnishing. The sizes varied from 880 to 2,680 cu ft, averaging 1550 cu ft. Figure 7-15 shows an optimum

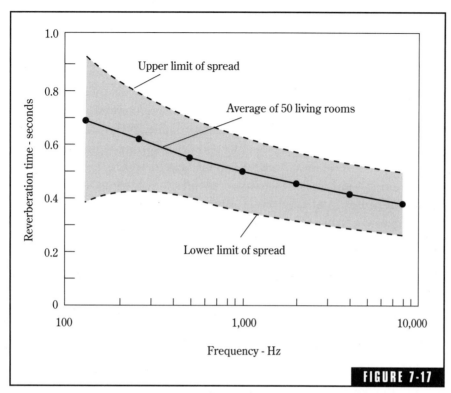

Average reverberation time for 50 living rooms. (After Jackson and Leventhall.[12])

reverberation time for speech for rooms of this size to be about 0.3 second. Only those living rooms near the lower limit approach this, and in them we would expect to find much heavy carpet and overstuffed furniture. These reverberation measurements tell us little or nothing about the possible presence of colorations. The BBC engineers checked for colorations and reported serious ones in a number of the living rooms studied.

Artificial Reverberation: The Past

Artificial reverberation is considered a necessity in audio-signal processing. Recordings of music played in "dry" (nonreverberant) studios lack the richness of the room effect contributed by the music hall. The addition of artificial reverberation to such recordings has become standard practice, and there is great demand for equipment that will provide natural-sounding artificial reverberation at a reasonable cost.

There are many ways of generating artificial reverberation, but the challenge is finding that method which mimics actual music halls and does not introduce colorations (frequency-response aberrations) into the signal. Historically, a dedicated reverberation room has been employed by the larger organizations. The program is played into this room, picked up by a microphone, and the reverberated signal mixed back into the original in the amount to achieve the desired effect. Small reverberation rooms are afflicted with serious coloration problems because of widely spaced modes. Large rooms are expensive. Even though the three-dimensional reverberation room approach has certain desirable qualities, the problems outweigh the advantages, and they are now a thing of the past.

Spring reverberators have been widely used in semiprofessional recording because of their modest cost. In this form of reverberator, the signal is coupled to one end of a spring, the sound traveling down the spring being picked up at the other end. Because of quality problems, spring reverberators are also rapidly passing from the scene.

The reverberation plate, such as the Gotham Audio Corporation's EMT-140, has been a professional standard for many years. It, too, is slipping into oblivion because of the more favorable cost/performance ratio of the newer digital devices.

Artificial Reverberation: The Future

An understanding of the digital reverberators now dominating the field can be achieved best by studying the basic principles of the old abandoned methods. What audiophile has not fed a signal from the playback head of his or her magnetic tape recorder back into the record head, and was enthralled by the repetitive "echo" effect? The secret lies in the delay resulting from the travel of the tape from one head to the other. Delay is the secret ingredient of every form of reverberation device. The delays associated with the return of successive echoes in a space is the secret ingredient of the natural reverberation.

This principle is illustrated in the simple signal flow schematic of Fig. 7-18. The incoming signal is delayed, and a portion of the delayed signal is fed back and mixed with the incoming signal, the mixture being delayed again, and so on.

Schroeder has found that approximately 1,000 echoes per second are required to avoid the flutter effect that dominates the above tape-

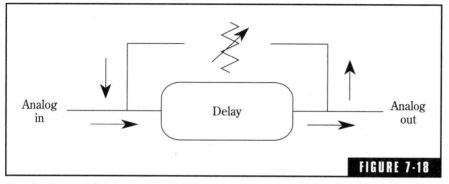

FIGURE 7-18

The simplest artificial reverberation device. The technique of utilizing the time delay between the record and playback heads of a magnetic tape recorder is of this type.

recorder experiment and to sound natural to the ear. With a 40-msec delay, only 1/0.04 = 25 echoes are produced each second, a far cry from the 1,000 per second desired. One solution is to arrange many of the simple reverberators of Fig. 7-18 in parallel. Four such simple reverberators, arranged in parallel as in Fig. 7-19, produce only 4 × 25 = 100 echoes per second. It would require 40 such reverberators in parallel to achieve the required echo density, and this is impractical.

One approach to producing the necessary echo density, and simultaneously, a flat frequency response is illustrated in Fig. 7-20. Here numerous delays feed back on themselves, combining to feed other delays, which in turn recirculate back to the first delay. The + signs in Fig. 7-20 indicate mixing (addition), and the × signs indicate gain (multiplication). Remember that multiplying by a fraction less than unity gives a gain of less than unity—in other words, attenuation. The digital reverberator of Fig. 7-20 only suggests how greater echo density along with good frequency response might be achieved. Actually, the better digital reverberators in use today are far more complicated than this.[13] The resulting artificial reverberation available today from the top of the line has far higher echo density, flatter frequency response, and a more natural sound than the best of the old mechanical devices. The less expensive digital units also show great improvement over those of the past.

Arrival Time Gap

There is one characteristic of natural reverberation in music halls that was revealed by Beranek's careful study of music halls around

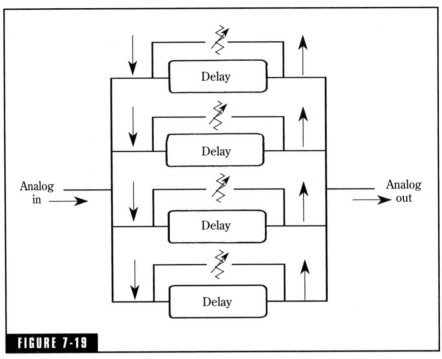

FIGURE 7-19

An artificial-reverberation device incorporating several different delays produces an effect far superior to that of the single delay in Fig. 7-18.

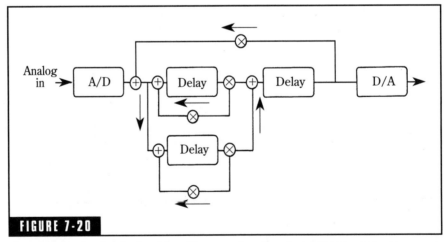

FIGURE 7-20

The required echo density is achieved in a reverberation device that incorporates numerous delays and recirculation of the signal. Actual artificial-reverberation devices available today are far more complicated than this one.

the world.[8,9] At a given seat, the direct sound arrives first because it follows the shortest path. Shortly after the direct sound, the reverberant sound arrives. The time between the two is called the *arrival time gap* or *early time gap* as shown in Fig. 7-21. If this gap is less than 40 or 50 msec, the ear integrates the direct and the reverberant sound successfully. This gap is important in recorded music because it is the cue that gives the ear information on the size of the hall. In addition to all of the delays responsible for achieving echo density, the initial time-delay gap is yet another important delay that must be included in digital reverberators.

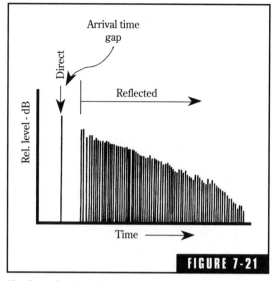

FIGURE 7-21

The introduction of an initial time-delay gap in the artificial reverberation adds to its lifelike character.

The Sabine Equation

Sabine's reverberation equation was developed at the turn of the century in a strictly empirical fashion. He had several rooms at his disposal and by adding or removing seat cushions of a uniform kind he established the following relationship (adapted from the metric units he used):

$$RT60 = \frac{0.049\ V}{Sa} \tag{7-1}$$

where

$RT60$ = reverberation time, seconds
V = volume of room, cu ft
S = total surface area of room, sq ft
a = average absorption coefficient of room surfaces
Sa = total absorption, sabins

Young[14] has pointed out that the absorption coefficients published by materials manufacturers (such as the list in the appendix) are Sabine coefficients and can be applied directly in the Sabine equation. After a thorough study of the historical development of the offshoots of Sabine's work, he recommends that Equation 7-1 be used for all engineering computations rather than the Eyring equation or the several derivatives thereof. Two unassailable reasons for this are simplicity and consistency. In spite of the fact that this simpler procedure was suggested as early as 1932, and Young's convincing arguments for it were given in 1959, many technical writings have continued to put forth the Eyring or other equations for studio use. Even though there was authoritative backing for using Eyring for more absorbent spaces, why continue if the commonly available coefficients apply only to Sabine? These are the reasons why we use only Equation 7-1 in this volume.

The total Sabine absorption in a room would be easy to get if all surfaces of the room were uniformly absorbing, but this condition rarely exists. Walls, floor, and ceiling may well be covered with quite different materials, and then there are the doors and windows. The total absorption, Sa, of Equation 7-1, can be found by considering the absorption contributed by each type of surface. For example, in our imaginary room, let us say that an area S_1 is covered with a material having an absorption coefficient a_1 as obtained from the table in the appendix. This area then contributes $(S_1)(a_1)$ absorption units, called *sabins*, to the room. Likewise, another area S_2 is covered with another kind of material with absorption coefficient a_2, and it contributes $(S_2)(a_2)$ sabins of absorption to the room. The total absorption in the room is $Sa = S_1a_1 + S_2a_2 + S_3a_3$ .... etc. With a figure for Sa in hand, it is a simple matter to go back to Equation 7-1 and calculate the reverberation time.

Reverberation Calculation: Example 1

A completely untreated room will first be taken to illustrate the implementation of Sabine's equation (Eq. 7-1). The dimensions of the room are assumed to be 23.3 × 16 × 10 ft. Other assumptions are that the room has a concrete floor and that the walls and the ceiling are of frame construction with ½ in gypsum board (drywall) covering. As a simplification the door and a window will be neglected as having minor effect. The tabulation of Fig. 7-22 illustrates the untreated condition. The concrete floor area of 373 sq ft and the gypsum board area of 1,159 sq ft are entered in the table. The appropriate absorption coefficients are entered from the table in the appendix for each material and for the six frequencies. Multiplying the concrete floor area of $S = 373$ sq ft by the coefficient $a = 0.01$ gives $Sa = 3.7$ sabins. This is

		125 Hz		250 Hz		500 Hz		1 kHz		2 kHz		4 kHz	
Material	S sq ft	a	Sa	a	Sa	a	Sa	a	Sa	a	Sa	a	Sa
Concrete	373	0.01	3.7	0.01	3.7	0.015	5.6	0.02	7.5	0.02	7.5	0.02	7.5
gypsum board	1,159	0.29	336.1	0.10	115.9	0.05	58.0	0.04	46.4	0.07	81.1	0.09	104.3
Total sabins		339.8		119.6		63.6		53.9		88.6		111.8	
Reverberation time (seconds)		0.54		1.53		2.87		3.39		2.06		1.63	

Size $23.3 \times 16 \times 10$ ft
Treatment   None
Floor   Concrete
Walls   Gypsum board, $1/2''$, on frame construction
Ceiling   Ditto
Volume   (23.3) (16) (10) = 3,728 cu ft

a = absorption coefficient for that material
and for that frequency (See Appendix)
Sa = S times a, absorption units, sabins

$$RT60 = \frac{(0.049)\,(3728)}{Sa} = \frac{182.7}{Sa}$$

Example: For 125 Hz, $RT60 = \dfrac{182.7}{339.8} = 0.54$ second

FIGURE 7-22

Room conditions and calculations for Example 1.

entered under Sa for 125 and 250 Hz. The absorption units (sabins) are then figured for both materials and for each frequency. The total number of sabins at each frequency is obtained by adding that of the concrete floor to that of the gypsum board. The reverberation time for each frequency is obtained by dividing 0.049 V = 182.7 by the total Sa product for each frequency.

To visualize the variation of reverberation time with frequency, the values are plotted in Fig. 7-23A. The peak reverberation time of 3.39 seconds at 1 kHz is excessive and would ensure quite poor sound conditions. Two persons separated 10 ft would have difficulty understanding each other as the reverberation of one word covers up the next word.

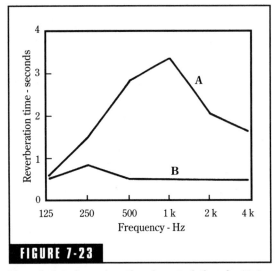

FIGURE 7-23

The calculated reverberation characteristics of a 23.3 × 16 × 10 ft room: (A) the "as found," untreated condition of Example 1, (B) treated condition of Example 2.

Reverberation Calculation: Example 2

The goal now is to correct the reverberation of curve *A* of Fig. 7-23. It is evident that much absorption is needed at midband frequencies, a modest amount at higher frequencies, and very little at lower frequencies. The need is for a material having an absorption characteristic shaped more or less like the reverberation curve *A*. Skipping the laborious thumbing through of handbooks, ³/₄-in acoustical tile seems to have the right shape. Giving no thought at this point to how it is to be distributed, what area of this tile is required to correct for Fig. 7-23A?

A new table, Fig. 7-24, is set up to organize the calculations. Everything is identical to Fig. 7-22 except that the ¾-in acoustical tile has been added with coefficients from the appendix. What area of tile is required? In Fig. 7-22 a total of 53.9 sabins is noted at the peak reverberation time at 1 kHz and 339.8 sabins at 125 Hz at which the reverberation time is a reasonable 0.54 sec. How much ¾-in acoustical tile would be required to add 286 sabins at 1 kHz? The absorption coefficient of this material is 0.84 at 1 kHz. To get 286 sabins at 1 kHz with this material would require 286/0.84 = 340 sq ft of the material. This is entered in Fig. 7-24 and the calculations extended. Plotting these reverberation time points gives the curve of Fig. 7-23B. Obtaining reverberation time this uniform across the band on the first trial is unusual, but satisfying. The overall precision of coefficients and measurements is so poor that the deviations of curve *B* from flatness are insignificant.

No carpet has been specified in this example and people usually demand it. The use of carpet would change everything as much absorption in the high frequencies is added.

Reverberant Field

In our 23.3 × 16 × 10 ft room the volume is 3,728 cu ft. The inner surface area is 1,533 sq ft. By statistical theory (geometrical ray acoustics

			125 Hz		250 Hz		500 Hz		1 kHz		2 kHz		4 kHz	
Material	S sq ft		a	Sa	a	Sa	a	Sa	a	Sa	a	Sa	a	Sa
Concrete	373		0.01	3.7	0.01	3.7	0.015	5.6	0.02	7.5	0.02	7.5	0.02	7.5
Gypsum board	1,159		0.29	336.1	0.10	115.9	0.05	58.0	0.04	46.4	0.07	81.1	0.09	104.3
Acoustical tile	340		0.09	30.6	0.28	95.2	0.78	265.2	0.84	285.6	0.73	248.2	0.64	217.6
Total sabins				370.4		214.8		328.8		339.5		336.8		329.4
Reverberation time (seconds)				0.49		0.85		0.56		0.54		0.54		0.55

Size   23.3 × 16 × 10 ft
Treatment   Acoustical tile
Floor   Concrete
Walls   Gypsum board, 1/2", on frame construction
Ceiling   Ditto
Volume   (23.3) (16) (10) = 3,728 cu ft

S = area of material
a = absorption coefficient for that material and for that frequency (See Appendix)
Sa = S times a, absorption units, sabins

$$RT60 = \frac{(0.049)(3728)}{Sa} = \frac{182.7}{Sa}$$

FIGURE 7-24

Room conditions and calculations for Example 2.

on which the Sabine equation is based) the mean free path (the average distance sound travels between reflections) is $4V/S$ or (4) (3,728)/1,533 = 9.7 ft. If the reverberation time is 0.3 second, there would be at least 35 reflections during the 60 dB decay. This would appear to be a fair involvement of all room surfaces.

In a small, relatively dead room such as the average studio, control room, and listening room, one never gets very far away from the direct influence of the source. A true reverberant field is often below the ambient noise level. The reverberation time equations have been derived for conditions that exist only in the reverberant field. In this sense, then, the concept of reverberation time is inapplicable to small, relatively dead rooms. And yet we measure something that looks very much like what is measured in large, more live spaces. What is it? What we measure is the decay rate of the normal modes of the room.

Each axial mode decays at its own rate determined by the absorbance of a pair of walls and their spacing. Each tangential and oblique mode has its own decay rate determined by distance traveled, the number of surfaces involved, the variation of the absorption coefficient of the surfaces with angle of incidence, etc. Whatever average decay rate is measured for an octave of random noise will surely be representative of the average decay rate at which that octave of speech or music signals would die away. Although the applicability of computing reverberation time from the equations based on reverberant field conditions might be questioned because of the lack of reverberant field, the measured decay rates (by whatever name you call them) most certainly apply to this space and to these signals.

Endnotes

[1]Beranek, Leo L., *Broadcast Studio Design*, J. Soc. Motion Picture and Television Eng., 64, Oct., 1955, 550-559.

[2]Schultz, Theodore J., *Problems in the Measurement of Reverberation Time*, J. Audio Eng. Soc., 11, 4, 1963, 307-317.

[3]Matsudaira, T. Ken, et al., *Fast Room Acoustic Analyzer (FRA) Using Public Telephone Line and Computer*, J. Audio Eng. Soc., 25, 3, 1977, 82-94.

[4]Mankovsky, V.S., *Acoustics of Studios and Auditoria*, London, Focal Press, 1971.

[5]Schroeder, M.R., *New Method of Measuring Reverberation Time*, J. Acous. Soc. Am., 37, 1965, 409-412.

[6]Peutz, V.M.A., *Articulation Loss of Consonants as a Criterion for Speech Transmission in a Room*, J. Audio Eng. Soc., 19, 11, 1971, 915-919.

[7]Klein, W., *Articulation Loss of Consonants as a Basis for the Design and Judgement of Sound Reinforcement Systems*, J. Audio Eng. Soc., 19, 11, 1971, 920-922.

[8]Beranek, L.L., *Music, Acoustics, and Architecture*, New York, John Wiley & Sons, 1962.

[9]Beranek, L.L., *Concert and Opera Halls—How They Sound*, Woodbury, N.Y., Acoustical Society of America, 1996.

[10]Balachandran, C.G., *Pitch Changes During Reverberation Decay*, J. Sound and Vibration, 48, 4, 1976, 559-560

[11]Spring, N.F. and K.E. Randall, *Permissible Bass Rise in Talks Studios*, BBC Engineering, 83, 1970, 29-34.

[12]Jackson, G.M. and H.G. Leventhall, *The Acoustics of Domestic Rooms*, Applied Acoustics, 5, 1972, 265-277

[13]Hall, Gary, *Digital Reverb—How It Works*, MIX, 9, 6, 1985, 32.

[14]Young, Robert W., *Sabine Equation and Sound Power Calculations*, J. Acous. Soc. Am., 31, 12, 1959, 1681.

Control of Interfering Noise

There are four basic approaches to reducing noise in a listening room or a recording studio:

■ Locating the room in a quiet place

■ Reducing the noise energy within the room

■ Reducing the noise output of the offending source

■ Interposing an insulating barrier between the noise and the room

Locating a sound-sensitive area away from outside interfering sounds is a luxury few can enjoy because of the many factors (other than acoustical) involved in site selection. If the site is a listening room, which is part of a residence, due consideration must be given to serving the other needs of the family—at least if some degree of peace is to prevail. If the room in question is a recording or broadcast studio, it is probably a part of a multipurpose complex and the noises originating from business machines, air conditioning equipment, or foot traffic within the same building, or even sounds from other studios, may dominate the situation.

Noise Sources and Some Solutions

Protecting a room from street traffic noise is becoming more difficult all the time. It is useful to remember that doubling the distance from a noisy street or other sound source reduces the level of airborne noise approximately 6 dB. Shrubbery and trees can help in shielding from street sounds; a cypress hedge 2 ft thick gives about a 4 dB reduction.

The level of noise that has invaded a room by one means or another can be reduced by introducing sound-absorbing material into the studio. For example, if a sound level meter registers a noise level of 45 dB inside a studio, this level might be reduced to 40 dB by covering the walls with great quantities of absorbing materials. Going far enough in this direction to reduce the noise significantly, however, would probably make the reverberation time too short. The control of reverberation must take priority. The amount of absorbent installed in the control of reverberation will reduce the noise level only slightly, and beyond this we must look to other methods for further noise reduction.

Reducing the noise output of the offending source, if accessible and if possible, is the most logical and profitable approach. Traffic noise on a nearby street or airplanes overhead may be beyond control, but the noise output of a ventilating fan might be reduced 20 dB by the installation of a pliant mounting or the separation of a metal air duct with a simple canvas collar. Installing a carpet in a hall might solve a foot traffic noise problem, or a felt pad might reduce a typewriter noise problem. In most cases working on the offending source and thus reducing its noise output is far more productive than corrective measures at or within the room in question.

As for terminology, a wall, for example, must offer a given *transmission loss* to sound transmitted through it, as shown in Fig. 8-1. An outside noise level of 80 dB would be reduced to 35 dB by a wall having a transmission loss of 45 dB. A 60 dB wall would reduce the same noise level to 20 dB if no "flanking" or bypassing of the wall by other paths is present. The wall "attenuates" the sound or it "insulates" the interior from the outside noise. The walls, floor, and ceiling of the sound-sensitive area must give the required transmission loss to outside noises, reducing them to tolerable levels inside the room.

Noise can invade a studio or other room in the following ways: airborne, transmitted by diaphragm action of large surfaces, transmitted through solid structures, or a combination of all three.

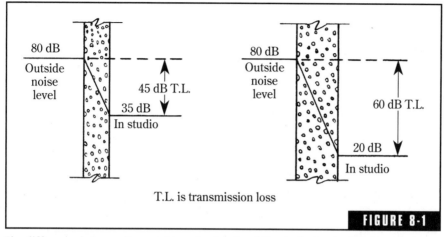

T.L. is transmission loss

FIGURE 8-1

The difference between the outside noise level and the desired noise level inside determines the required transmission loss of the wall.

Airborne Noise

A heavy metal plate with holes to the extent of 13 percent of the total area can transmit as much as 97 percent of the sound impinging on it. The amount of sound that can pass through a small crack or aperture in an otherwise solid wall is astounding. A crack under a door or loosely fitting electrical service box can compromise the insulating properties of an otherwise excellent structure. Air-tightness is especially necessary to insulate against airborne noises.

Noise Carried by Structure

Unwanted sounds can invade an enclosure by mechanical transmission through solid structural members of wood, steel, concrete, or masonry. Air conditioner noises can be transmitted to a room by the air in the ducts, by the metal of the ducts themselves, or both. Water pipes and plumbing fixtures have excellent sound-carrying capabilities.

It is very difficult to make a solid structure vibrate by airborne noise falling upon it because of the inefficient transfer of energy from tenuous air to a dense solid. On the other hand, a motor bolted to a floor, a slammed door, or an office machine on a table with legs on the bare floor can cause the structure to vibrate very significantly. These vibrations can travel great distances through solid structure with little loss. With wood, concrete, or brick beams, longitudinal vibrations are attenuated only about 2 dB in 100 ft. Sound travels in steel about 20 times as far for the same loss! Although joints and cross-bracing members increase the transmission loss, it is still very low in common structural configurations.

Noise Transmitted by Diaphragm Action

Although very little airborne sound energy is transmitted directly to a rigid structure, airborne sound can set a wall to vibrating as a diaphragm and the wall, in turn, can transmit the sound through the interconnected solid structure. Such structure-borne sound might then cause another wall at some distance to vibrate, radiating noise into the space we are interested in protecting. Thus two walls interconnected by solid structure can serve as a coupling agent between exterior airborne noise and the interior of the listening room or studio itself.

Sound-Insulating Walls

For insulating against outside airborne sounds, the general rule is the heavier the wall the better. The more massive the wall, the more difficult it is for sound waves in air to move it to and fro. Figure 8-2 shows how the transmission loss of a rigid, solid wall is related to the density of the wall. The wall weight in Fig. 8-2 is expressed as so many pounds per square foot of surface, sometimes called the *surface density*. For example, if a 10 × 10 ft concrete block wall weighs 2,000 lb, the "wall weight" would be 2,000 lb per 100 sq ft, or 20 lb per sq ft. The thickness of the wall is not directly considered.

From Fig. 8-2 you can see that the higher the frequency, the greater the transmission loss, or in other words, the better the wall is as a barrier to outside noises. The line for 500 Hz is made heavier than the lines for other frequencies as it is common to use this frequency for casual comparisons of walls of different materials. However, don't forget that below 500 Hz the wall is less effective and for frequencies greater than 500 Hz it is more effective as a sound barrier.

The transmission losses indicated in Fig. 8-2 are based on the mass of the material rather than the kind of material. The transmission loss through a layer of lead of certain thickness can be matched by a plywood layer about 95 times thicker. But doubling the thickness of a concrete wall, for instance, would increase the transmission loss only about 5 dB.

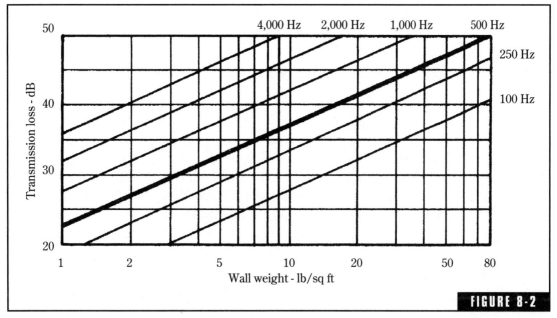

The mass of the material in a barrier rather than the kind of material determines the transmission loss of sound going through the barrier. The transmission loss is also dependent on frequency although values at 500 Hz are commonly used in casual estimates. The wall weight is expressed in pounds per square foot of wall surface.

A discontinuous structure such as bricks set in lime mortar conducts sound less efficiently than a more homogeneous material like concrete or steel. Unbridged air cavities between walls are very effective in sound reduction, but completely unbridged cavities are unattainable. Only in the case of two separate structures, each on its own foundation, is this unbridged condition approached.

Porous Materials

Porous materials such as fiberglass (rock wool, mineral fiber) are excellent sound absorbers and good heat insulators, but they are of limited value in insulating against sound. Using fiberglass to reduce sound transmission will help to a certain extent, but only moderately. The transmission loss for porous materials is directly proportional to the thickness traversed by the sound. This loss is about 1 dB (100 Hz) to 4 dB (3,000 Hz) per inch of thickness for a dense, porous material (rock wool, density 5 lb/cu ft) and less for lighter material. This direct dependence of transmission loss on thickness for porous materials is in contrast to the transmission loss for solid, rigid walls, which is approximately 5 dB for each doubling of the thickness.

Sound Transmission Classification (STC)

The solid line of Fig. 8-3 is simply a replotting of data from the mass law graphs of Fig. 8-2 for a wall weight of 10 lb per sq ft. If the mass law were perfectly followed, we would expect the transmission loss of a practical wall of this density to vary with frequency, as shown by the solid line. Unfortunately, things are not this simple and actual measurements of transmission loss on this wall might be more like the broken line of Fig. 8-3. These deviations reflect resonance and other effects in the wall panel, which are not included in the simple mass law concept.

Because of such commonly occurring irregularities, it would be of great practical value to agree on some arbitrary procedure of arriving

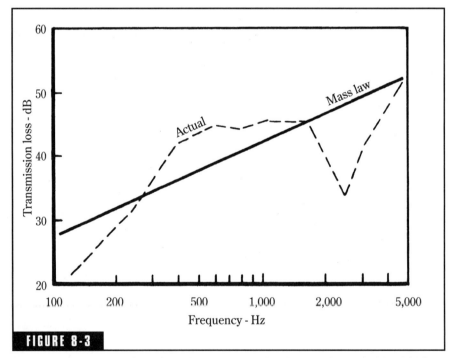

FIGURE 8-3

Actual measurements of transmission loss in walls often deviate considerably from the mass law (Fig. 8-2) because of resonances and other effects.

at a single number that would give a reasonably accurate indication of the sound transmission loss characteristics of a wall. This has been done in a procedure specified by the American Society for Testing and Materials in which the measured graph of a wall would be placed in a certain Sound Transmission Class (STC) by comparison to a reference graph (STC contour). The details of this procedure are beyond the scope of this book, but the results of such classification have been applied to walls of various types to be described for ready comparison. An STC rating of 50 dB for a wall would mean that it is better in insulating against sound than a wall of STC 40 dB. It is not proper to call STC ratings "averages" but the whole procedure is to escape the pitfalls of averaging dB transmission losses at various frequencies.[1]

Comparison of Wall Structures

Figure 8-4 gives the measured performance of a 4 in concrete block wall as a sound barrier. It is interesting to note that plastering both sides increases the transmission loss of the wall from STC 40 to 48. Figure 8-5 shows a considerable improvement in doubling the thickness of the concrete block wall. In this case the STC 45 is improved 11 dB by plastering both sides. In Fig. 8-6 is illustrated the very common 2 × 4 frame construction with ⅝-in gypsum board covering. The STC 34 without fiberglass between is improved only 2 dB by filling the cavity with fiberglass material, a meager improvement that would probably not justify the added cost.

Figure 8-7 describes a very useful and inexpensive type of wall of staggered stud construction. Here the inherently low coupling between the two independent wall diaphragms is further reduced by filling the space with fiberglass building material. Attaining the full STC 52 rating would require careful construction to ensure that the two wall surfaces are truly independent and not "shorted out" by electrical conduits, outlet boxes, etc.

The last wall structure to be described is the double wall construction of Fig. 8-8. The two walls are entirely separate, each having its own 2 × 4 plate.[2] Without fiberglass this wall is only 1 dB better than the staggered stud wall of Fig. 8-7 but by filling the inner space with building insulation, STC ratings up to 58 dB are possible.

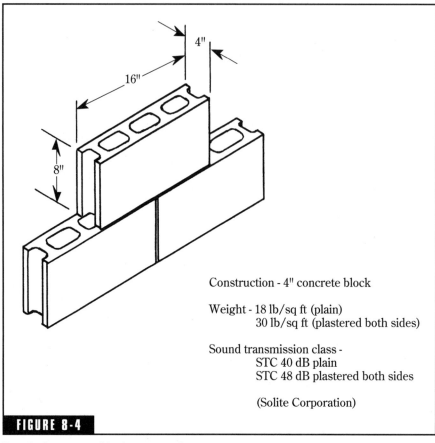

4"

16"

8"

Construction - 4" concrete block

Weight - 18 lb/sq ft (plain)
30 lb/sq ft (plastered both sides)

Sound transmission class -
STC 40 dB plain
STC 48 dB plastered both sides

(Solite Corporation)

FIGURE 8-4

Four-inch concrete block construction.

It was stated earlier that porous sound absorbing materials are of limited value in insulating against sound. This is true when normal transmission loss is considered, but in structures as those in Figs. 8-7 and 8-8, such porous materials have a new contribution to make in absorbing sound energy in the cavity. This can improve the transmission loss in some wall structures by as much as 15 dB, principally by reducing resonances in the space between the walls, while in others the effect is negligible. The low-density mineral fiber batts commonly used in building construction are as effective as the high-density boards, and they are much cheaper. Mineral fiber batts within a wall may also meet certain fire-blocking requirements in building codes.

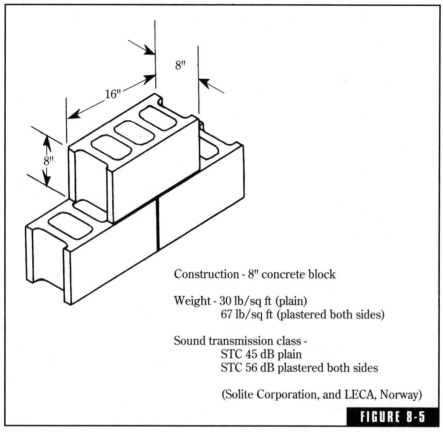

Construction - 8" concrete block

Weight - 30 lb/sq ft (plain)
　　　　67 lb/sq ft (plastered both sides)

Sound transmission class -
　　　　STC 45 dB plain
　　　　STC 56 dB plastered both sides

(Solite Corporation, and LECA, Norway)

FIGURE 8-5

Eight-inch concrete block construction.

The staggered stud wall and the double wall, on the basis of mass alone, would yield a transmission loss of only about 35 dB (Fig. 8-2). The isolation of the inner and the outer walls from each other and the use of insulation within have increased the wall effectiveness by 10 or 15 dB.

Double Windows

Between the control room and the studio a window is quite necessary, and its sound transmission loss should be comparable to that of the wall itself. A well-built staggered stud or double wall might have an STC of 50 dB. To approach this performance with a window requires very careful design and installation.[3]

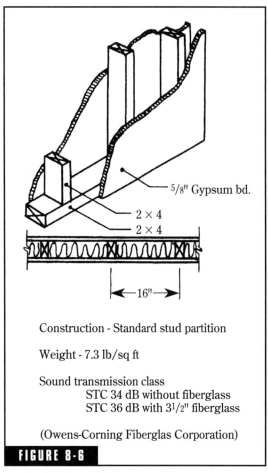

5/8" Gypsum bd.

2 × 4

2 × 4

16"

Construction - Standard stud partition

Weight - 7.3 lb/sq ft

Sound transmission class
STC 34 dB without fiberglass
STC 36 dB with 3¹/₂" fiberglass

(Owens-Corning Fiberglas Corporation)

FIGURE 8-6

Standard stud partition.

A double window is most certainly indicated; a triple window adds little more. The mounting must minimize coupling from one wall to the other. One source of coupling is the window frame, another is the stiffness of the air between the glass panels. The plan of Fig. 8-9 is a practical solution to the double-window problem for concrete block walls. Figure 8-9B is an adaptation to the staggered stud construction. In the latter there are, in effect, two entirely separate frames—one fixed to the inner and the other to the outer staggered stud walls. A felt strip may be inserted between them to ensure against accidental contact.

Heavy plate glass should be used, the heavier the better. There is a slight advantage in having two panes of different thickness. If desired, one glass can be inclined to the other to control light or external sound reflections, but this will have negligible effect on the transmission loss of the window itself. The glass should be isolated from the frame by rubber or other pliable strips. The spacing between the two glass panels has its effect—the greater the spacing the greater the loss—but there is little gain in going beyond 8 inches, nor serious loss in dropping down to 4 or 5 inches.

The absorbent material between the panes in the design of Fig. 8-9 discourages resonances in the air space. This adds significantly to the overall insulation efficiency of the double window, and it should extend completely around the periphery of the window. If the double window of Fig. 8-9 is carefully constructed, sound insulation should approach that of an STC 50 wall but will probably not quite reach it. For the staggered stud wall in which a double window is to be placed, the use of a 2 × 8 plate instead of the 2 × 6 plate will simplify mounting of the inner and outer window frames.

Prefabricated double glass windows are available commercially, one of which is rated at STC 49 dB.

Sound-Insulating Doors

The transmission loss of a door is determined by its mass, stiffness, and air-tightness. An ordinary household panel door hung in the usual way might offer less than 20 dB sound insulation. Increasing the weight and taking reasonable precautions on seals might gain another 10 dB, but a door to match a 50 dB wall requires great care in design, construction, and maintenance. Steel doors or patented acoustical doors giving specified values of transmission loss are available commercially but they are quite expensive. To avoid the expense of doors having high transmission loss, *sound locks* are commonly used. These small vestibules with two doors of medium transmission loss are very effective and convenient.

Doors with good insulating properties can be constructed if the requirements of mass, stiffness, and airtightness are met. Figure 8-10 suggests one inexpensive approach to the mass requirement, filling a hollow door with sand. Heavy plywood (¾ in) is used for the door panels.

Achieving a good seal around a "sound-proof" door can be very difficult. Great force is necessary to seal a heavy door. Wear and tear on pliant sealing strips can destroy their effectiveness, especially at the floor where foot-wear is a problem. The detail of Fig. 8-10 shows one approach to the sound leakage problem in which a very absorbent edge built around the periphery of the door serves as a trap for sound traversing the crack between door and jamb. This absorbent trap could also

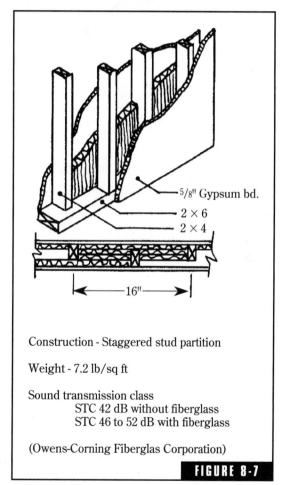

⁵⁄₈" Gypsum bd.

2 × 6

2 × 4

16"

Construction - Staggered stud partition

Weight - 7.2 lb/sq ft

Sound transmission class
STC 42 dB without fiberglass
STC 46 to 52 dB with fiberglass

(Owens-Corning Fiberglas Corporation)

FIGURE 8-7

Staggered stud partition.

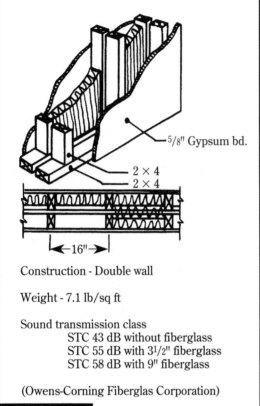

5/8" Gypsum bd.

2 × 4

2 × 4

←16"→

Construction - Double wall

Weight - 7.1 lb/sq ft

Sound transmission class
    STC 43 dB without fiberglass
    STC 55 dB with 3^1/$_2$" fiberglass
    STC 58 dB with 9" fiberglass

(Owens-Corning Fiberglas Corporation)

FIGURE 8-8

Double wall partition.

be embedded in the door jamb. Such a soft trap could also be used in conjunction with one of the several types of seals.

Figure 8-11 shows a do-it-yourself door seal that has proved reasonably satisfactory. The heart of this seal is a rubber or plastic tubing an inch or less in outside diameter with a wall thickness of about ^{3}/$_{32}$ in. The wooden nailing strips hold the tubing to the door frame by means of a canvas wrapper. A raised sill is required at the floor if the tubing method is to be used all around the door (or another type of seal such as weatherstripping could be utilized at the bottom of the door). An advantage of tubing seal is that the degree of compression of the tubing upon which the sealing properties depend is available for inspection.

A complete door plan patterned after BBC practice is shown in Fig. 8-12. It is based upon a 2-in-thick solid slab door and utilizes a magnetic seal such as used on refrigerator doors. The magnetic material is barium ferrite in a PVC (polyvinyl chloride) rod. In pulling toward the mild steel strip, a good seal is achieved. The aluminum strip "C" decreases sound leakage around the periphery of the door.

It is possible to obtain a very slight acoustical improvement and, to some at least, an improvement in appearance by padding both sides of a door. A plastic fabric over 1-in foam rubber sheet can be "quilted" with upholstery tacks.

Noise and Room Resonances

Room resonances can affect the problem of outside noise in a studio. Any prominent modes persisting in spite of acoustic treatment make a room very susceptible to interfering noises having appreciable energy at these frequencies. In such a case a feeble interfering sound could be augmented by the resonance effect to a very disturbing level.

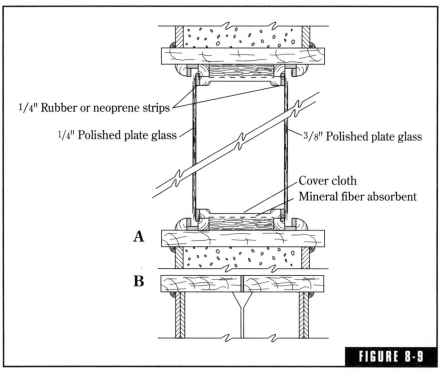

1/4" Rubber or neoprene strips

1/4" Polished plate glass

3/8" Polished plate glass

Cover cloth
Mineral fiber absorbent

A

B

FIGURE 8-9

Four-inch concrete block construction.

Active Noise Control

Many have had the idea of canceling noise by radiating a replica of the noise in inverse phase. It sounds simple but is very difficult to implement. It has been made to work fairly well in the immediate vicinity of a telephone in an industrial area of heavy noise. Very serious investigations are in progress to apply the principle to automobiles and other such controlled spaces. The prospects of active noise control being applied in sound-sensitive areas such as home listening rooms, recording studios, or control rooms are rather remote but new digital sound processing techniques might change that.

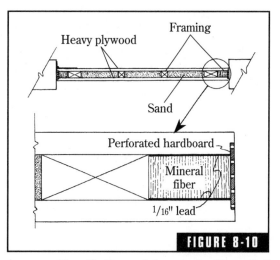

Framing

Heavy plywood

Sand

Perforated hardboard

Mineral fiber

1/16" lead

FIGURE 8-10

A reasonably effective and inexpensive "acoustic" door. Dry sand between the plywood faces adds to the mass and thus the transmission loss. Sound traveling between the door and jamb tends to be absorbed by the absorbent door edge.

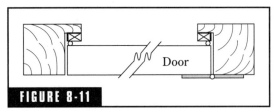

A door can be sealed by compressible rubber or plastic tubing held in place by a fabric wrapper.

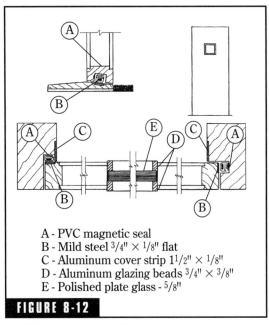

A - PVC magnetic seal
B - Mild steel $3/4'' \times 1/8''$ flat
C - Aluminum cover strip $11/2'' \times 1/8''$
D - Aluminum glazing beads $3/4'' \times 3/8''$
E - Polished plate glass - $5/8''$

A BBC door design utilizing magnetic seals of the type used on refrigerator doors.

Endnotes

[1]Jones, R.E., *How to Design Walls of Desired STC Ratings*, Sound and Vibration, 12, 8, 1978, 14-17.

[2]Green, D.W. and C.W. Sherry, *Sound Transmission Loss of Gypsum Wallboard Partitions, Report No. 3, 2x4 Wood Stud Partitions*, J. Acous. Soc. Am., 71, 4, 1982, 908-914.

[3]Everest, F.A., *Glass in the Studio*, dB The Sound. Eng. Mag. Part I, 18, 3 April 1984, 28-33. Part II, 18, 4 May 1984, 41-44.

Absorption of Sound

The law of the conservation of energy states that energy can neither be created nor destroyed, but it can be changed from one form to another. If we have some sound energy in a room to get rid of, how can it be done? Sound is the vibratory energy of air particles, and it can be dissipated in the form of heat. If it takes the sound energy of a million people talking to brew a cup of tea, we must give up any idea of heating our home with sound from the high-fidelity loud speakers.

Dissipation of Sound Energy

When sound wave S hits a wall (such as in Fig. 9-1), what happens to the energy it contains? If the sound wave is traveling in air and it strikes a concrete block wall covered with an acoustical material, there is first a reflected component A returned to the air from the surface of the acoustical material. Of course, there is a certain heat loss E in the air that is appreciable only at the higher audio frequencies.

Some of the sound penetrates the acoustical material represented by the shaded layer in Fig. 9-1. The direction of travel of the sound is refracted downward because the acoustical material is denser than air. There is heat lost (F) by the frictional resistance the acoustical material offers to the vibration of air particles. As the sound ray

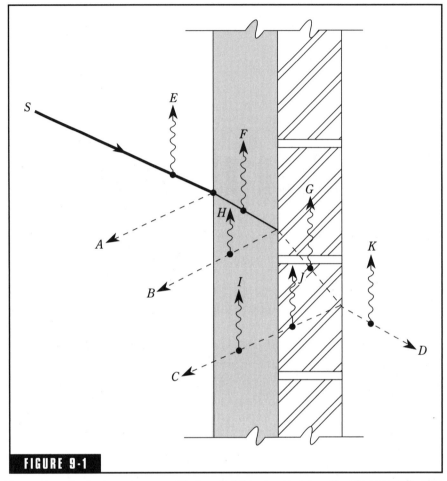

FIGURE 9-1

A sound ray impinging on an acoustical material on a masonry wall undergoes reflection from three different surfaces and absorption in the air and two different materials, with different degrees of refraction at each interface. In this chapter, the absorbed component is of chief interest.

strikes the surface of the concrete blocks, two things happen: a component is reflected (*B*), and the ray is bent strongly downward as it enters the much denser concrete blocks. Of course, there is further heat loss (*G*) within the concrete blocks. As the ray travels on, getting weaker all the time, it strikes the concrete-air boundary and undergoes another reflection (*C*) and refraction (*D*) with heat lost (*I*, *J*, and *K*) in three media.

The sound ray *S* of Fig. 9-1 experiences many rather complex events during its odyssey through this barrier, and every reflection and passage through air or acoustical material dissipates some of its original energy. The refractions bend the ray but do not necessarily dissipate heat. Fortunately, all this minutia is not involved in practical absorption problems.

Evaluation of Sound Absorption

The *absorption coefficient* is a measure of the efficiency of a surface or material in absorbing sound. If 55 percent of the incident sound energy is absorbed, the absorption coefficient is said to be 0.55. One square foot of this material gives 0.55 absorption units (sabins). An open window is considered a perfect absorber because sound passing through it never returns to the room. It would have an absorption coefficient of 1.0. Ten square feet of open window would give 10 sabins of absorbance.

The absorption coefficient of a material varies with frequency and with the angle at which the sound wave or ray impinges upon the material. In an established sound field in a room, sound is traveling in every imaginable direction. What we need in our calculations are sound absorption coefficients averaged over all possible angles of incidence.

Young has pointed out a long-standing and widespread confusion in the field of acoustics concerning the sound absorption coefficient.[1] There really are two kinds, one based on the arithmetic mean reflection coefficient of the several sound absorbing surfaces, *a*, and the other the geometric mean reflection coefficient, α, which are related by:

$$a = -\log_e (1 - \alpha) \qquad \textbf{(9-1)}$$

in which
   a = Sabine absorption coefficient
   α = energy absorption coefficient

We can skirt this problem by concentrating our attention on the Sabine coefficient, *a*, which is actually what is measured and published in various tables.

Reverberation Chamber Method

The reverberation chamber method of determining the absorption coefficients of absorbing materials automatically measures the average value. This chamber is a large room with highly reflective walls, ceiling, and floor. The reverberation time of such a room is very long, and the longer it is, the more accurate the measurement. A standard sample of the material to be tested, 8 × 9 ft in size, is laid on the floor and the reverberation time measured. Comparing this time with the known reverberation time of the empty room yields the number of absorption units the sample adds to the room. From this the absorption attributed to each square foot of material is determined, giving the equivalent of the absorption coefficient.

This description is a highly simplified view of the reverberation chamber method. The construction of the chamber is very important to ensure many modal frequencies and to equalize mode spacing as much as possible. The position of the sound source and the number and position of the measuring microphones must be carefully worked out. It is common to use large rotating vanes to ensure adequate diffusion of sound. All absorption coefficients supplied by manufacturers for use in architectural acoustic calculations are measured by the reverberation chamber method.

If the open window is the perfect absorber, what happens if chamber measurements show absorption coefficients greater than 1? This is a regular occurrence. The diffraction of sound from the edges of the standard sample makes the sample appear, acoustically, of greater area than it really is. There is no standard method of making adjustments for this artifact. Some manufacturers publish the actual measured values if greater than unity; others arbitrarily adjust the values down to unity or to 0.99.

Sound absorption coefficients vary with frequency. It is standard practice to publish coefficients and make calculations for the following six frequencies: 125, 250, 500, 1,000, 2,000, and 4,000 Hz.

Impedance Tube Method

The Kundt tube has been applied to the measurement of the absorption coefficient of materials. Used in this way it is commonly a *standing-*

wave tube or an *impedance tube.* No matter what name it bears, it is a very handy device for quickly and accurately determining coefficients. It also has the advantage of small size, modest demands in terms of supporting equipment, and it requires only a small sample. This method is primarily used for porous absorbers because it is not suited to those absorbers that depend on area for their effect such as vibrating panels and large slat absorbers.

The construction and operation of the impedance tube are illustrated in Fig. 9-2. The tube usually has a circular cross section with rigid walls. The sample to be tested is cut to fit snugly into the tube. If the sample is intended to be used while mounted on a solid surface, it is placed in contact with the heavy backing plate. If the material is to be used with a space behind it, it is mounted an appropriate distance from the backing plate.

At the other end of the tube is a small loudspeaker with a hole drilled through its magnet to accommodate a long, slender probe tube coupled to a microphone. Energizing the loudspeaker at a given frequency sets up standing waves due to the interaction of the outgoing wave with the wave reflected from the sample. The form of this standing wave gives important information on the absorbance of the material under test.

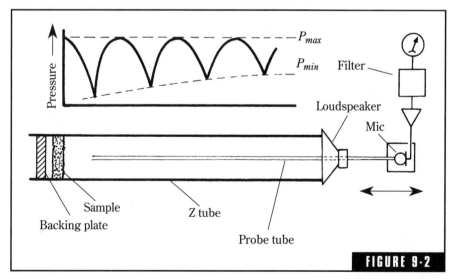

FIGURE 9-2

The standing-wave tube method of measuring the absorption coefficient of absorbing materials at normal incidence.

The sound pressure is maximal at the surface of the sample. As the microphone probe tube is moved away from the sample, the sound pressure falls to the first minimum. Successive, alternating maxima and minima will be detected as the probe tube is further withdrawn. If n is the ratio of the maximum sound pressure to its adjacent minimum, the normal absorption coefficient a_n is equal to:

$$a_n = \frac{4}{n + \dfrac{1}{n} + 2} \qquad \text{(9-2)}$$

which is plotted in Fig. 9-3.[2]

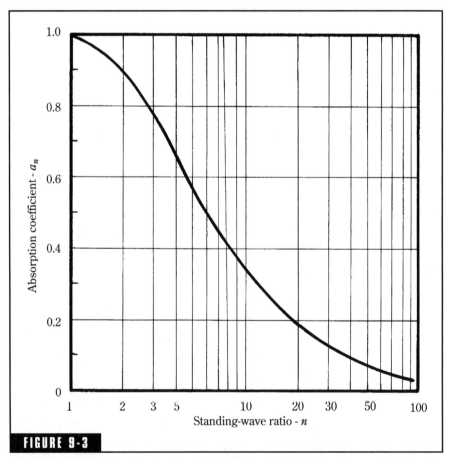

FIGURE 9-3

A graph for interpreting the standing-wave ratio in terms of absorption coefficient. The standing-wave ratio can be found by dividing any pressure maximum by its adjacent pressure minimum (see Fig. 9-2).

The advantages of the impedance tube method are offset by the disadvantage that the absorption coefficient so determined is truly only for normal (perpendicular) incidence. In a room, sound impinges on the surface of a material at all angles. Figure 9-4 is a graph for obtaining the random-incidence coefficient from the absorption coefficient for normal incidence as measured by the standing-wave method. The random-incidence coefficients are always higher than the coefficients for normal incidence.

Tone-Burst Method

The utilization of short pulses of sound has made it possible to perform anechoic acoustical measurements in ordinary rooms. It takes time for

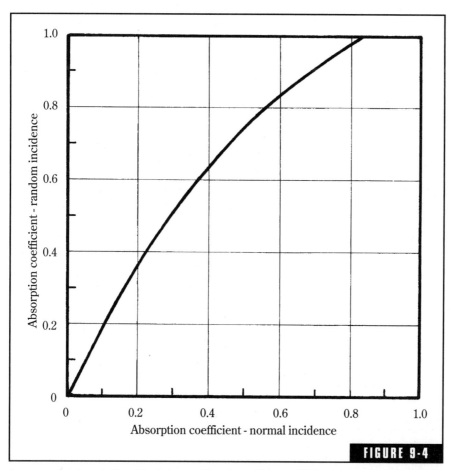

FIGURE 9-4

An approximate relationship between the absorption coefficients at normal incidence and those with random incidence.

bothersome reflections from walls and other surfaces to arrive at the measuring position. If the pulse is short enough, the time gate can be opened only for the desired sound pulse, shutting out the interfering pulses. This *tone-burst method* can be used to measure the sound-absorption coefficient of a material at any desired angle of incidence.

Such an arrangement is illustrated in principle in Fig. 9-5. The source-microphone system is calibrated at distance x as shown in Fig. 9-5A. The geometry of Fig. 9-5B is then arranged so that the total path of the pulse reflected from the material to be tested is equal to this same distance x. The strength of the reflected pulse is then compared to that of the unreflected pulse at distance x to determine the absorption coefficient of the sample.

A recent surge of interest in the influence of individual reflections on the timbre of sound is a new and promising development in acoustics. In this new field specific normal reflections, called "early sound," are of special interest. Although random-incidence coefficients are still of interest in room reverberation calculations, for these image control problems normal-incident reflection coefficients are generally required. Thus, interest may be returning to normal (right-angle) coefficients obtained by the resonance-tube method. There may even be a renewed interest in the old "quarter-wavelength rule" in which the porous absorber for normal incidence must be at least a quarter wavelength thick at the frequency of interest. For example, for a frequency of 1 kHz, the minimum absorber thickness should be about 3.4".

Mounting of Absorbents

The method of mounting the test sample on the reverberation chamber floor is intended to mimic the way the material is actually used in practice. Table 9-1 lists the standard mountings, both in the old form and in the ASTM form that will be used in the future.

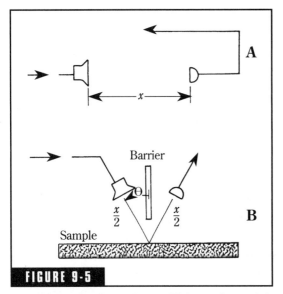

FIGURE 9-5

Determining the absorption coefficients of materials by a tone-burst method. The source-microphone system is calibrated at distance X as shown in (A).

Table 9-1 Mountings commonly used in sound-absorption measurements.

New mounting designation*		Old mounting designation**
A	Material directly on hard surface	#4
B	Material cemented to plasterboard	#1
C-20	Material with perforated, expanded or other open facing furred out 20 mm (3/4″)	#5
C-40	Ditto, furred out 40 mm (1^1/2″)	#8
D-20	Material furred out 20 mm (^{3}/4″)	#2
E-405	Material spaced 405 mm (16″) from hard surface	#7

*ASTM designation: E 795-83.

**Mountings formerly listed by Acoustical and Board Products Manufacturers Association, ABPMA (formerly the Acoustical and Insulating Materials Association, AIMA).

The cooperation of Riverbank Acoustical Laboratory in providing information on current practice is gratefully acknowledged.

The mounting has a major effect on the absorption characteristics of the material. For example, the absorption of porous materials is much greater with an airspace between the material and the wall. Tables of absorption coefficients should always identify the standard mounting or include a description of the way the material was mounted during the measurements, or the coefficients are of little value. Mounting *A* with no air space between the sound absorbing material and the wall is widely used. Another one commonly used is mounting *E-405,* which is at least an approximation to the varying spaces encountered in suspended ceilings (Fig. 9-6).

Mid/High-Frequency Absorption by Porosity

The key word in this discussion of porous sound absorbers is *interstices.* It is simply the space between two things. If a sound wave strikes a wad of cotton batting, the sound energy sets the cotton fibers vibrating. The fiber amplitude will never be as great as the air particle amplitude of the sound wave because of frictional resistance. Some sound energy is changed to frictional heat as fibers are set in motion, restricted as this motion is. The sound penetrates more and more into the interstices of the cotton, losing more and more energy as more and

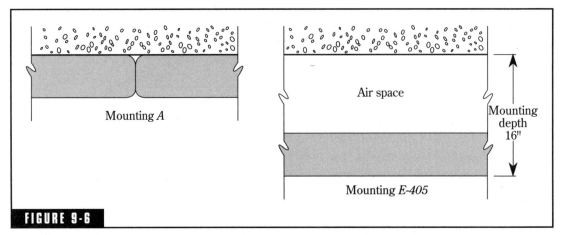

FIGURE 9-6

Commonly used standard mountings associated with listings of absorption coefficients. With Mounting A the material is flat against the backing. Mounting B (not shown) is similar to Mounting A but with a $^1/_{16}''$ air-space resulting when acoustical materials are cemented to a surface. Mountings B and A are essentially the same. Mounting E-405 applies to suspended ceilings with lay-in panels. (See Table 9-1.)

more fibers are vibrated. Cotton is an excellent sound absorber that has been specified in studio treatment in Africa where it was plentiful and cheap, and because imported materials were out of the question.

Porous absorptive materials most commonly used as sound absorbers are usually fuzzy, fibrous materials in the form of boards, foams, fabrics, carpets, cushions, etc. If the fibers are too loosely packed, there will be little energy lost as heat. On the other hand, if they are packed too densely, penetration suffers and the air motion cannot generate enough friction to be effective. Between these extremes are many materials that are very good absorbers of sound. These are commonly composed of cellulose or mineral fiber. Their effectiveness depends on the thickness of the material, the airspace, and the density of the material.

The absorption efficiency of materials depending on the trapping and dissipating of sound energy in tiny pores can be seriously impaired if the surface pores are filled so that penetration is limited. Coarse concrete block, for example, has many such pores and is a fair absorber of sound. Painting that block fills the surface pores and greatly reduces sound penetration, and thus absorption. However, if spray painted with a nonbridging paint, the absorption may be reduced very modestly. Acoustical tile painted at the factory minimizes the problem of reduced absorption. Under certain conditions, a painted surface can reduce porosity but act as a diaphragm that might actually become a fair absorber on a different principle, that of a damped vibrating diaphragm.

In the first radio broadcasting studios, the acoustical treatment was an overuse of carpeted floors and drapes, which emphasized a serious shortcoming of most porous absorbers—that of poor low frequency absorption. Tiles of cellulose fiber with perforated faces became the next style of treatment, but they were also deficient in low-frequency absorption. Overly enthusiastic use of porous absorbers, not only during the early days but even today, causes overabsorption of the high-frequency sound energy, without touching a major problem of room acoustics, low-frequency standing waves.

Glass Fiber: Building Insulation

Great quantities of glass fiber materials are used in the acoustical treatment of recording studios, control rooms, and public gathering spaces. These glass fibers can consist of both special, high-density materials, and ordinary building insulation. In wood or steel stud single frame walls, double walls, and staggered stud walls thermal insulating batts are commonly used. This material usually has a density of about 1 lb/cu ft. Such material is often identified as R-11, R-19, or other such numbers. These R-prefix designations have to do with thermal insulating qualities, but are related to thickness. Thus R-8 is 2.5″ thick, R-11 is 3.5″, and R-19 is 6″.[3]

Building insulation installed within a wall increases its transmission loss a modest amount, primarily by reducing cavity resonance that would tend to couple the two wall faces at the resonance frequency of the cavity. A certain increase in the transmission loss of the wall can also be attributed to attenuation of sound in passing through the glass fiber material, but this loss is small because of the low density of the material. Considering all mechanisms, the transmission loss of a staggered stud wall with a layer of gypsum board on each side can be increased about 7 dB by adding 3.5″ of building insulation. A double wall might show as much as a 12 dB increase by adding 3.5″ and 15 dB with 9″ of insulation. As far as wall transmission loss is concerned, using the denser, more expensive glass fiber between wall faces offers only a slight advantage over ordinary building insulation.

Glass Fiber: Boards

This type of glass fiber usually used in the acoustical treatment of audio rooms is in the form of semi-rigid boards of greater density than building insulation. Typical of such materials are Owens-Corning Type 703 Fiberglas and Johns-Manville 1000 Series Spin-Glass, both of 3 lb/cu ft density. Other densities are available, for example, Type 701 has a density of 1.58 lb/cu ft and Type 705 a density of 6 lb/cu ft. The Type 703 density, however, is widely applied in studios.

These semi-rigid boards of glass fiber do not excel cosmetically, hence they are usually covered with fabric. They do excel in sound absorption.

Acoustical Tile

During the 1960s and 1970s many top-line manufacturers of acoustical materials offered their competitive lines of 12″- × -12″-acoustical tiles. Surface treatments of the tiles included even-spaced holes, random holes, slots, or fissured or other special textures. They continue to be available from local building material suppliers. Such tiles are reputable products for noise and reverberation control as long as they are used with full knowledge of their limitations. One of the problems of using acoustical tile in critical situations is that absorption coefficients are rarely available for the specific tile obtainable. Going back into the earlier literature, the average of the coefficients for eight cellulose and mineral fiber tiles of $3/4''$ thickness is shown in Fig. 9-7. The range of the coefficients is indicated by the vertical lines. The average points could be used for $3/4''$ tile for which no coefficients are available. Coefficients 20% lower would be a fair estimate for $1/2''$ tiles.

Effect of Thickness of Absorbent

It is logical to expect greater sound absorption from thicker materials, but this logic holds primarily for the lower frequencies. Figure 9-8 shows the effect of varying absorbent thickness where the absorbent is mounted directly on a solid surface (mounting A). In Fig. 9-8, there is little difference above 500 Hz in increasing the absorbent thickness from 2 inches to 4 inches, but there is considerable improvement below 500 Hz as thickness is increased. There is also a proportionally greater gain in overall absorption in a 1-inch increase of thickness in

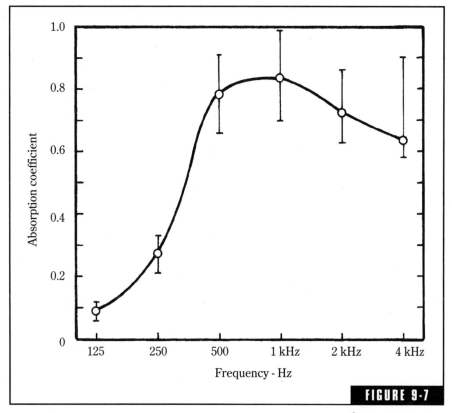

The average absorption characteristics of 8 acoustical tile brands of $^3/_4$" thickness. The vertical lines show the spread of the data.

going from 1 to 2 inches than going from 2 to 3 inches or 3 to 4 inches. A 4-inch thickness of glass fiber material of 3-lb/cu ft density has essentially perfect absorption over the 125-Hz to 4-kHz region.

Effect of Airspace Behind Absorbent

Low-frequency absorption can also be improved by spacing the absorbent out from the wall. This is an inexpensive way to get improved performance—within limits. Figure 9-9 shows the effect on the absorption coefficient of furring 1-inch glass-fiber wallboard out from a solid wall. Spacing 1-inch material out 3 inches makes its absorption approach that of the 2-inch material of Fig. 9-8 mounted directly on the wall.

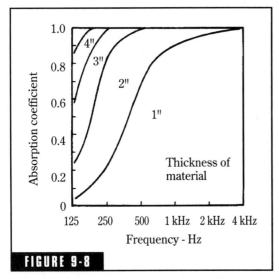

FIGURE 9-8

The thickness of glass-fiber sound-absorbing material determines the low-frequency absorption (density, 3 lb/cu ft). The material is mounted directly on a hard surface.

Effect of Density of Absorbent

Glass fiber and other materials come in various densities from the flimsy thermal insulation batts to the semi-rigid and rigid boards used widely in industry. All of these have their proper place in acoustical treatment of spaces, but the question right now is, "What effect does density—the packing of the fibers—have on sound absorption coefficient?" In other words, is the sound able to penetrate the interstices of the high-density, harder surface material as well as one of the flimsy kind? The answer appears in Fig. 9-10, which shows relatively little difference in absorption coefficient as the density is varied over a range of almost 4 to 1. For very low densities the fibers are so widely spaced that absorption suffers. For extremely dense boards, the surface reflection is high and sound penetration low.

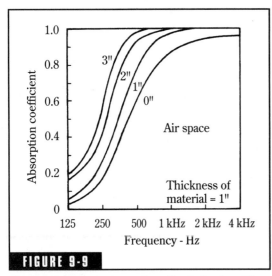

FIGURE 9-9

The low-frequency absorption of 1-inch glass fiberboard is improved materially by spacing it out from the solid wall.

Open-Cell Foams

Flexible polyurethane foams are widely used in noise quieting of automobiles, machinery, aircraft, and in various industrial applications. After a slow start, foams are finding some application as sound absorbers in architectural applications, including recording studios and home listening rooms. Figure 9-11 is a photograph of one form of Sonex,[4] a foam product contoured to simulate the wedges used in anechoic rooms. These are shaped in male and female molds and come in meshed pairs. This material can be cemented or stapled to the surface to be treated.

The sound absorption coefficients of Sonex for thicknesses of 2″, 3″, and 4″ are shown in Fig. 9-12 for Mounting A. The 2″ glass fiber of Fig. 9-8 is considerably superior acoustically to the 2″ Sonex but a few things should be considered in this comparison. These are:

- The Type 703 has a density of 3 lb/cu ft while Sonex is 2 lb/cu ft.

- The 2″ Sonex is the wedge height and the average thickness is far less, while the 703 thickness prevails throughout.

- Comparing the two products is, in a sense, specious because the much higher cost of Sonex is justified in the minds of many by appearance and ease of mounting rather than straight acoustical considerations.

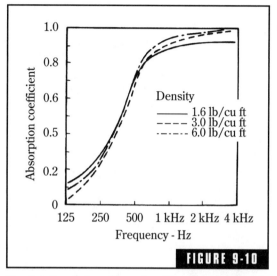

FIGURE 9-10

The density of glass-fiber absorbing material has relatively little effect on absorption in the range of 1.6 to 6 lb/cu ft. The material is mounted directly on solid wall.

Drapes as Sound Absorbers

Drapes are a porous type of sound absorber because air can flow through the fabric under pressure. Variables affecting absorbency include weight of material, degree of drape, and distance from the wall. Data are scarce, but Fig. 9-13 compares the absorption of 10, 14, and 18 oz/sq yd velour hung straight and presumably at some distance from the wall. One intuitively expects greater absorption with heavier material. However, the greater absorption in going from 14 to 18 oz/sq yd than in going from 10 to 14 oz/sq yd is difficult to explain. The effect, whatever it is, is concentrated in the 500- to 1-kHz region.

The amount of fullness of the drape has a great effect as shown in Fig. 9-14. The "draped to 7/8 area" means that the entire 8/8 area is drawn in only slightly (1/8) from the flat condition. The deeper the drape fold, the greater the absorption.

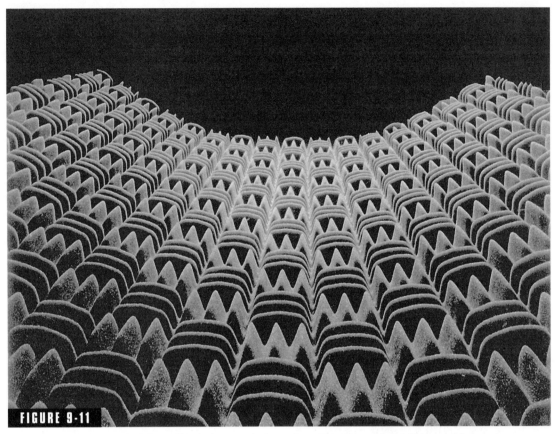

FIGURE 9-11

Sonex contoured acoustical foam simulating anechoic wedges. This is an open-cell type of foam.

The distance a drape is hung from a reflecting surface can have a great effect on its absorption efficiency. This is best explained by Fig. 9-15. In Fig. 9-15A a drape or other porous material is hung parallel to a solid wall, and the distance d between the two is varied. The frequency of the sound impinging on the porous material is held constant at 1,000 Hz. If the sound absorption provided by the porous material is measured, we find that it varies greatly as the distance d from the wall is changed. Looking at the situation closely reveals that the wavelength of the sound is related to maxima and minima of absorption. The wavelength of sound is the speed of sound divided by frequency, which in the case of 1,000 Hz, is $1,130/1,000 = 1.13$ feet or about 13.6 inches. A quarter wavelength is 3.4 inches, and a half wavelength is 6.8 inches. We note that there are absorption peaks at $1/4$ wavelength, and if we carry it further than indicated in Fig. 9-15A at each odd multiple of quarter wavelengths. Absorption minima occur at even multiples of quarter wavelengths.

These effects are explained by reflections of the sound from the solid wall. At the wall surface, pressure will be highest, but air particle velocity will be zero because the sound waves cannot supply enough energy to shake the wall. At a quarter wavelength from the wall, however, pressure is zero, and air particle velocity is maximum. By placing the porous material, such as a drape, a quarter wavelength from the wall, it will have maximum absorbing effect because the particle velocity is maximum. At half wavelength, particle velocity is at a minimum, hence absorption is minimum.

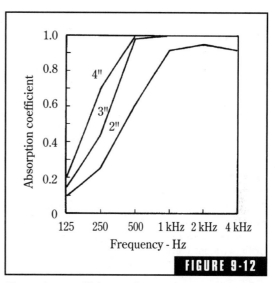

FIGURE 9-12

Absorption coefficients of Sonex contoured foam of various thicknesses.[4]

In Fig. 9-15B, the spacing of the drape from the wall is held constant at 12 inches as the absorption is measured at different frequencies. The same variation of absorption is observed, maximum when the spacing is at odd quarter wavelengths and minimum at even quarter wavelengths from the wall. At this particular spacing of 12 inches, a wavelength of spacing occurs at $1{,}130/1 = 1{,}130$ Hz, a quarter wavelength at 276 Hz, and a half wavelength at 565 Hz, etc.

Figure 9-16 shows actual reverberation-chamber measurements of the absorption of 19 oz/sq yd velour. The solid graph is presumably for a drape well removed from all walls. The other graphs, very close together, are for the same material spaced 10 cm (about 4″) and 20 cm (about 8″) from the wall. The 10-cm distance is one wavelength at 3,444 Hz, the 20-cm distance is at

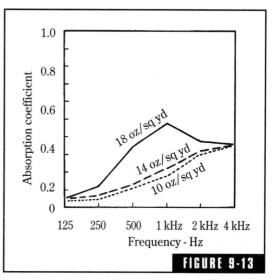

FIGURE 9-13

The sound absorption of velour hung straight for three different weights of fabric. (After Beranek.[18])

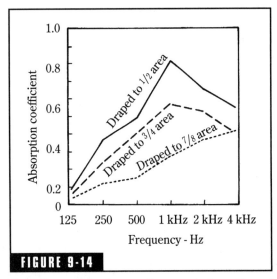

FIGURE 9-14

The effect of draping on the sound absorption of drapes. "Draped to $^1/_2$ area" means that folds are introduced until the resulting drape area is half that of the straight fabric. (After Mankovsky.[9])

1,722 Hz. The odd multiples of both the 10-cm and the 20-cm quarter wavelengths are spotted on the upper part of Fig. 9-16.

The absorption of the velour is greater when spaced from the wall, and the effect is greatest in the 250-Hz to 1-kHz region. At 125 Hz, the 10 and 20 cm spacing adds practically nothing to the drape absorption because at 125 Hz, the quarter wavelength spacing is 2.26 feet. When referring to quarter wavelengths, sine waves are inferred. Absorption measurements are invariably made with bands of random noise. Hence we must expect the wiggles of Fig. 9-15B to be averaged out by the use of such bands.

Carpet as Sound Absorber

Carpet commonly dominates the acoustical picture in spaces as diverse as living rooms, recording studios, and churches. It is the one amenity the owner often specifies in advance and the reason is more often comfort and appearance than acoustic. For example, the owner of a recording studio with a floor area of 1,000 square feet specified carpet. He was also interested in a reverberation time of about 0.5 second, which requires 1,060 sabins of absorption. At the higher audio frequencies, this heavy carpet and pad with an absorption coefficient of around 0.6 gives 600 sabins of absorption at 4 kHz or 57% of the required absorption for the entire room before the absorption needs of walls and ceiling are even considered. The acoustical design is almost frozen before it is started.

There is another, more serious problem. This high absorbance of carpet is only at the higher audio frequencies. Carpet having an absorption coefficient of 0.60 at 4 kHz offers only 0.05 at 125 Hz. In other words, the 1,000 sq ft of carpet introduces 600 sabins at 4 kHz but only 50 sabins at 125 Hz! This is a major problem encountered in many acoustical treatment jobs. The unbalanced absorption of carpet

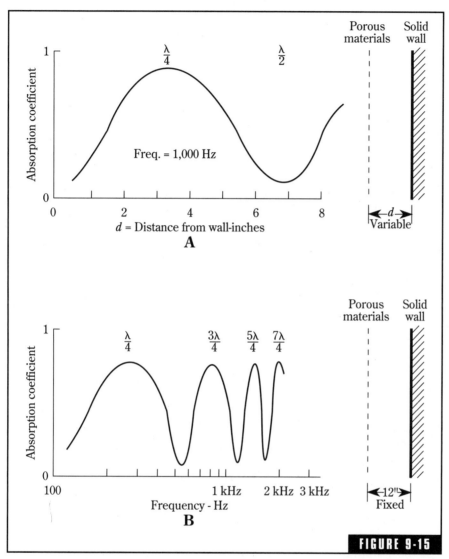

(A) The sound absorption of porous material such as a drape varies with the distance from a hard wall. The maximum absorption is achieved when the drape is one-quarter wavelength from the wall, the minimum at a half wavelength. (B) The sound absorption of porous material hung at a fixed distance from a wall will show maxima at spacings of a quarter wavelength and odd multiples of a quarter wavelength as the frequency is varied.

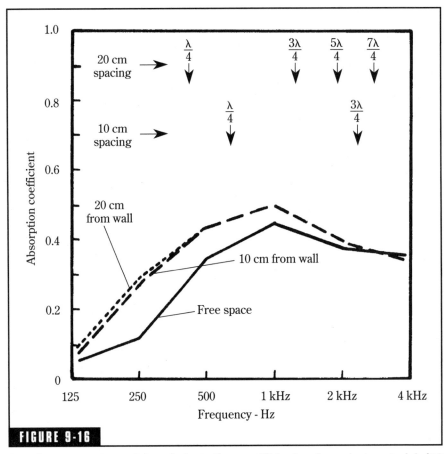

FIGURE 9-16

Actual measurements of sound-absorption coefficients of a velvet material (19 oz/sq yd) in free space and 10 cm and 20 cm from a solid wall. The point at which the increase in absorption due to wall reflection is to be expected are indicated. (After Mankovsky.[9])

can be compensated in other ways, principally with resonant-type, low-frequency absorbers.

To compound the problem of unbalanced absorption of carpet, dependable absorption coefficients are hard to come by. A bewildering assortment of types of carpet and variables in underlay add to the uncertainty. Unfortunately, reverberation chamber measurements of random-incidence absorption coefficients for specific samples of carpet are involved and expensive, and generally unavailable to the acoustical designer. Therefore, it is well to be informed on the factors affecting the absorption of carpets and make judgments on what coefficients to use for the specific carpet at hand.

Effect of Carpet Type on Absorbance

What variation in sound absorption should one expect between types of carpet? Figure 9-17 shows the difference between a heavy Wilton carpet and a velvet carpet with and without a latex back. The latex backing seems to increase absorption materially above 500 Hz and to decrease it a modest amount below 500 Hz.

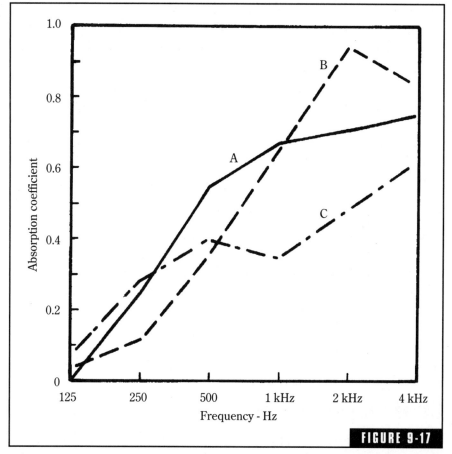

FIGURE 9-17

A comparison of the sound-absorption characteristics of three different types of carpet. (A) Wilton, pile height 0.29", 92.6 oz/sq yd. (B) Velvet, latex backed, pile height 0.25", 76.2 oz/sq yd. (C) The same velvet without latex backing, 37.3 oz/sq yd, all with 40-oz hair felt underlay. (After Harris.[5])

Effect of Carpet Underlay on Absorbance

Hair felt was formerly used almost exclusively as the padding under the carpet. It is interesting that in the early days of this century hair felt was used for general acoustical treatment until it was displaced by numerous proprietary materials. Today foam rubber, sponge rubber, felts, polyurethane, or combinations have replaced hair felt. Foam rubber is made by whipping a latex water dispersion, adding a gelling agent, and pouring into molds. The result is always open-celled. Sponge rubber, on the other hand, formed by chemically generated gas bubbles, can yield either open or closed cells. Open cells provide the interstices required for good sound absorption while closed cells do not.

The influence of underlay on carpet absorption is very great. Figure 9-18 shows chamber measurements of absorption coefficients for a single Axminster type of carpet with different underlay conditions. Graphs A and C show the effect of hair felt of 80 and 40 oz/sq yd weight. Graph B shows an intermediate combination of hair felt and foam. While these three graphs differ considerably, they all stand in stark contrast to graph D for the carpet laid directly on bare concrete. Conclusion: the padding underneath the carpet contributes markedly to overall carpet absorption.

Carpet Absorption Coefficients

The absorption coefficients plotted in Figs. 9-17 and 9-18 are taken from Harris's 1957 paper,[5] perhaps the most exhaustive study of carpet characteristics available. In general, these coefficients are higher than those in currently published tables. In Fig. 9-19 the coefficients listed in a widely used publication[6] and included in the appendix are plotted for comparison with Figs. 9-17 and 9-18. Carpets vary widely, which can account for some of the great variability with which any designer of acoustical systems is confronted.

Sound Absorption by People

People absorb sound too. Just how much of this is due to absorption by flesh and how much by hair and clothing has yet to be reported. The important thing is that the people making up an audience account for a significant part of the sound absorption of the room. It also makes an

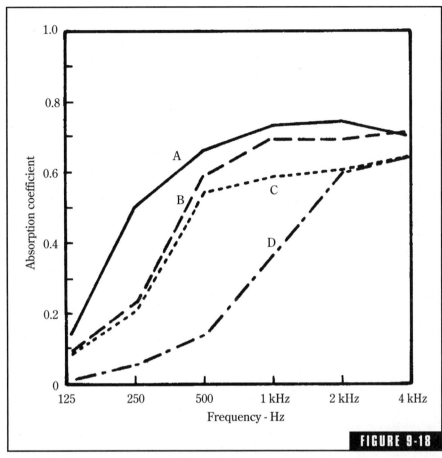

FIGURE 9-18

Sound-absorption characteristics of the same Axminster carpet with different underlay. (A) 80-oz hair felt. (B) Hair felt and foam. (C) 40-oz hair felt. (D) No underlay, on bare concrete. (After Harris.[5])

acoustical difference whether one or ten people are in a small monitoring room. The problem is how to rate human absorption and how to involve it in calculations. The usual method of multiplying a human absorption coefficient by the area of a human has its problems. The easy way is to determine the absorption units (sabins) a human presents at each frequency and add them to the sabins of the carpet, drapes, and other absorbers in the room at each frequency. Table 9-2 lists the absorption of informally dressed college students in a classroom along with a range of absorption for more formally dressed people in an auditorium environment.

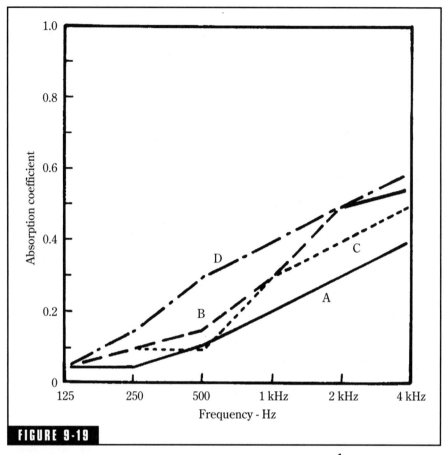

Carpet absorption coefficients from a commonly used table. (A) $^1/_8''$ pile height. (B) $^1/_4''$ pile height. (C) $^3/16''$ combined pile and foam. (D) $^5/16''$ combined pile and foam. Compare these graphs with those of Figs. 9-16 and 9-17.[6]

For 1 kHz and higher, the absorption offered by college students in informal attire in the Spartan furnishings of a classroom falls at the lower edge of the range of a more average audience. The low-frequency absorption of the students, however, is considerably lower than that of the more formally dressed people.

Sound propagated across rows of people, as in an auditorium or music hall, is subjected to an unusual type of attenuation. In addition to the normal decrease in sound with distance from the stage, there is an additional dip of up to 15 or 20 dB around 150 Hz and spreading over the 100- to 400-Hz region.[7] In fact, this is not strictly an audience effect because it prevails even when the seats are empty. A similar dip

Table 9-2 Sound absorption by people (Sabins per person).

	Frequency, Hz					
	125	250	500	1 kHz	2 kHz	4 kHz
College students informally dressed seated in tablet arm chairs[6]	—	2.5	2.9	5.0	5.2	5.0
Audience seated, depending on spacing and upholstery of seats[22]	2.5–4.0	3.5–5.0	4.0–5.5	4.5–6.5	5.0–7.0	4.5–7.0

in sound-pressure level affects those most important first reflections from the sidewalls. All of this apparently results from interference.

Absorption of Sound in Air

For frequencies 1 kHz and above and for very large auditoriums, the absorption of sound by the air in the space becomes important. For example, a church seating 2,000 has a volume of about 500,000 cubic feet.

Frequency (Hz)	Absorption (sabins per 1,000 cu ft)
1,000	0.9
2,000	2.3
4,000	7.2

Notice that for 50 percent relative humidity the absorption is 7.2 sabins per 1,000 cubic feet or a total of (500) (7.2) = 3,600 sabins at 4 kHz. This is equivalent to 3,600 square feet of perfect absorber.

This could be 20 percent to 25 percent of the total absorption in the space, and there is nothing that can be done about it other than to take it into consideration and taking consolation in at least knowing why the treble reverberation time falls off so much!

Low-Frequency Absorption by Resonance

The concept of wall reflection, graphically portrayed for drapes in Fig. 9-15, applies as well to *bass traps*. This phrase is applied to many kinds of low-frequency sound absorbers, such as panel absorbers, but perhaps it should really be reserved for a special type of reactive cavity absorber that has been widely used in sound recording circles.[8] A true bass trap is shown in Fig. 9-20. It is simply a box or cavity of critical depth but with a mouth opening of size to suit particular purposes. This is a tuned cavity with a depth of a quarter wavelength at the design frequency at which maximum absorption is desired. Sound absorption at the lowest octave or two of the audible spectrum is often difficult to achieve. The bass trap is commonly used in recording studio control rooms to reduce standing waves at these bass frequencies.

The sound pressure at the bottom of the cavity is maximum at the quarter-wavelength design frequency. The air particle velocity is zero at the bottom. At the mouth the pressure is zero and the particle velocity is maximum, which results in two interesting phenomena. First, a glass fiber semi-rigid board across the opening offers great friction to the rapidly vibrating air particles resulting in maximum absorption at this frequency. In addition, the zero pressure at the opening consti-

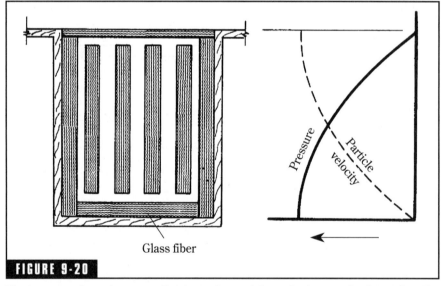

Glass fiber

FIGURE 9-20

The bass trap depends upon reflections of sound from the bottom for its action. The pressure for the frequency at which the depth is a quarter wavelength is maximum at the bottom and the particle velocity is zero at the bottom. At the mouth, the pressure is zero (or very low) and the particle velocity is maximum. Absorbent placed where the particle velocity is maximum will absorb sound very effectively. The same action occurs at odd multiples of the quarter wavelength.

tutes a vacuum that tends to suck sound energy into this "sound sink" from surrounding areas. The bass trap's effect, then, is greater than its opening area would suggest because of this "vacuum cleaner" effect.

The bass trap effect, like the drape spaced from a reflective wall, occurs not only at a quarter-wavelength depth, but also at odd multiples of a quarter wavelength. Great trap depths are required for very low bass frequencies. For example, a quarter wavelength for 40 Hz is 7 ft. Unused spaces above control room ceilings and between inner walls and outer shells are often used for trap space.

Diaphragmatic Absorbers

The absorption of sound at the lower audible frequencies can be achieved by porous absorbers or by resonant (or reactive) absorbers. Glass fiber and acoustical tiles are common forms of porous absorption in which the sound energy is dissipated as heat in the interstices of the fibers. The absorption of commercial forms of glass fiber and other fibrous absorbers at low audio frequencies, however, is quite poor. To absorb well, the thickness of the porous material must be comparable to the wavelength of the sound. At 100 Hz, the wavelength is 11.3 ft, and using any porous absorber approaching this thickness would be impractical. For this reason we turn our attention to the resonant type of absorber to obtain absorption at low frequencies.

Some of the great chamber music rooms owe their acoustical excellence to the low-frequency absorption offered by extensive paneled walls. Plywood or tongue-and-groove flooring or subflooring vibrates as a diaphragm and contributes to low-frequency absorption. Drywall construction on walls and the ceiling does the same thing. All such components of absorption must be included in the acoustical design of a room, large or small.

Drywall or gypsum board plays a very important part in the construction of homes, studios, control rooms, and other spaces. It also plays an important part in the absorption of low-frequency sound in these spaces. Usually, such low-frequency absorption is welcome, but in larger spaces designed for music, drywall surfaces can absorb so much low-frequency sound as to prevent the achievement of the desired reverberatory conditions. Drywall of 1/2-inch thickness on studs spaced 16 inches offers an absorption coefficient of 0.29 at 125 Hz and even higher at 63 Hz (which would be of interest in music recording studios). Drywall absorption in small audio rooms is free; you just have to recognize its existence and remember to include its low-frequency absorption in calculations.

The simplest resonant type of absorber utilizes a diaphragm vibrating in response to sound and absorbing some of that sound by frictional heat losses in the fibers as it flexes.

A piece of $1/4''$ plywood is an excellent example. Assume that it is spaced out from the wall on two-by-fours, which gives close to $3^3/4''$ airspace behind. The frequency of resonance of this structure can be calculated from the expression:

$$f_o = \frac{170}{\sqrt{(m)\,(d)}} \qquad \text{(9-3)}$$

where

f_o = frequency of resonance, Hz
m = surface density of the panel, lb/sq ft of panel surface
d = depth of airspace, inches.

The surface density of $1/4''$ plywood, 0.74 lb/sq ft, can be measured or found in the books. Substituting in Eq. 9-3 we get:

$$f_o = \frac{170}{\sqrt{(0.74)\,(3.75)}}$$

$$f_o = 102 \text{ Hz}$$

Figure 9-21 is a graphical solution of Eq. 9-3 for maximum convenience. Knowing only the thickness of the plywood and the depth of the space behind the plywood, the frequency of resonance can be read off the diagonal lines. Equation 9-3 applies to membranes and diaphragms of materials other than plywood such as masonite, fiberboard, or even Kraft paper. For other than plywood, the surface density must be determined. The surface density is easily found by weighing a piece of the material of known area.

How accurate are Eq. 9-3 and Fig. 9-21? Actual measurements on three plywood membrane absorbers are shown in Fig. 9-22. Such calculations of the frequency of peak absorption at resonance are not perfect, but they are a good first approximation of sufficient accuracy for most purposes.

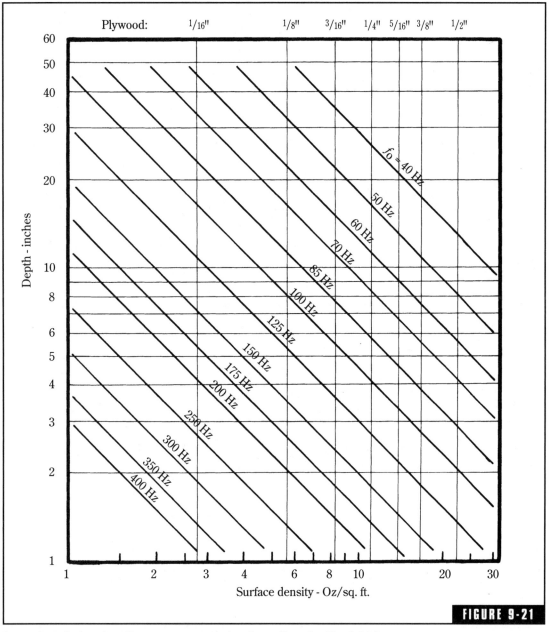

Convenient design chart for resonant panel absorbers. (See also Fig. 9-34.)

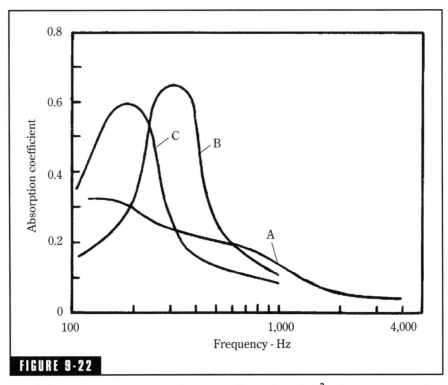

Actual absorption measurements of three panel absorbers. (A) [3]/16″ plywood with 2″ air space.[19] (B) [1]/16″ plywood with 1″ mineral wool and [1]/4″ air space. (C) The same as (B) but for [1]/8″ panel.[20]

Panel sound absorbers are quite simple to build. An example of a panel absorber to be mounted on a flat wall or ceiling surface is shown in Fig. 9-23. A ¼″ or ¹/16″ plywood panel is fastened to a wooden framework to give the desired spacing from the wall. A glass or mineral fiber blanket of 1″ to 1¹/2″ is glued to the wall surface. An airspace of ¹/4″ or ¹/2″ should be maintained between the absorbent and the rear surface of the plywood panel.

A corner panel absorber is shown in Fig. 9-24. For computations, an average depth is used. Depths greater and smaller than the average simply mean that the peak of absorbance is broader than that of an absorber with uniform depth. Spacing the absorber ¹/4″ to ¹/2″ from the rear of the plywood panel is simple if a mineral fiber board such as Tectum is used. Using a flexible blanket of glass fiber requires support by hardware cloth, open-weave fabric, or even expanded metal. For applications in which reflected mid/high-frequency sound from the panel absorber might create problems, a facing of glass fiber board would not interfere with the low-frequency absorbing action if it was spaced to avoid damping of the vibration of the plywood panel. Chapter 15 emphasizes that all room modes terminate in the corners of a room. A corner panel absorber could be used to control such modes.

Polycylindrical Absorbers

Flat paneling in a room might brighten an interior decorator's eye and do some good acoustically, but wrapping a plywood or hardboard skin around some semicylindrical bulkheads can provide some very attractive features. These polycylindrical elements (polys) are coming back into fashion. A few are used in top-flight recording studios today. Visually, they are rather overpowering, which can be good or bad depending on the effect one wants to achieve. With polys it is acoustically possible to achieve a good diffuse field along with liveness and brilliance, factors tending to oppose each other in rooms with flat surfaces.

One of the problems of using polys has been the scarcity of published absorption coefficients. The Russian acoustician, V. S. Mankovsky, has taken care of that in his book.[9] As expected, the larger the chord dimension, the better the bass absorption. Above 500 Hz there is little significant difference between the polys of different sizes.

The overall length of polys is rather immaterial, ranging in actual installations from the length of a sheet of plywood to the entire length, width, or height of a

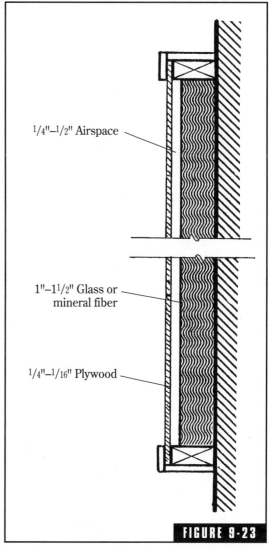

1/4"–1/2" Airspace

1"–1 1/2" Glass or mineral fiber

1/4"–1/16" Plywood

FIGURE 9-23

Typical resonant panel absorber for wall mounting.

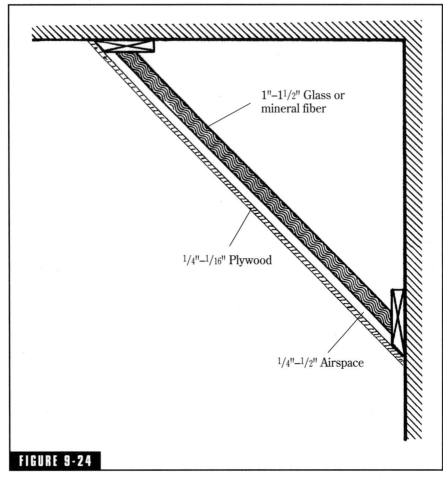

1"–1¹/₂" Glass or
mineral fiber

¹/₄"–¹/₁₆" Plywood

¹/₄"–¹/₂" Airspace

FIGURE 9-24

Typical resonant panel absorber for either vertical or horizontal corner mounting.

studio. It is advisable, however, to break up the cavity behind the poly skin with randomly spaced bulkheads. The polys of Fig. 9-25 incorporate such bulkheads.

Should the polys be empty or filled with something? Mankovsky again comes to our rescue and shows us the effect of filling the cavities with absorbent material. Figures 9-25D and 9-25C show the increase in bass absorption resulting from filling the cavities with absorbent. If needed, this increased bass absorption can be easily achieved by simply filling the polys with glass fiber. If the bass absorption is not needed the polys can be used empty. The great value of this adjustable

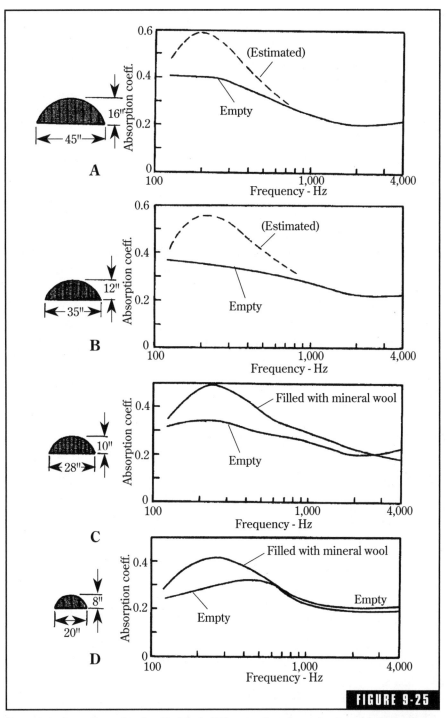

FIGURE 9-25

Measured absorption of polycylindrical diffusors of various chord and height dimensions. In C and D, graphs are shown for both empty polys and for polys filled with mineral wool. In A and B only empty poly data are available; the broken lines are estimated absorption when filled with mineral wool. (After Mankovsky.[9])

feature will become more apparent as the actual acoustical design of listening rooms and studios is approached.

Poly Construction

The construction of polycylindrical diffusers is reasonably simple. A framework for vertical polys is shown in Fig. 9-26 mounted above a structure intended for a low-frequency slat absorber. In this photograph the variable chord dimensions are apparent, and also the random placement of bulkheads so that cavities will be of various volumes resulting in different natural cavity frequencies. It is desirable that each cavity be essentially airtight, isolated from adjoining cavities by well-fitted bulkheads and framework. Irregularities in the wall can be sealed with a nonhardening acoustical sealant. The bulkheads of each poly are carefully cut to the same radius on a bandsaw. Sponge rubber weatherstripping with an adhesive on one side is struck to the edge of each bulkhead to ensure a tight seal against the plywood or hardboard cover. If such precautions are not taken, annoying rattles and coupling between cavities can result.

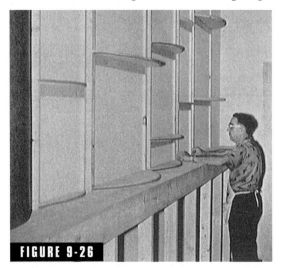

FIGURE 9-26

The construction of polys in a motion picture sound mixing studio. Note the foam rubber anti-rattle strip on the edge of each bulkhead. Also note the random spacing of the bulkheads. *(Moody Institute of Science photo, reprinted with permission of the Journal of the Audio Engineering Society.[21])*

The polys of Fig. 9-26 use $1/8''$ tempered Masonite as the poly skin. A few hints can simplify the job of stretching this skin. In Fig. 9-27 slots of a width to fit the Masonite snugly are carefully cut along the entire length of strips 1 and 2 with a radial saw. Let us assume that poly A is already mounted and held in place by strip 1, which is nailed or screwed to the wall. Working from left to right, the next job is to mount poly B. First the left edge of Masonite sheet B is inserted in the remaining slot of strip 1. The right edge of Masonite sheet B is then inserted in the left slot of strip 2. If all measurements and cuts have been accurately made, swinging strip 2 against the wall should make a tight seal over the bulkheads 3 and weatherstripping 4. Securing strip 2 to the wall

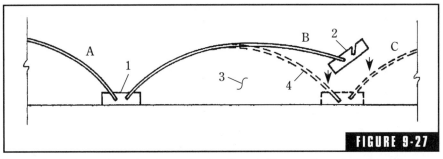

FIGURE 9-27

The method of stretching the plywood or hardboard skin over the poly bulkheads shown in Fig. 9-26.

completes poly B. Poly C is mounted in a similar fashion and so on to the end of the series of polys. The end result is shown in Fig. 9-28. Notice that the axes of symmetry of the polys on the side wall are perpendicular to those on the rear wall. If polys were used on the ceiling, their axes should be perpendicular to both the others.

It is quite practical and acceptable to construct each poly as an entirely independent structure rather than building them on the wall. Such independent polys can be spaced at will.

Membrane Absorbers

Building insulation commonly comes with a kraft paper backing. Between walls this paper has no significant effect, but if building insulation is to be used as a sound absorber on walls, perhaps behind a fabric facing, the paper becomes significant. Figure 9-29 compares the sound absorption efficiency of R-19 (6 inch) and R-11 (3.5 inch) with the kraft paper backing exposed and with the glass fiber exposed to the incident sound. When the paper is exposed it shields the glass fiber from sound above 500 Hz but has little effect below 500 Hz. The net effect is an absorption peak at 250 Hz (R-19) and 500 Hz (R-11), which may be important in

FIGURE 9-28

Finished poly array of Fig. 9-26 mounted on the wall above a low-frequency absorber structure. Note that the axes of the polys on the rear wall are perpendicular to the axes of the polys on the other wall. *(Moody institute of Science photo, reprinted with permission of Journal of the Audio Engineering Society.[21])*

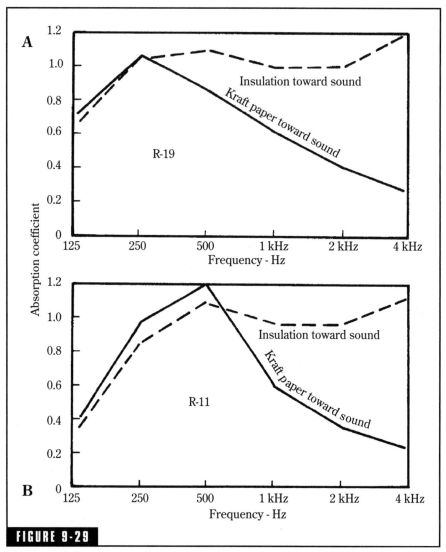

FIGURE 9-29

When ordinary building insulation is used as a wall treatment (perhaps with fabric facing) the position of the kraft paper backing becomes important. (A) R-19 Fiberglass building insulation, (B) R-11 Fiberglass building insulation, Mounting A.

room treatment. With insulation exposed there is essentially perfect absorption above 250 Hz (R-19) or 500 Hz (R-11).

Building insulation has not caught on as an inexpensive absorbing material. One reason is that some sort of cosmetic and protective cover is required, but this is often true of denser materials as well. Fabric, expanded metal, metal lath, hardware cloth, or even perforated vinyl wall covering can be used as a cover. Do not be surprised by absorption coefficients greater than 1.0. This manufacturer elected to publish the coefficients actually obtained from laboratory measurements rather than arbitrarily reducing those over 1.0 to 0.99 or 1.0. The greater absorption of the standard 8 × 9 ft sample results from edge diffraction and other effects that make the sample appear larger acoustically than it really is.

Helmholtz Resonators

The Helmholtz type of resonator is widely used to achieve adequate absorption at lower audio frequencies. There is nothing particularly mysterious about such resonators; in fact they pop up in various forms in everyday life. Blowing across the mouth of any bottle or jug produces a tone at its natural frequency of resonance. The air in the cavity is springy, and the mass of the air in the neck of the jug reacts with this springiness to form a resonating system, much as a weight on a spring vibrating at its natural period. Change the volume of the air cavity, or the length or diameter of the neck, and you change the frequency of resonance. Such a Helmholtz resonator has some very interesting characteristics. For instance, sound is absorbed at the frequency of resonance and at nearby frequencies. The width of this absorption band depends on the friction of the system. A glass jug offers little friction to the vibrating air and would have a very narrow absorption band. Adding a bit of gauze across the mouth of the jug or stuffing a wisp of cotton into the neck, the amplitude of vibration is reduced and the width of the absorption band is increased.

The sound impinging on a Helmholtz resonator that is not absorbed is reradiated. As the sound is reradiated from the resonator opening, it tends to be radiated in a hemisphere. This means that unabsorbed energy is diffused, and diffusion of sound is a very desirable thing in a studio or listening room.

Bottles and jugs are not appropriate forms of a Helmholtz resonator with which to apply the resonance principle in studios. An interesting experiment conducted many years ago at Riverbank Acoustical Laboratories bears this out.[10] To demonstrate the effectiveness of a continuously swept narrow-band technique of measuring sound absorption coefficients, the idea was conceived to measure the absorption of Coca Cola bottles. A tight array of 1,152 empty 10-oz bottles was arranged in a standard 8 × 9 ft space on the concrete floor of the reverberation chamber. It was determined that a single, well-isolated bottle has an absorption of 5.9 sabins at its resonance frequency of 185 Hz, but with a bandwidth (between −3dB points) of only 0.67 Hz! Absorption of 5.9 sabins is an astounding amount of absorption for a Coke bottle! This is about what a person, normally clothed, would absorb at 1,000 Hz, or what 5.9 sq ft of glass fiber (2 in thick, 3 lb/cu ft density) would absorb at midband. The sharpness of this absorption characteristic is even more amazing. This would correspond to a Q of 185/0.67 = 276! As interesting as these data are, they tell us that leaving an empty Coke bottle in a studio will not devastate the acoustics of the room, but it might have a tiny effect at 185 Hz.

In Helmholtz resonators, we have acoustical artifacts that far ante-date Helmholtz himself. Resonators in the form of large pots were used in ancient times by the Greeks and Romans in their open-air theaters. Apparently they were used to provide some reverberation in this nonreverberant outdoor setting. Some of the larger pots that have survived to modern times have reverberation times of from 0.5 to 2 seconds. These would also absorb sound at the lower frequencies. Groupings of smaller pots supplied sound absorption at the higher frequencies.

More recently (that is, in medieval times) such resonators were used in a number of churches in Sweden and Denmark.[11] Pots like those of Fig. 9-30 were embedded in the walls, presumably to reduce low-frequency reverberation that is often a problem in churches. Ashes have been found in some of the pots, undoubtedly introduced to "kill the Q" of the ceramic pot and to broaden the frequency of its effectiveness.

If bottles and ceramic pots are not suitable forms of Helmholtz resonators for a studio, what is? Figure 9-31 shows a conveniently idealized square bottle with a tubular neck. This bottle alone would produce its characteristic tone if one were to blow across the opening. Stacking these bottles does not detract from the resonator action, but rather enhances it. It is a small step to a box of length L, width W, and depth H that has a lid of thickness equal to the length of necks of the bottles. In this lid are drilled holes having the same

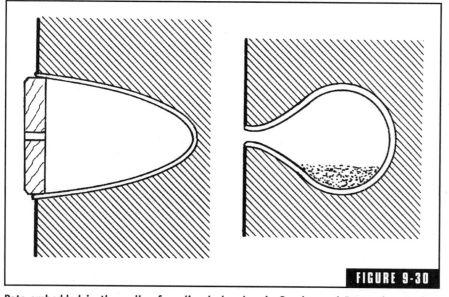

FIGURE 9-30

Pots embedded in the walls of medieval churches in Sweden and Denmark served as Helmholtz resonators, absorbing sound. Ashes, found in some of the pots, apparently served as a dissipative agent. (After Brüel.[11])

diameter as the holes in the neck. It is just a bit harder to realize that partitions between each segment can be removed without greatly affecting the Helmholtz action. In this way, a Helmholtz resonator of the perforated face type can be related to oddly shaped bottles, giving something of a visual picture of how perforated face resonators perform.

In a similar way, Fig. 9-32 illustrates another bottle with an elongated slit neck. These, too, can be stacked, even in multiple rows. It is but a short step to a slot-type resonator. The separating walls in the air cavity can also be eliminated without destroying the resonator action. A word of caution is in order, however. Subdividing the airspace can improve the action of perforated face or slit resonators but only because this reduces

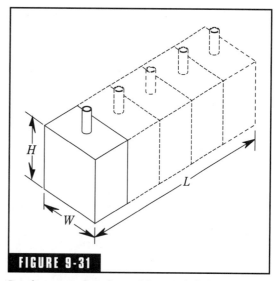

FIGURE 9-31

Development of perforated-face Helmholtz resonator from a single rectangular-battle resonator.

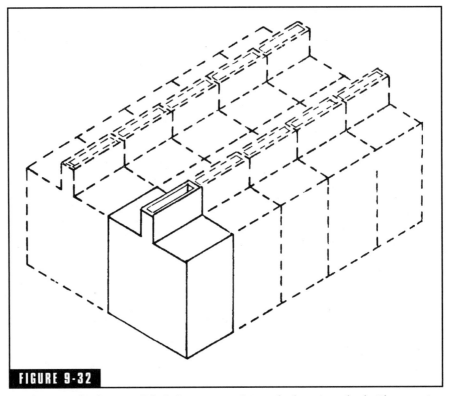

FIGURE 9-32

Development of a slot-type Helmholtz resonator from a single rectangular-bottle resonator.

spurious, unwanted modes of vibration being set up within the air cavity.

Perforated Panel Absorbers

Perforated hardboard or plywood panels spaced from the wall constitute a resonant type of sound absorber.[12–15] Each hole acts as the neck of a Helmholtz resonator, and the share of the cavity behind "belonging" to that hole is comparable to the cavity of the Helmholtz resonator. In fact, we can view this structure as a host of coupled resonators. If sound arrives perpendicular to the face of the perforated panel, all the tiny resonators are in phase. For sound waves striking the perforated board at an angle, the absorption efficiency is somewhat decreased. This loss can be minimized by sectionalizing the cavity behind the perforated face with an egg crate type of divider of wood or corrugated paper.

The frequency of resonance of perforated panel absorbers backed by a subdivided air space is given approximately by:

$$f_o = 200 \sqrt{\frac{p}{(d)\,(t)}} \qquad\qquad \textbf{(9-4)}$$

in which

f_o = frequency, Hz
p = perforation percentage,
  = hole area divided by panel area $\times$ 100
t = effective hole length, inches, with correction factor applied,
  = (panel thickness) + (0.8) (hole diameter)
d = depth of air space, inches

There is a certain amount of confusion in the literature concerning p, the perforation percentage. Some writers use the decimal ratio of hole area to panel area, rather than the percentage of hole area to panel area, introducing an uncertainty factor of 100. This perforation percentage is easily calculated by reference to Fig. 9-33.

Equation 9-4 is true only for circular holes. This information is presented in graphical form in Fig. 9-34 for a panel thickness of 3/16″. Common pegboard with holes 3/16″ in diameter spaced 1″ on centers with the square configuration of Fig. 9-33 has 2.75% of the area in holes. If this pegboard is spaced out from the wall by 2 × 4s on edge, the system resonates at about 420 Hz and the peak absorption appears near this frequency.

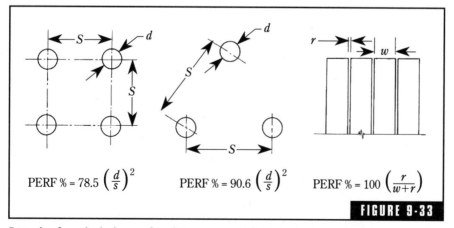

$$\text{PERF \%} = 78.5 \left(\frac{d}{s}\right)^2 \qquad \text{PERF \%} = 90.6 \left(\frac{d}{s}\right)^2 \qquad \text{PERF \%} = 100 \left(\frac{r}{w+r}\right)$$

FIGURE 9-33

Formulas for calculating perforation percentage for Helmholtz resonators.

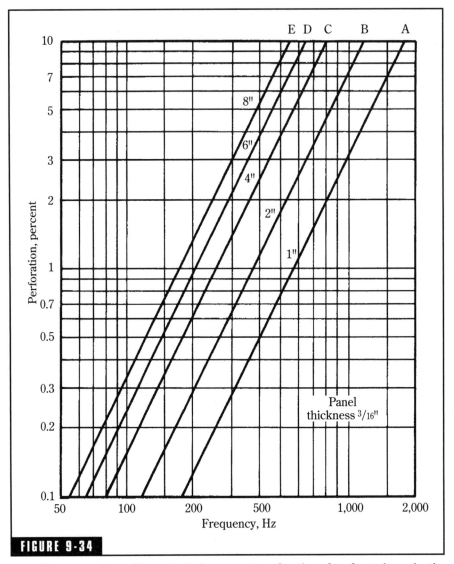

FIGURE 9-34

A graphic presentation of Eq. 9-4 relating percent perforation of perforated panels, the depth of air space, and the frequency of resonance. The graphs are for a panel of $^3/16''$ thickness. (See also Fig. 9-21.) (A) For 1″ furring lumber. The lines are drawn to correspond to furring lumber which is furnished, e.g., the line for 8″ is actually 7 $^3/4$ airspace. (B) For 2″ furring lumber. (C) For 4″ furring lumber. (D) For 6″ furring lumber. (E) For 8″ furring lumber.

In commonly available perforated materials, such as pegboard, the holes are so numerous that resonances at only the higher frequencies can be obtained with practical air spaces. To obtain much needed low-frequency absorption, the holes can be drilled by hand. Drilling $7/32''$ holes $6''$ on centers gives a perforation percentage of about 1.0%.

Figure 9-35 shows the effect of varying hole area from 0.18% to 8.7% in a structure of otherwise fixed dimensions. The plywood is $5/32''$ thick perforated with $3/16''$ holes, except for the 8.7% case in which the hole diameter is about $3/4''$. The perforated plywood sheet is spaced $4''$ from the wall and the cavity is half filled with glass fiber and half is air space.

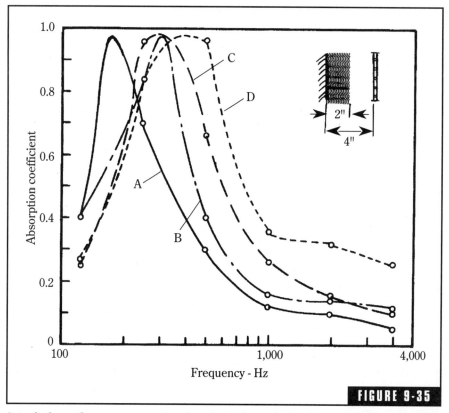

FIGURE 9-35

Actual absorption measurements of perforated panel absorbers of 4" air space, half filled with mineral wool and for panel thickness of $5/32''$. (A) Perforation 0.18%. (B) Perforation 0.79%. (C) Perforation 1.4%. (D) Perforation 8.7%. Note that the presence of the mineral wool shifts the frequency of resonance considerably from the theoretical values of Eq. 9-4 and Fig. 9-34. (Data from Mankovsky.[9])

Figure 9-36 is identical to Fig. 9-35 except that the perforated plywood is spaced 8″ and glass fiber of 4″ thickness is mounted in the cavity. The general effect of these changes is a substantial broadening of the absorption curve.

It would be unusual to employ such perforated panel absorbers without acoustic resistance in the cavity in the form of glass fiber batts or boards. Without such resistance the graph is very sharp. One possible use of such sharply tuned absorbers would be to control specific troublesome room modes or isolated groups of modes with otherwise minimum effect on the signal and overall room acoustics.

Table 9-3 includes the calculated frequency of resonance of 48 different combinations of airspace depth, hole diameter, panel thickness,

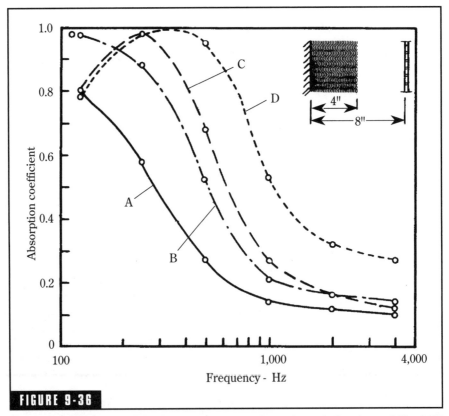

FIGURE 9-36

Actual absorption measurements on the same perforated panel absorbers of Fig. 9-35 except that the air space is increased to 8″, half of which is taken up with mineral fiber. Panel thickness is ⁵/32″. (A) Perforation 0.18%. (B) Perforation 0.79%. (C) Perforation 1.4%. (D) Perforation 8.7%. (Data from Mankovsky.[9])

Table 9-3 Helmholtz low-frequency absorber perforated-face type.

Depth of airspace	Hole dia.	Panel x dia.	% Perf.	Hole spacing	Freq. of resonance
3⁵/₈″	¹/₈″	¹/₈″	0.25%	2.22″	110 Hz
			0.50	1.57	157
			0.75	1.28	192
			1.00	1.11	221
			1.25	0.991	248
			1.50	0.905	271
			2.00	0.783	313
			3.00	0.640	384
3⁵/₈″	¹/₈″	¹/₄″	0.25%	2.22″	89 Hz
			0.50	1.57	126
			0.75	1.28	154
			1.00	1.11	178
			1.25	0.991	199
			1.50	0.905	217
			2.00	0.783	251
			3.00	0.640	308
3⁵/₈″	¹/₄″	¹/₄″	0.25%	4.43″	89 Hz
			0.50	3.13	126
			0.75	2.56	154
			1.00	2.22	178
			1.25	1.98	199
			1.50	1.81	217
			2.00	1.57	251
			3.00	1.28	308
5⁵/₈″	¹/₈″	¹/₈″	0.25%	2.22″	89 Hz
			0.50	1.57	126
			0.75	1.28	154
			1.00	1.11	178
			1.25	0.991	199
			1.50	0.905	218
			2.00	0.783	251
			3.00	0.640	308

Table 9-3 Helmholtz low-frequency absorber perforated-face type. (*Cont.*)

Depth of airspace	Hole dia.	Panel x dia.	% Perf.	Hole spacing	Freq. of resonance
5⅝″	⅛″	¼″	0.25%	2.22″	74 Hz
			0.50	1.57	105
			0.75	1.28	128
			1.00	1.11	148
			1.25	0.991	165
			1.50	0.905	181
			2.00	0.783	209
			3.00	0.640	256
5⅝″	¼″	¼″	0.25%	4.43″	63 Hz
			0.50	3.13	89
			0.75	2.56	109
			1.00	2.22	126
			1.25	1.98	141
			1.50	1.81	154
			2.00	1.57	178
			3.00	1.28	218

and hole spacing. This convenient listing should assist in approximating the desired condition.

Slat Absorbers

Another type of resonant absorber is that utilizing closely spaced slats over a cavity. The mass of the air in the slots between the slats reacts with the springiness of air in the cavity to form a resonant system, again comparable to the Helmholtz resonator. The glass fiber board usually introduced behind the slots acts as a resistance, broadening the peak of absorption. The narrower the slots and the deeper the cavity, the lower the frequency of maximum absorption.

The resonance frequency of the slat absorber can be estimated from the statement[16]:

$$f_o = 216 \sqrt{\frac{p}{(d)\,(D)}} \qquad (9\text{-}5)$$

in which

f_o = frequency of resonance, Hz

p = percent perforation (see Fig. 9-33)

D = airspace depth, inches

d = thickness of slat, inches

Suggestion: If slats are mounted vertically, it is recommended that they be finished in a dark color conforming to the shadows of the slots to avoid some very disturbing "picket fence" optical effects!

Placement of Materials

The application of sound-absorbing materials in random patches has already been mentioned as an important contribution to diffusion. Other factors than diffusion might influence placement. If several types of absorbers are used, it is desirable to place some of each type on ends, sides, and ceiling so that all three axial modes (longitudinal, transverse, and vertical) will come under their influence. In rectangular rooms it has been demonstrated that absorbing material placed near corners and along edges of room surfaces is most effective. In speech studios, some absorbent effective at the higher audio frequencies should be applied at head height on the walls. In fact, material applied to the lower portions of high walls can be as much as twice as effective as the same material placed elsewhere. Untreated surfaces should never face each other.

Winston Churchill once remarked that as long as he had to wear spectacles he intended to get maximum cosmetic benefit from them. So it is with placement of acoustical materials. After the demands of acoustical function have been met, every effort should be made to arrange the resulting patterns, textures, and protuberances into esthetically pleasing arrangements, but do not reverse priorities!

Reverberation Time of Helmholtz Resonators

Some concern has been expressed about the possibility of acoustically resonant devices, such as Helmholtz absorbers, "ringing" with a "reverberation time" of their very own adding coloration to the voice

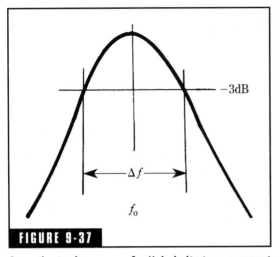

FIGURE 9-37

Once the tuning curve of a Helmholtz type resonant absorber has been determined, its Q-factor can be found from the expression, $f_0/\Delta f$. The "reverberation time" of such absorbers is very short for Qs normally encountered.

and music signals. It is true that any resonant system, electronic or acoustical, has a certain time constant associated with it. The Q-factor (quality factor) describes the sharpness of tuning of the Helmholtz resonator as shown in Fig. 9-37. Once the tuning curve has been obtained experimentally, the width of the tuning curve at the −3 dB points gives Δf. The Q of the system is then $Q = f_0/\Delta f$, where f_0 is the frequency to which the system is tuned. Measurements on a number of perforated and slat Helmholtz absorbers gave Qs around 1 or 2 but some as high as 5. Table 9-4 shows how the decay rate of resonant absorbers of several Qs relates to reverberation time.

With resonant absorber Qs of 100, real problems would be encountered in a room having a reverberation time of, say, 0.5 second as the absorbers tailed off sound for several seconds. However, Helmholtz absorbers with such Qs would be very special devices, made of ceramic, perhaps. Absorbers made of wood with glass fiber to broaden the absorption curve have Qs so low that their sound dies away much faster than the studio or listening room itself.

Taming Room Modes

The following example of taming a troublesome room mode is based on Acoustisoft's ETF 5 room acoustics analyzer program.[23] The ETF

Table 9-4 Sound decay in resonant absorbers.

Q	f_0	"Reverberation time" (seconds)
100	100 Hz	2.2
5	100	0.11
1	100	0.022

stands for energy, time, frequency. By means of this program the detailed low-frequency modal structure of the room is revealed in Fig. 9-38A and B. The mode that caused an audible distortion in room sound is the one with the pronounced reverberant "tail" at 47 Hz on the extreme left of Fig. 9-38A. Once the culprit is identified, steps can be taken to reduce its activity so that it will behave as the other modes of the room.

The solution rests in the placement of a highly tuned absorber in the room at some point of high pressure of the 47-Hz mode. The

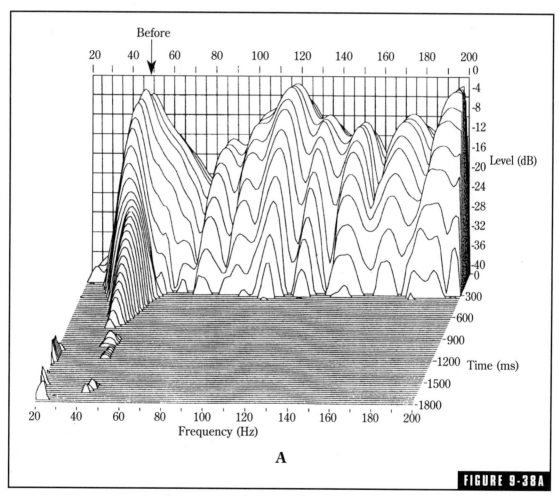

A

FIGURE 9-38A

Low-frequency model structure of the sound field of a small room *before* introduction of the tuned Helmholtz resonator absorber.

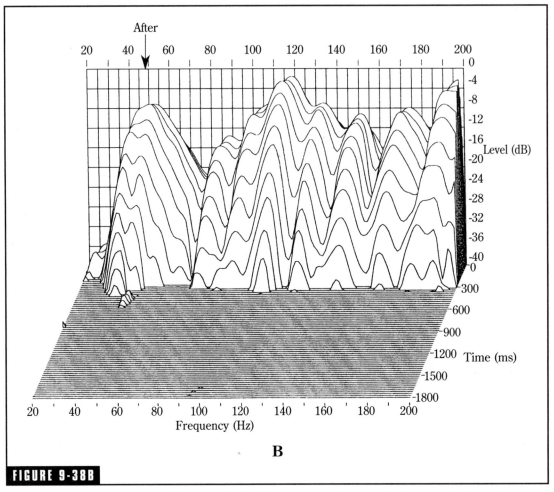

FIGURE 9-38B

Low-frequency model structure of the sound field of the same small room *after* the introduction of the tuned Helmholtz resonator absorber.

locating of high-pressure points of the 47-Hz mode is found simply by energizing the room with a 47-Hz sine wave on the loudspeaker and exploring with a sound-level meter. A spot both convenient and effective will probably be found in a corner.

The finished Helmholtz resonator is shown in Fig. 9-39A. Constructional details are obtained from resonator design instruction included in the ETF 5 program. The resonator is made from a concrete-forming tube available in hardware stores. Laminated wood covers are tightly fitted into both ends of the tube.

The length of the PVC pipe, Fig. 9-39(A), is varied to tune the resonator to specific frequencies. An absorbent partially fills the resonator. The Q-factor of the finished resonator was measured at $Q = 1$.

Increasing Reverberation Time

Low-Q Helmholtz resonators are capable of shortening reverberation time by increasing absorption. High-Q resonators can increase reverberation time through storage of energy as described by Gilford.[2] To achieve the high Qs necessary, plywood, particleboard, masonite, and other such materials must be abandoned and ceramics, plaster, concrete, etc., used in resonator construction. By proper tuning of the resonators, the increase in reverberation time can be placed where needed in regard to frequency.

Modules

The British Broadcasting Corporation has pioneered a modular approach to the acoustical treatment of their numerous small voice studios, which is very interesting.[17] Because they have applied it in several hundred such studios economically and with very satisfactory acoustical results, it deserves our critical attention. Basically, the idea is to cover the walls with standard-sized modules, say 2 × 3 ft, having a maximum depth of perhaps 8″. These can be framed on the walls to give a flush surface appearing very much like an ordinary room, or they can be made into boxes with

A

FIGURE 9-39A

Typical Helmholtz resonator made of readily available materials.

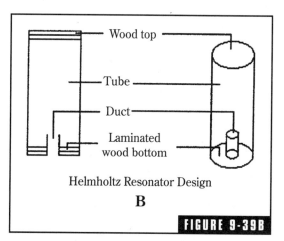

Helmholtz Resonator Design

B

FIGURE 9-39B

Details of Helmholtz resonator design.

grill cloth covers mounted on the walls in regular patterns. All modules can be made to appear identical, but the similarity is only skin deep.

There are commonly three, or perhaps four, different types of modules, each having its own distinctive contribution to make acoustically. Figure 9-40 shows the radically different absorption characteristics obtained by merely changing the covers of the standard module. This is for a 2×3 ft module having a 7″ air space and a 1″ semirigid glass fiber board of 3 lb/cu ft density inside. The wideband absorber has a highly perforated cover (25% or more perforation percentage) or no cover at all, yielding essentially complete absorption down to about 200 Hz. Even better low-frequency absorption is possible by breaking up the air space with egg-crate type dividers of corrugated paper to discourage unwanted resonance modes. A cover $^1/_4$″ thick with a 5% perforation percentage peaks in the 300–400 Hz range. A true bass absorber is obtained with a low-perforation cover (0.5% perforation). If essentially

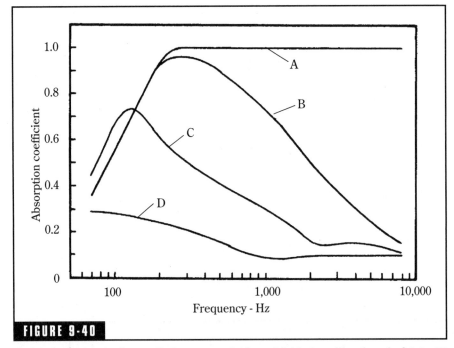

FIGURE 9-40

Modular absorber having a 7″ air space and 1″ semirigid glass fiber board of 9 to 10 lb/cu ft density behind the perforated cover. (A) No perforated cover at all, or at least more than 25% perforation. (B) 5% perforated cover. (C) 0.5% perforated cover. (D) $^3/_4$″ plywood cover, essentially to neutralize the module. (Data from Brown.[17])

neutral modules are desired, they can be covered with $^3/_8''$ and $^1/_4''$ plywood, which would give relatively low absorption with a peak around 70 Hz. Using these three or four modules as acoustical building blocks, the desired effect can be designed into a studio by specifying the number and distribution of each of the basic types.

Figure 9-41 shows an adaptation from BBC practice where the wall is used as the "bottom" of the module box. In this case the module size is 2 × 4 ft. The modules are fastened to the 2 × 2″ mounting strips, that in turn are fastened to the wall. A studio wall 10 ft high

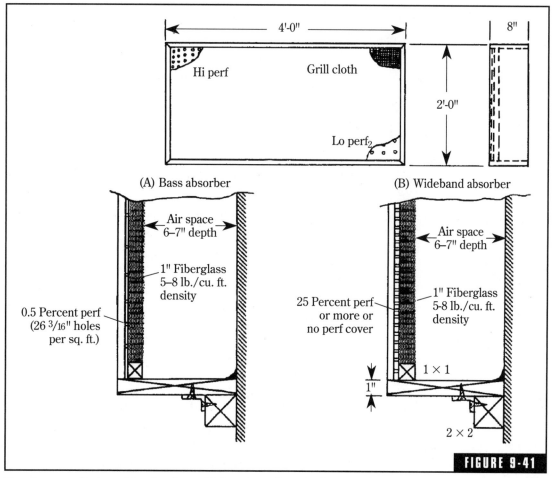

(A) Bass absorber

(B) Wideband absorber

FIGURE 9-41

Plan for a practical module absorber utilizing the wall as the bottom of the module. (Left) Bass absorber. (Right) Wideband absorber.

and 23 or 24 ft long might use 20 modules of distributed types, four modules high and five long. It is good practice to have acoustically dissimilar modules opposing each other on opposite walls.

The question that comes to mind is, "How about diffusion of sound with such modular treatment?" BBC experience has shown that careful distribution of the different types of modules results in adequate diffusion.

Endnotes

[1]Young, Robert W., *Sabine Reverberation and Sound Power Calculations,* J. Acous. Soc. Am., 31, 7 (July 1959) 912–921. See also by the same author, *On Naming Reverberation Equations,* 31, 12 (Dec 1959) 1681.

[2]Gilford, Christofer, *Acoustics for Radio and Television Studios,* London, Peter Peregrinus, Ltd. (1972), Chapter 8.

[3]Anon., *Noise Control Manual,* Owens-Corning Fiberglas Corp. (1980), publication No. 5-BMG-8277-A.

[4]Sonex is manufactured by Illbruck USA, 3600 Washington Ave., North, Minneapolis, MN 55412.

[5]Harris, Cyril M., *Acoustical Properties of Carpet,* J. Acous. Soc. Am., 27, 6 (Nov 1955) 1077–1082.

[6]Anon., *Acoustical Ceilings: Use and Practice,* Published by the Ceiling and Interior Systems Contractors Association (1978). See Appendix.

[7]Schultz, Theodore J. and B. G. Watters, *Propagation of Sound Across Audience Seating,* J. Acous. Soc. Am. 36, 5 (May 1964) 885–896.

[8]Rettinger, Michael, *Bass Traps,* Recording Eng./Prod., 11, 4 (Aug 1980) 46–51.

[9]Mankovsky, V. S., *Acoustics of Studios and Auditoria,* London, Focal Press, Ltd., (1971).

[10]Siekman, William, private communication. (Mr. Seikman was manager of Riverbank Acoustical Laboratories at time of these measurements, which were reported to the Acous. Soc. Am. April 1969.)

[11]Brüel, Per V., *Sound Insulation and Room Acoustics,* London, Chapman and Hall (1951).

[12]Callaway, D. B. and L. G. Ramer, *The Use of Perforated Facings in Designing Low-Frequency Resonant Absorbers,* J. Acous. Soc. Am., 24, 3 (May 1952), 309–312.

[13]Ingard, U. and R. H. Bolt, *Absorption Characteristics of Acoustic Materials with Perforated Facings,* J. Acous. Soc. Am., 23, 5 (Sept 1951), 533–540.

[14]Davern, W. A., *Perforated Facings Backed with Porous Materials as Sound Absorbers—An Experimental Study,* Applied Acoustics, 10 (1977), 85–112.

[15]Rettinger, Michael, *Low Frequency Sound Absorbers,* db the Sound Engineering Magazine, 4, 4 (April 1970), 44–46.

[16]Rettinger, Michael, *Low-Frequency Slot Absorbers,* db the Sound Engineering Magazine, 10, 6 (June 1976) 40–43.

[17]Brown, Sandy, *Acoustic Design of Broadcasting Studios,* J. Sound and Vibration, I (3) (1964) 239–257.

[18]Beranek, Leo L., Acoustics, New York, McGraw-Hill Book Co. (1954).

[19]Evans, E. J. and E. N. Bazley, *Sound Absorbing Materials,* London, National Physical Laboratories (1960).

[20]Sabine, Paul E. and L. G. Ramer, *Absorption-Frequency Characteristics of Plywood Panels,* J. Acous. Soc. Am., 20, 3 (May 1948) 267–270.

[21]Everest, F. Alton, *The Acoustic Treatment of Three Small Studios,* J. Audio Eng. Soc., 15, 3 (July 1968) 307–313.

[22]Kingsbury, H. F. and W. J. Wallace, *Acoustic Absorption Characteristics of People,* Sound and Vibration, 2, 2 (Dec. 1968) 15,16.

[23]Acoustisoft, Inc., 53 King St., Peterborough, Ontario, Canada K9J-2T1. e-mail: *doug@etfacoustic.com,* telephone 800-301-1429, fax (705) 745-4955. Internet site, www.etfacoustic.com.

Reflection of Sound

If a sound is activated in a room, sound travels radially in all directions. As the sound waves encounter obstacles or surfaces, such as walls, their direction of travel is changed, i.e., they are reflected.

Reflections from Flat Surfaces

Figure 10-1 illustrates the reflection of waves from a sound source from a rigid, plane wall surface. The spherical wavefronts (solid lines) strike the wall and the reflected wavefronts (broken lines) are returned toward the source.

Like the light/mirror analogy, the reflected wavefronts act as though they originated from a *sound image*. This image source is located the same distance behind the wall as the real source is in front of the wall. This is the simple case—a single reflecting surface. In a rectangular room, there are six surfaces and the source has an image in all six sending energy back to the receiver. In addition to this, images of the images exist, and so on, resulting in a more complex situation. However, in computing the total sound intensity at a given receiving point, the contributions of all these images must be taken into consideration.

Sound is reflected from objects that are large compared to the wavelength of the impinging sound. This book would be a good

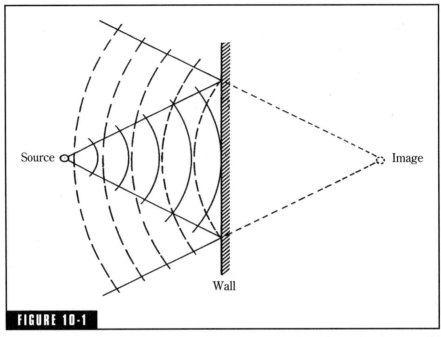

FIGURE 10-1

Reflection of sound from a point source from a flat surface (incident sound, solid lines; reflected sound, broken lines). The reflected sound appears to be from a virtual image source.

reflector for 10 kHz sound (wavelength about an inch). At the low end of the audible spectrum, 20 Hz sound (wavelength about 56 ft) would sweep past the book and the person holding it as though they did not exist, and without appreciable *shadows*.

Below 300–400 Hz, sound is best considered as waves (chapter 15 expounds on this). Sound above 300–400 Hz is best considered as traveling in rays. A ray of sound may undergo many reflections as it bounces around a room. The energy lost at each reflection results in the eventual demise of that ray. Even the ray concept is an oversimplification: Each ray should really be considered as a "pencil" of diverging sound with a spherical wavefront to which the inverse square law applies.

The mid/high audible frequencies have been called the *specular* frequencies because sound in this range acts like light rays on a mirror. Sound follows the same rule as light: The angle of incidence is equal to the angle of reflection, as in Fig. 10-2.

Doubling of Pressure at Reflection

The sound pressure on a surface normal to the incident waves is equal to the energy-density of the radiation in front of the surface. If the surface is a perfect absorber, the pressure equals the energy-density of the incident radiation. If the surface is a perfect reflector, the pressure equals the energy-density of both the incident and the reflected radiation. Thus the pressure at the face of a perfectly reflecting surface is twice that of a perfectly absorbing surface. At this point, this is only an interesting sidelight. In the study of standing waves in Chap. 15, however, this pressure doubling takes on greater significance.

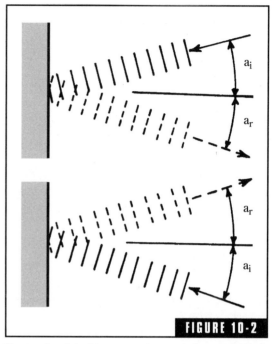

FIGURE 10-2

The angle of incidence, a_i, is equal to the angle of reflection, a_r.

Reflections from Convex Surfaces

Spherical wavefronts from a point source tend to become plane waves at greater distance from the source. For this reason impinging sound on the various surfaces to be considered will be thought of as plane wavefronts. Reflection of plane wavefronts of sound from a solid convex surface tends to scatter the sound energy in many directions as shown in Fig. 10-3. This amounts to a diffusion of the impinging sound.

The polycylindrical sound-absorbing system described in the previous chapter both absorbs sound and contributes to much-needed diffusion in the room by reflection from the cylindrically shaped surface.

Reflections from Concave Surfaces

Plane wavefronts of sound striking a concave surface tend to be focussed to a point as illustrated on Fig. 10-4. The precision with which sound is focussed to a point is determined by the shape of the

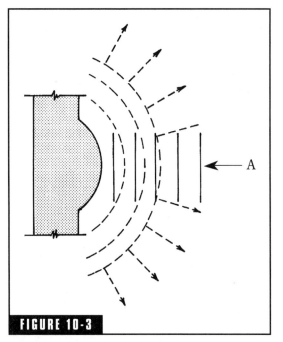

FIGURE 10-3

Plane sound waves impinging on a convex irregularity tend to be dispersed through a wide angle if the size of the irregularity is large compared to the wavelength of the sound.

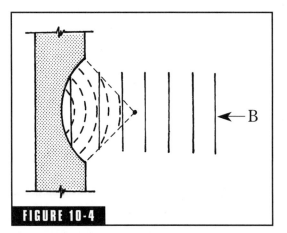

FIGURE 10-4

Plane sound waves impinging on a concave irregularity tend to be focussed if the size of the irregularity is large compared to the wavelength of the sound.

concave surface. Spherical concave surfaces are common because they are readily formed. They are often used to make a microphone highly directional by placing it at the focal point. Such microphones are frequently used to pick up field sounds at sporting events or in recording songbirds or other animal sounds in nature. In the early days of broadcasting sporting events in Hong Kong, a resourceful technician saved the day by using an ordinary Chinese wok, or cooking pan, as a reflector. Aiming the microphone into the reflector at the focal point provided an emergency directional pickup. Concave surfaces in churches or auditoriums can be the source of serious problems as they produce concentrations of sound in direct opposition to the goal of uniform distribution of sound.

The effectiveness of reflectors for microphones depends on the size of the reflector with respect to the wavelength of sound. A 3-ft-diameter spherical reflector will give good directivity at 1 kHz (wavelength about 1 ft), but it is practically nondirectional at 200 Hz (wavelength about 5.5 ft).

Reflections from Parabolic Surfaces

A parabola has the characteristic of focusing sound precisely to a point (Fig. 10-5). It is generated by the simple equation $y = x^2$. A very "deep" parabolic surface, such as that of Fig. 10-5, exhibits far better directional properties than a shallow one. Again, the directional properties depend on the size of the opening in terms of wavelengths. Figure 10-5 shows the parabola used as a directional sound source with a small, ultrasonic Galton Whistle pointed inward at the focal point.

Plane waves striking such a reflector would be brought to a focus at the focal point. Conversely, sound emitted at the focal point of the parabolic reflector generates plane wavefronts. This is demonstrated in the photographs of Figs. 10-6 and 10-7 in which standing waves are produced by reflections from a heavy glass plate. The force exerted by the vibration of the air particles on either side of a node is sufficient to hold slivers of cork in levitation.

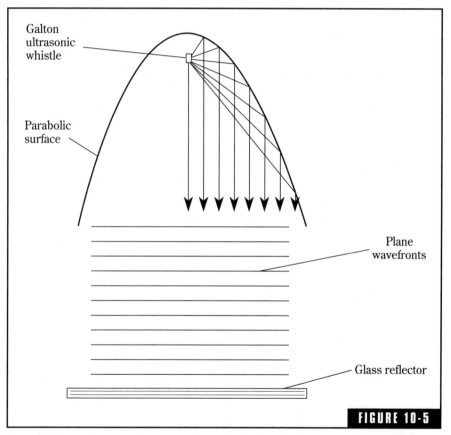

FIGURE 10-5

A parabolic surface can focus sound precisely at a focal point or, the converse, a sound source placed at the focal point can produce plane, parallel wavefronts. In this case, the source is an ultrasonic Galton Whistle blown by compressed air with the results shown in Figs. 10-6 and 10-7.

CHAPTER TEN

Reflections Inside a Cylinder

St. Paul's Cathedral in London boasts a *whispering gallery*. The way this whispering gallery works is explained in the diagram of Fig. 10-8. Reflections from the exterior surfaces of cylindrical shapes have been mentioned in the treatment of "polys." In this case the source and receiver are both inside a mammoth, hard-surfaced cylindrical room.

At the source, a whisper directed tangentially to the surface is clearly heard on the receiver side. The phenomenon is assisted by the fact that the walls are dome-shaped. This means that upward-directed components of the whispered sounds tend to be reflected downward and conserved rather than lost above.

Standing Waves

The concept of standing waves is directly dependent on the reflection of sound as emphasized in Chap. 15. Assume two flat, solid parallel walls separated a given distance. A sound source between them radiates sound of a specific frequency. The wavefront striking the right wall is reflected back toward the source, striking the left wall where it is again reflected back toward the right wall, and so on. One wave travels to the right, the other toward the left. The two traveling waves interact to form a standing wave. Only the standing wave, the interaction of the two, is stationary. The frequency of the radiated sound is such as to establish this resonant condition between the wavelength of the sound and the distance between the two surfaces. The pertinent point at the moment is that this phenomenon is entirely dependent on the reflection of sound at the two parallel surfaces.

Reflection of Sound from Impedance Irregularities

The television repairman is concerned about matching the electrical impedance of the television receiver to that of the transmission line, and matching the transmission line to the impedance of the antenna (or cable). Mismatches of impedance give rise to reflections, which cause numerous undesirable effects.

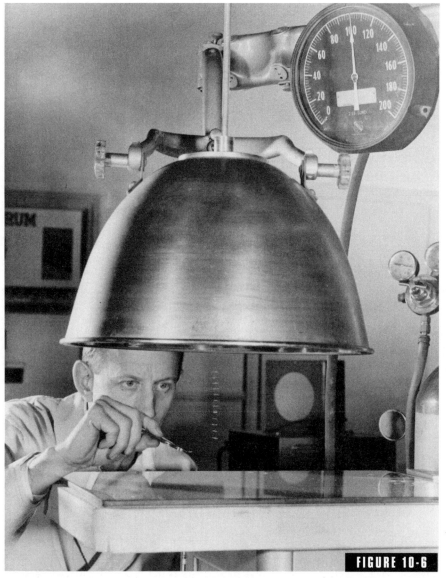

A parabolic reflector following the equation $y=x^2$ reflects sound from an ultrasonic Galton Whistle to form a stable standing wave system capable of levitating bits of cork.

FIGURE 10-7

Close-up of the levitated cork chips of Fig. 10-6. If the cork chips are spaced about ½ inch, the frequency of the sounds emitted by the Galton Whistle is about 30 kHz.

A similar situation prevails in an air-conditioning duct. A sound wave (noise) traveling in the duct suddenly encounters the large open space of the room. This discontinuity (impedance mismatch) reflects a significant portion of the sound (fan noise, etc.) back toward the source. This is an example of a benevolent mismatch as the air-conditioner noise is reduced in the room.

The Corner Reflector

In an art museum with large Dutch paintings on display, the eyes of certain subjects seem to follow as one walks by. Corner reflectors are like that. There seems to be no way of escaping their pernicious effect. The corner reflector of Fig. 10-9, receiving sound from the source *S*, sends a reflection directly back toward the source. If the angles of incidence and reflection are carefully noted, a source at *B* will also send a direct, double-surface reflection returning to the source. A source at *C*, on the opposite of *B*, is subject to the same effect.

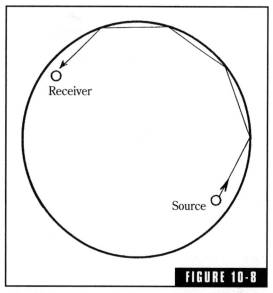

FIGURE 10-8

Graphic explanation of the "Whispering Gallery" of St. Paul's Cathedral, London. A whisper directed tangentially to the cylinderical surface is readily heard by the receiver on the far side of the room.

You might be instinctively aware of perpendicular (normal) reflections from surrounding walls, but now consider reflections from the four corners of the room that follow the source around the room. Corner reflections suffer losses at two surfaces, tending to make them somewhat less intense than normal reflections at the same distance.

The corner reflector of Fig. 10-9 involves only two surfaces. How about the four upper tri-corners of the room formed by ceiling and walls and another four formed by floor and wall surfaces? The same follow-the-source principle applies. In fact, sonar and radar people have long employed targets made of three circular plates of reflecting material assembled so that each is perpendicular to the others.

Echo-Sounding

Objects can be located by sending out a pulse of sound and noting the time it takes for the reflected echo to return. Directional sources of sound make possible the determination of both the azimuth angle and the distance to the reflecting object. This principle has been widely applied in water depth sounders, sonar on submarines, etc. All

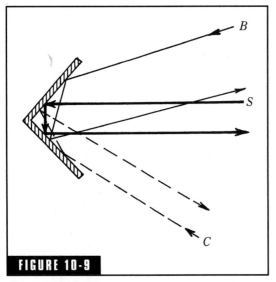

FIGURE 10-9

The corner reflector has the property of reflecting sound back toward the source from any direction.

depend on the reflection of sound from the bottom of the ocean, or enemy target.

Perceptive Effects of Reflections

In the reproduction of sound in a high-fidelity listening room or control room of a recording studio, the sound of the loudspeakers falling on the ear of the listener is very much affected by reflections from the surfaces of the room. This is another manifestation of sound reflection. A comprehensive consideration of human perception of such reflections is included in Chap. 16.

Diffraction of Sound

It is well known that sound travels around corners and around obstacles. Music reproduced in one room of a home can be heard down the hall and in other rooms. Diffraction is one of the mechanisms involved in this. The character of the music heard in distant parts of the house is different. In distant rooms the bass notes are more prominent because their longer wavelengths are readily diffracted around corners and obstacles.

Rectilinear Propagation

Wavefronts of sound travel in straight lines. Sound rays, a concept applicable at mid/high audible frequencies, are considered to be pencils of sound that travel in straight lines perpendicular to the wavefront. Sound wavefronts and sound rays travel in straight lines, except when something gets in the way. Obstacles can cause sound to be changed in its direction from its original rectilinear path. The process by which this change of direction takes place is called *diffraction*.

Alexander Wood, the early Cambridge acoustician, recalled Newton's pondering over the relative merits of the corpuscular and wave theories of light. Newton finally decided that the corpuscular theory was the correct one because light is propagated rectilinearly. Later it was demonstrated that light is not always propagated rectilinearly, that diffraction can cause light to change its direction of travel. In fact, all types of wave motion, including sound, are subject to diffraction.

The shorter the wavelength (the higher the frequency), the less dominant is the phenomenon of diffraction. Diffraction is less noticeable for light than it is for sound because of the extremely short wavelengths of light. Obstacles capable of diffracting (bending) sound must be large compared to the wavelength of the sound involved. The well-worn example of ocean waves is still one of the best. Ocean waves sweep past a piling of a dock with scarcely a disturbance. Ocean waves, however, are bent around an end of an island.

Diffraction and Wavelength

The effectiveness of an obstacle in diffracting sound is determined by the acoustical size of the obstacle. Acoustical size is measured in terms of the wavelength of the sound. One way of looking at the illustration shown later, Fig. 11-3, is that the obstacle in B is the same physical size as that of A, but the frequency of the sound of A is one tenth that of B. If the obstacle in B is 1 ft long and that of A 0.1 ft long, the frequency of the sound in A could well be 1,000 Hz (wavelength 1.13 ft), and that of B could be 100 Hz (wavelength 11.3 ft). The same drawing could be used if the obstacle of A were 0.01 ft long with a frequency of 10,000 Hz (wavelength 0.113 ft) and the obstacle of B were 0.1 ft long with a frequency of 1,000 Hz (wavelength 1.13 ft).

In Fig. 11-1, two types of obstructions to plane wavefronts of sound are depicted. In Fig. 11-1A a heavy brick wall is the obstacle. The sound waves are reflected from the

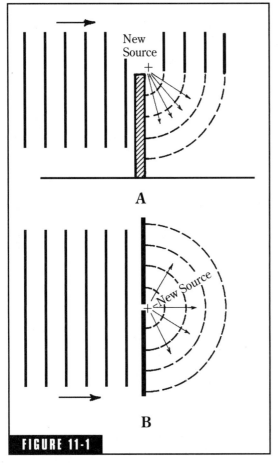

FIGURE 11-1

(A) If the brick wall is large in terms of the wavelength of the sound, the edge acts as a new source, radiating sound into the shadow zone. (B) Plane waves of sound impinging on the heavy plate with a small hole in it sets up spherical wavefronts on the other side due to diffraction of sound.

face of the wall, as expected. The upper edge of the wall acts as a new, virtual source sending sound energy into the "shadow" zone behind the wall by diffraction. The mechanism of this effect will be considered in more detail later in this chapter.

In Fig. 11-1B the plane wavefronts of sound strike a solid barrier with a small hole in it. Most of the sound energy is reflected from the wall surface, but that tiny portion going through the hole acts as a virtual point source, radiating a hemisphere of sound into the "shadow" zone on the other side.

Diffraction of Sound by Large and Small Apertures

Figure 11-2A illustrates the diffraction of sound by an aperture that is many wavelengths wide. The wavefronts of sound strike the heavy obstacle: some of it is reflected, some goes right on through the wide aperture. The arrows indicate that some of the energy in the main beam is diverted into the shadow zone. By what mechanism is this diversion accomplished?

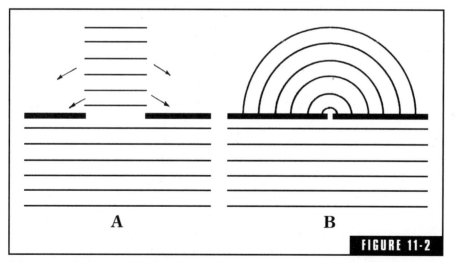

FIGURE 11-2

(A) An aperature large in terms of wavelength of sound allows wavefronts to go through with little disturbance. These wavefronts act as lines of new sources radiating sound energy into the shadow zone. (B) If the aperture is small compared to the wavelength of the sound, the small wavefronts which do penetrate the hole act almost as point sources, radiating a hemispherical field of sound into the shadow zone.

For an answer, the work of Huygens is consulted.[1] He enunciated a principle that is the basis of very difficult mathematical analyses of diffraction. The same principle also gives a simple explanation of how sound energy is diverted from the main beam into the shadow zone. Huygens' principle can be paraphrased as:

Every point on the wavefronts of sound that has passed through an aperture or passed a diffracting edge is considered a point source radiating energy back into the shadow zone.

The sound energy at any point in the shadow zone can mathematically be obtained by summing the contributions of all of these point sources on the wavefronts.

In Fig. 11-2A, each wavefront passing through the aperture becomes a row of point sources radiating diffracted sound into the shadow zone. The same principle holds for Fig. 11-2B except that the aperture is very small and only a small amount of energy passes through it. The points on the limited wavefront going through the hole are so close together that their radiations take the form of a hemisphere.

Diffraction of Sound by Obstacles

In Fig. 11-3A the obstacle is so small compared to the wavelength of the sound that it has no appreciable effect on the passage of sound. In Fig. 11-3B, however, the obstacle is many wavelengths long and it has a definite effect in casting a shadow behind the obstacle. Each wavefront passing the obstacle becomes a line of new point sources radiating sound into the shadow zone.

A very common example of an obstacle large compared to the wavelength of the impinging sound is the highway noise barrier shown in Fig. 11-4. If the wavelength of the impinging sound is indicated by the spacing of the spherical wavefronts hitting the barrier, the barrier size is acoustically great. At higher frequencies the barrier becomes even *larger*, and at lower frequencies it becomes acoustically *smaller*. First, the sound reflected from the wall must be noted. It is as though the sound were radiated from a virtual image on the far side of the wall. The wavefronts passing the top edge of the wall can be considered as lines of point sources radiating sound. This is the source of the sound penetrating the shadow zone.

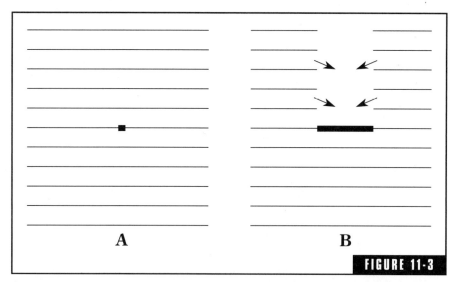

FIGURE 11-3

(A) An obstacle very much smaller than the wavelength of sound allows the wavefronts to pass essentially undisturbed. (B) An obstacle large compared to the wavelength of sound casts a shadow that tends to be irradiated from sources on the wavefronts of sound that go past the obstacle.

Figure 11-5 gives some idea of the effectiveness of highway barriers and of the intensity of the sound in the shadow of a high, massive wall. The center of the highway is taken to be 30 ft from the wall on one side, and the home or other sensitive area is considered to be 30 ft on the other side of the wall (the shadow side). A wall 20 ft high yields something like 25 dB of protection from the highway noise at 1,000 Hz. At 100 Hz, the attenuation of the highway noise is only about 15 dB. At the higher audible frequencies, the wall is more effective than at lower frequencies. The shadow zone behind the wall tends to be shielded from the high-frequency components of the highway noise. The low-frequency components penetrate the shadow zone by diffraction.

Diffraction of Sound by a Slit

Figure 11-6 diagrams a classical experiment performed by Pohl in acoustical antiquity and described by Wood[1] in somewhat more recent antiquity. One must admire the precise results obtained with crude measuring instruments (high-pitched whistle, sound radiometer). The

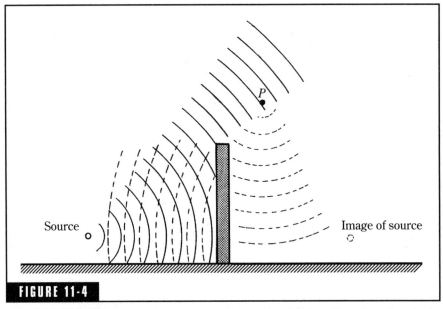

FIGURE 11-4

The classic sound barrier case. The sound striking the wall is reflected as though the sound is radiated from a virtual image of the source. That sound passing the top edge of the wall acts as though the wavefronts are lines of sources radiating sound energy into the shadow zone.

equipment layout of Fig. 11-6A is very approximate. Actually the source/slit arrangement rotated about the center of the slit and the measuring radiometer was at a distance of 8 meters. The slit width was 11.5 cm wide, the wavelength of the measuring sound was 1.45 cm (23.7 kHz). The graph of Fig. 11-6B shows the intensity of the sound versus the angle of deviation. The dimension B indicates the geometrical boundaries of the ray. Anything wider that B is caused by diffraction of the beam by the slit. A narrower slit would yield correspondingly more diffraction and a greater width of the beam. The increase in width of the beam is the striking feature of this experiment.

Diffraction by the Zone Plate

The zone plate can be considered an acoustic lens. It consists of a circular plate with a set of concentric, annular slits of cunningly devised radii. If the focal point is at a distance of r from the plate,

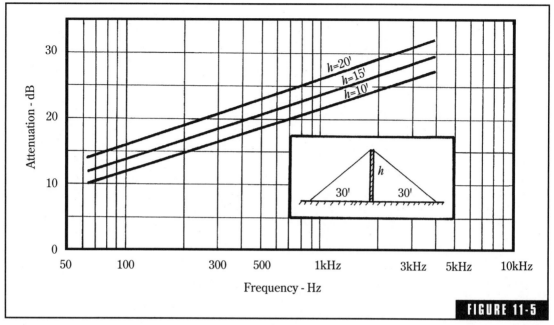

FIGURE 11-5

An estimation of the effectiveness of a sound barrier in terms of sound (or noise) attentuation as a function of frequency and barrier height. *(After Rettinger.[4])*

the next longer path must be $r + \lambda/2$ where λ is the wavelength of the sound falling on the plate from the source. Successive path lengths are $r + \lambda$, $r + 3/2\lambda$, and $r + 2\lambda$. These path lengths differ by $\lambda/2$, which means that the sound through all the slits will arrive at the focal point in phase which, in turn, means that they add constructively, intensifying the sound.[2] See Fig. 11-7.

Diffraction around the Human Head

Figure 11-8 illustrates the diffraction caused by a sphere roughly the size of the human head. This diffraction by the head as well as reflections and diffractions from the shoulders and the upper torso influences human perception of sound. In general, for sound of frequency 1–6 kHz arriving from the front, head diffraction tends to increase the sound pressure in front and decrease it behind the head. For frequencies in the lower range the directional pattern tends to become circular.[2,3]

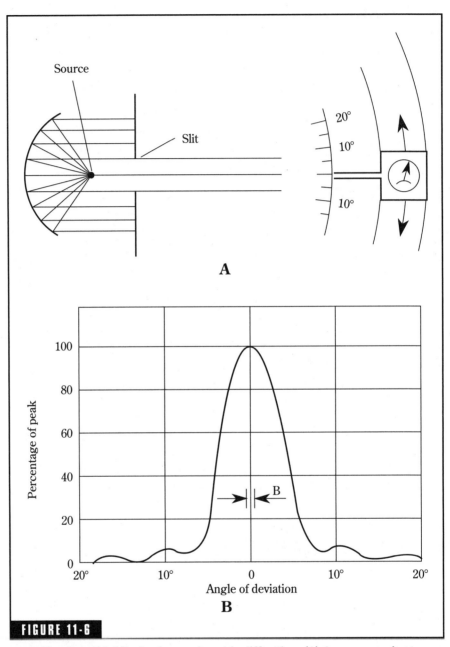

FIGURE 11-6

A consideration of Pohl's classic experiment in diffraction. (A) A very approximate suggestion of the equipment arrangement (see text). (B) The broadening of the beam B by diffraction. The narrower the slit the greater this broadening of the beam. *(After Wood.[1])*

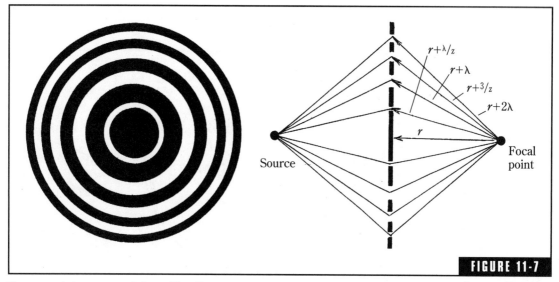

$$r+\lambda/_2$$
$$r+\lambda$$
$$r+{}^3/_2$$
$$r+2\lambda$$
$$r$$

Source

Focal point

The zone plate or acoustic lens. The slits are so arranged that the several path lengths differ by multiples of a half wavelength of the sound so that all diffracted rays arrive at the focal point in phase, combining constructively. *(After Olson.[2])*

Diffraction by Loudspeaker Cabinet Edges

Loudspeaker cabinets are notorious for diffraction effects. If a loudspeaker is mounted near a wall and aimed away from the wall, the wall is still illuminated with sound diffracted from the corners of the box. Reflections of this sound can affect the quality of the sound at the listener's position. Measurements of this effect have been scarce, but Vanderkooy[5] and Kessel[6] have recently computed the magnitude of loudspeaker cabinet edge diffraction. The computations were made on a box loudspeaker with front baffle having the dimensions 15.7×25.2 in and depth of 12.6 in (Fig. 11-9). A point source of sound was located symmetrically at the top of the baffle. The sound from this point source was computed at a distance from the box. The sound arriving at the observation point is the combination of the direct sound plus the edge diffraction. This combination is shown in Fig. 11-10. Fluctuations due to edge diffraction for this particular typical situation approached plus or minus 5 dB. This is a significant change in overall frequency response of a reproduction system.

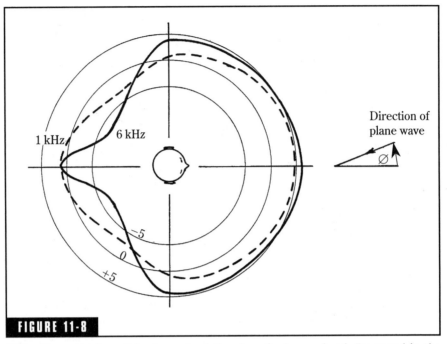

FIGURE 11-8

Diffraction around a solid sphere about the size of a human head. For sound in the 1-6 kHz range, sound pressure is generally increased in the front hemisphere and generally reduced in the rear. *(After Muller, Black and Davis, as reported by Olson.[2])*

This effect can be controlled (eliminated?) by setting the loudspeaker box face flush in a much larger baffling surface. There is also the possibility of rounding edges and the use of foam or "fuzz".[7]

Diffraction by Various Objects

Sound level meters were, in early days, boxes with a microphone protruding. Diffraction from the edges and corners of the box seriously affected the calibration of the microphones. Modern sound level meters have carefully rounded contours with the microphone mounted on a smooth, slender, rounded neck.

Diffraction from the casing of a microphone can cause deviations from the desired flat sensitivity.

In the measurement of sound absorption in large reverberation chambers, the common practice is to place the material to be measured

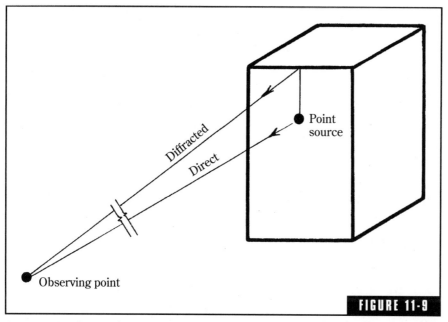

FIGURE 11-9

Arrangement for Vanderkooy's calculation of loudspeaker cabinet edge diffraction, shown in Fig. 11-10.

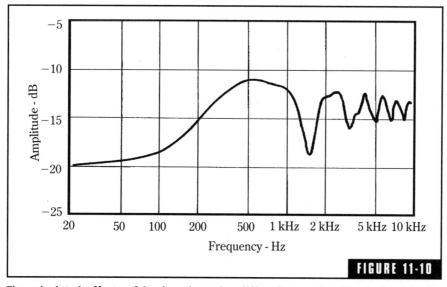

FIGURE 11-10

The calculated effects of loudspeaker edge diffraction on the direct signal in the arrangement of Fig.11-9. *(After Vanderkooy,[5] and Kessel.[6])*

in a 8 × 9-ft frame on the floor. Diffraction from the edges of this frame often result in absorption coefficients greater than unity. In other words, diffraction of sound makes the sample appear larger than it really is.

Small cracks around observation windows or back-to-back microphone or electrical service boxes in partitions can destroy the hoped-for isolation between studios or between studio and control room. The sound emerging on the other side of the hole or slit is spread in all directions by diffraction.

In summary, diffraction causes sound, which normally travels rectilinearly, to travel in other directions.

Endnotes

[1]Wood, Alexander, *Acoustics*, New York, Interscience Publishers, Inc. (1941).

[2]Olson, Harry F., *Elements of Acoustical Engineering*, New York, D. Van Nostrand Co. (1940).

[3]Muller, C.G., R. Black, and T.E. Davis, *The Diffraction Produced by Cylinders and Cubical Obstacles and by Circular and Square Plates*, J. Acous. Soc. Am., 10, 1 (1938) p. 6.

[4]Rettinger, M., *Acoustic Design and Noise Control*, Chemical Publishing Co. (1973).

[5]Vanderkooy, John, *A Simple Theory of Cabinet Edge Diffraction*, J. Audio Eng. Soc., 39, 12 (1991) 923-933.

[6]Kessel, R.T., *Predicting Far-Field Pressures from Near-Field Loudspeaker Measurements*, J. Audio Eng. Soc., Abstract, Vol. 36, p.1,026 (Dec 1988), preprint 2729.

[7]Kaufman, Richard J., *With a Little Help from My Friends*, AUDIO, 76, 9 (Sept 1992) 42-46.

Refraction of Sound

About the turn of this century Lord Rayleigh was puzzled because some very powerful sound sources, such as cannon fire, could be heard only short distances some times and very great distances at other times. He set up a powerful siren that required 600 hp to maintain it. He calculated that if all this power were converted into energy as sound waves and spread uniformly over a hemisphere, how far could it be heard? Knowing the minimum audible intensity (10^{-16} watts per sq cm), his calculations indicated that the sound should be audible to a distance of 166,000 miles, more than 6 times the circumference of the earth!

It is indeed fortunate that such sound propagation is never experienced and that a range of a few miles is considered tops. There are numerous reasons why sound is not heard over greater distances. For one thing, the efficiency of sound radiators is usually quite low; not much of that 600 hp was actually radiated as sound. Energy is also lost as wavefronts drag across the rough surface of the earth. Another loss is dissipation in the atmosphere, but this is known to be very small. The result of such calculations and early experiments that fell far short of expectations served only to accelerate research on the effects of temperature and wind gradients on the transmission of sound.

Refraction of Sound

Refraction changes the direction of travel of the sound by differences in the velocity of propagation. *Diffraction* is changing the direction of travel of sound by encountering sharp edges and physical obstructions (chapter 11). Most people find it easy to distinguish between *absorption* and *reflection* of sound, but there is often confusion between *diffraction* and *refraction* (and possibly *diffusion*, the subject of the next chapter). The similarity of the sound of the words might be one cause for this confusion, but the major reason is the perceived greater difficulty of understanding diffraction, refraction, and diffusion compared to absorption and reflection. Hopefully Chaps. 9, 10, 11, 13, and this chapter will help to equalize and advance understanding of these five important effects.

Figure 12-1 recalls a very common observation of the apparent bending of a stick as one end touches the water surface or is actually immersed. This is an illustration of refraction of light. As the present subject is refraction of sound, which is another wave phenomenon, the relative refractive indices of air and water will be passed over.

Refraction of Sound in Solids

Figure 12-2 illustrates sound passing from a dense solid medium to a less dense medium. The sound speed in the denser medium is greater

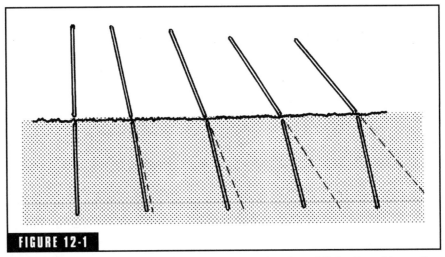

FIGURE 12-1

Touching a stick to the water surface illustrates refraction of light. Sound is another wave phenomenon that is also refracted by changes in media sound speed.

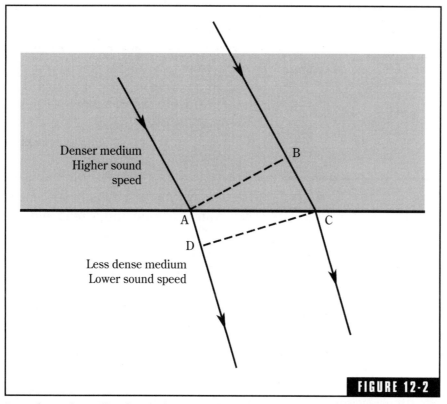

Denser medium
Higher sound
speed

B

A

D

C

Less dense medium
Lower sound speed

FIGURE 12-2

Rays of sound traveling from a denser medium having a certain sound speed into a less dense medium having a lower sound speed. The wavefront *AB* is not parallel to wavefront *DC* because the direction of the wave is changed due to refraction.

than that in the less dense one (Table 12-1). As one ray reaches the boundary between the two media at A, the other still has some distance to go. In the time it takes one ray to travel from B to C, the other ray has traveled a shorter distance from A to D in the new medium. Wavefront A-B represents one instant of time as does wavefront D-C an instant later. But these two wavefronts are no longer parallel. The rays of sound have been refracted at the interface of the two media having unlike sound speeds.

An analogy may assist memory and logic. Assume that the shaded area is paved and that the lower-density area is ploughed. Assume also that the wavefront A-B is a line of soldiers. The line of soldiers A-B, marching in military order, has been making good progress on the

Table 12-1. Speed of sound.

| | Speed of sound | |
Medium	Ft/sec	Meters/sec
Air	1,130	344
Sea water	4,900	1,500
Wood, fir	12,500	3,800
Steel bar	16,600	5,050
Gypsum board	22,300	6,800

pavement. As soldier A reaches the ploughed ground he or she slows down and begins plodding over the rough surface. Soldier A travels to D on the ploughed surface in the same time that soldier B travels the distance BC on the pavement. This tilts the wavefront off in a new direction, which is the definition of refraction. In any homogeneous medium, sound travels rectilinearly (in the same direction). If a medium of another density is encountered, the sound is refracted.

Refraction of Sound in the Atmosphere

The atmosphere is anything but a stable, uniform medium for the propagation of sound. Sometimes the air near the earth is warmer than the air at greater heights, sometimes it is colder. Horizontal changes are taking place at the same time this vertical layering exists. All is a wondrously intricate and dynamic system, challenging the meteorologists (as well as acousticians) to make sense of it.

In the absence of thermal gradients, a sound ray may be propagated rectilinearly as shown in Fig. 12-3A. The sound ray concept is helpful in considering direction of propagation. Rays of sound are always perpendicular to sound wavefronts.

In Fig. 12-3B a thermal gradient exists between the cool air near the surface of the earth and the warmer air above. This affects the wavefronts of the sound. Sound travels faster in warm air than in cool air causing the tops of the wavefronts to go faster than the lower parts. The tilting of the wavefronts is such as to direct the sound rays downward. Under such conditions, sound from the source is bent down toward the surface of the earth and can be heard at relatively great distances.

The thermal gradient of Fig. 12-3C is reversed from that of Fig. 12-3B as the air near the surface of the earth is warmer than the air higher up. In this case the bottom parts of the wavefronts travel faster than the tops, resulting in an upward refraction of the sound rays. The same sound energy from the source S would now be dissipated in the upper reaches of the atmosphere, reducing the chances of it being heard at any great distance at the surface of the earth.

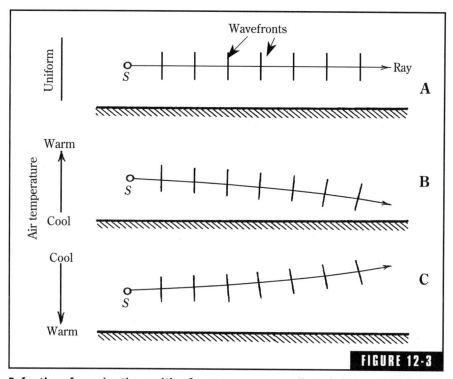

FIGURE 12-3

Refraction of sound paths resulting from temperature gradients in the atmosphere; (A) air temperature constant with height, (B) cool air near the surface of the earth and warmer air above, (C) warm air near the earth and cooler air above.

Figure 12-4A presents a distant view of the downward refraction situation of Fig. 12-3B. Sound traveling directly upward from the source S penetrates the temperature gradient at right angles and would not be refracted. It would speed up and slow down slightly as it penetrates the warmer and cooler layers, but would still travel in the vertical direction.

All rays of sound except the vertical would be refracted downward. The amount of this refraction varies materially: the rays closer to the vertical are refracted much less than those more or less parallel to the surface of the earth.

Figure 12-4B is a distant view of the upward refraction situation of Fig.12-3C. Shadow zones are to be expected in this case. Again, the vertical ray is the only one escaping refractive effects.

It is a common experience to hear sound better downwind than upwind. Air is the medium for the sound. If wind moves the air at a

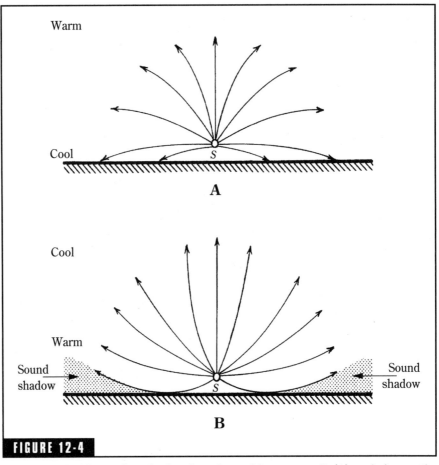

Comprehensive illustration of refraction of sound from source S; (A) cool air near the ground and warmer air above, (B) warm air near the ground and cooler air above. In (B) note that sound shadow areas result from the upward refraction.

certain speed, it is to be expected that the speed of sound will be affected. If sound travels 1,130 ft/sec and a 10 mi/hour (about 15 ft/sec) wind prevails, what will be the effect of the wind on the sound? Upwind the sound speed with respect to the earth would be increased about 1%, and downwind it would be decreased the same amount. This seems like a very small change but it is enough to affect refraction materially. Figure 12-5 illustrates the effect of wind on the downward refraction case of Fig. 12-4A. A downwind shadow is created and upwind sound is refracted downward.

Wind speed near the surface of the earth is usually less than that at greater heights. A wind gradient exists in such a case that has its effect on propagation of sound. This is not a true refraction but the effect is the same. Plane waves from a distant source traveling with the wind would bend the sound down toward the earth. Plane waves traveling against the wind will be bent upward.

It is possible, under unusual circumstances, that sound traveling upwind may actually be favored. For instance, upwind sound is kept above the surface of the ground, minimizing losses at the ground surface. After all, does not the sportsman approach his prey upwind? Doing so keeps footstep noises from being heard by the prey until the sportsman is quite close.

Refraction of Sound in the Ocean

In 1960 some oceanographers devised an ambitious plan to see how far underwater sound could be detected.[1,2] Charges of 600 lb were discharged at various depths in the ocean off Perth, Australia. Sounds from these discharges were detected near Bermuda. The great circle path the sound presumably followed is shown in Fig. 12-6. Even though sound in sea water travels 4.3 times faster than in air, it took 13,364 seconds (3.71 hr) for the sound to make the trip. This distance is over 12,000 miles, close to half the circumference of the earth. Interesting, but what has this to do with refraction? Everything!

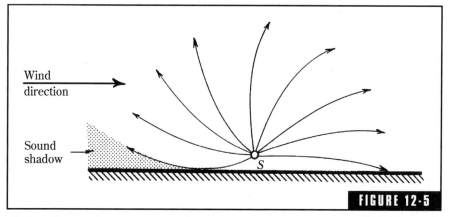

FIGURE 12-5

Wind gradients refract (not a true refraction) sound. A shadow sound is created upwind and good listening conditions downwind.

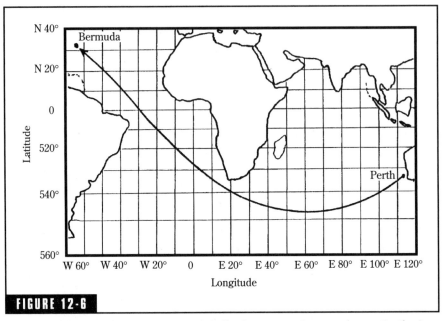

FIGURE 12-6

Refraction of sound in the ocean. A 600-lb charge was detonated near Perth, Australia, and the sound was recorded at Bermuda, over 12,000 miles away. The secret lies in the fact that the sound was confined to a sound channel by refraction that reduced losses. The sound took 3.71 hours to travel almost half way around the world. Such long-distance transmission of sound in the sea is being used to study long-range warming effects of the ocean. (After Heaney et al.[2])

An explanation is found in Fig. 12-7. The depth of the oceanic abyss is 5,000 or more fathoms (30,000 ft). At about 700 fathoms (4,200 ft) a very interesting effect takes place. The sound speed profile shown in Fig. 12-7A is very approximate to illustrate a principle. In the upper reaches of the ocean the speed of sound decreases with depth because temperature decreases. At greater depths the pressure effect prevails causing sound speed to increase with depth because of the increase in density. The "V" change-over from one effect to the other occurs near the 700 fathom (4,200 ft) depth.

A sound channel is created by this V-shaped sound-speed profile. A sound emitted in this channel tends to spread out in all directions. Any ray traveling upward will be refracted downward, any ray traveling downward will be refracted upward. Sound energy in this channel is propagated great distances with modest losses.

Refraction in the vertical plane is very prominent because of the vertical temperature/pressure gradient of Fig. 12-7A. There is relatively little horizontal sound speed gradient and therefore very little horizontal refraction. Sound tends to be spread out in a thin sheet in this sound channel at about 700 fathom depth. Spherical divergence in three dimensions is changed to two-dimensional propagation at this special depth.

These long-distance sound channel experiments have suggested that such measurements can be used to check on the "warming of the planet" by detecting changes in the average temperature of the oceans. The speed of sound is a function of the temperature of the ocean. Accurate measures of time of transit over a given course yield information on the temperature of that ocean.[3]

Refraction of Sound in Enclosed Spaces

Refraction is an important effect on a world-sized scale, how about enclosed spaces? Consider a multi-use gymnasium that serves as an

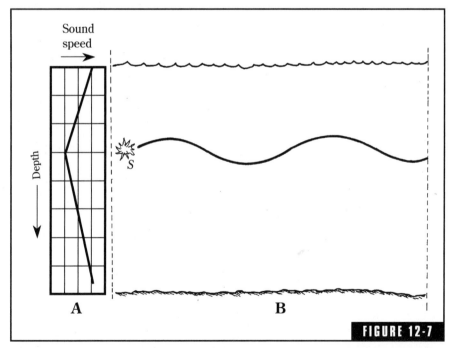

FIGURE 12-7

How the long range sound experiment of Fig. 12-6 was accomplished; (A) sound speed decreases with depth in the upper reaches of the ocean (temperature effect) and increases at greater depths (pressure effect) creating a sound channel at the inversion depth (about 700 fathoms). (B) A ray of sound is kept in this sound channel by refraction. Sound travels great distances in this channel because of the lower losses.

auditorium at times. With a normal heating and air-conditioning system, great efforts are made to avoid large horizontal or vertical temperature gradients. The goals of temperature uniformity and no troublesome drafts have reduced sound refraction effects to inconsequential levels.

Consider the same gymnasium used as an auditorium but with less sophisticated air conditioning. In this case a large ceiling-mounted heater near the rear acts as a space heater. Working against gravity, the unit produces copious hot air near the ceiling, relying on slow convection currents to move some of the heat down to the audience level.

This reservoir of hot air near the ceiling and cooler air below can have a minor effect on the transmission of sound from the sound system and on the acoustics of the space. The feedback point of the sound system might shift. The standing waves of the room might change slightly as longitudinal and transverse sound paths are increased in length because of their curvature due to refraction. Flutter echo paths are also shifted. With a sound radiating system mounted high at one end of the room, lengthwise sound paths would be curved downward. Such downward curvature might actually improve audience coverage, depending somewhat on the directivity of the radiating system.

Endnotes

[1]Shockley, R.C., J. Northrop, P.G. Hansen, and C. Hartdegen, *SOFAR Propagation Paths from Australia to Bermuda*, J. Acous. Soc. Am., 71, 51 (1982).

[2]Heaney, K.D., W.A. Kuperman, and B.E. McDonald, *Perth-Bermuda Sound Propagation (1960): Adiabatic Mode Interpretation*, J. Acous. Soc. Am., 90, 5 (Nov 1991) 2586-2594.

[3]Spiesberger, John, Kent Metzger, and John A. Ferguson, *Listening for Climatic Temperature Changes in the Northeast Pacific 1983-1989*, J. Acous. Soc. Am., 92, 1 (July 1992) 384-396.

Diffusion of Sound

Diffusion problems are most troublesome in smaller rooms and at the lower audio frequencies. The problem with small spaces such as the average recording studio, control room, or music listening room is that modal spacings below 300 Hz guarantee a sound field far from diffuse (Chap. 15).

The Perfectly Diffuse Sound Field

Even though unattainable, it is instructive to consider the characteristics of a diffuse sound field. Randall and Ward[1] have given us a list of these:

- The frequency and spatial irregularities obtained from steady-state measurements must be negligible.

- Beats in the decay characteristic must be negligible.

- Decays must be perfectly exponential, i.e., they must be straight lines on a logarithmic scale.

- Reverberation time will be the same at all positions in the room.

- The character of the decay will be essentially the same for different frequencies.

■ The character of the decay will be independent of the directional characteristics of the measuring microphone.

These six factors are observation oriented. A professional physicist specializing in acoustics might stress fundamental and basic factors in his definition of a diffuse sound field such as energy density, energy flow, superposition of an infinite number of plane progressive waves, and so on. The six characteristics suggested by Randall and Ward point us to practical ways of obtaining solid evidence for judging the diffuseness of the sound field of a given room.

Evaluating Diffusion in a Room

There is nothing quite as upsetting as viewing one's first attempt at measuring the "frequency response" of a room. To obtain the frequency response of an amplifier, a variable-frequency signal is put in the front end and the output observed to see how flat the response is. The same general approach can be applied to a room by injecting the variable frequency signal into "the front end" by means of a loudspeaker and noting the "output" picked up by a microphone located elsewhere in the room.

Steady-State Measurements

Figure 13-1 is a graphic-level recorder tracing of the steady-state response of a studio having a volume of 12,000 cubic feet. In this case, the loudspeaker was in one lower tricorner of the room, and the microphone was at the upper diagonal tricorner about one foot from each of the three surfaces. These positions were chosen because all room modes terminate in the corners and all modes should be represented in the trace. The fluctuations in this response cover a range of about 35 dB over the linear 30- to 250-Hz sweep. The nulls are very narrow, and the narrow peaks show evidence of being single modes because the mode bandwidth of this room is close to 4 Hz. The wider peaks are the combined effect of several adjacent modes. The rise from 30 to 50 Hz is due primarily to loudspeaker response and the 9-dB peak between 50 and 150 Hz (due to radiating into ¼ space) should not be charged against the room. The rest is primarily room effect.

The response of Fig. 13-1 is typical of even the best studios. Such variations in response are, of course, evidence of a sound field that is

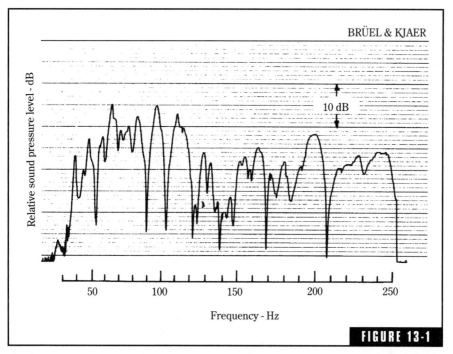

Slowly swept sine-wave sound-transmission response of a 12,000-cu ft. video studio. Fluctuations of this magnitude, which characterize the best of studios, are evidence of nondiffusive conditions.

not perfectly diffused. A steady-state response such as this taken in an anechoic room would still show variations, but of lower amplitude. A very live room, such as a reverberation chamber, would show even greater variations.

Figure 13-1 illustrates one way to obtain the steady-state response of a room. Another is to traverse the microphone while holding the loud-speaker frequency constant. Both methods reveal the same deviations from a truly homogeneous sound field. Thus, we see that Randall and Ward's criteria of negligible frequency and spatial irregularities are not met in the studio of Fig. 13-1 or, in fact, in any practical recording studio.

Decay Beats

By referring to Chap. 7, Fig. 7-10, we can compare the smoothness of the reverberation decay for the eight octaves from 63 Hz to 8 kHz. In general, the smoothness of the decay increases as frequency is increased. The reason for this, as explained in Chap. 7, is that the number of modes within an octave span increases greatly with frequency, and the greater the mode density, the smoother their average effect. Beats in the decay are greatest at 63 Hz and 125 Hz. The decays of Fig. 7-10 indicate that the diffusion of sound in this particular studio is about as good as can be achieved by traditional means. It is the beat information on the low-frequency reverberation decay that makes possible a judgment on the degree of diffusion prevailing. Reverberation-time measuring devices that yield information only on the average slope and not the shape of the decay pass over information that most consultants consider important in evaluating the diffuseness of a space.

Exponential Decay

A truly exponential decay is a straight line on a level vs. time plot, and the slope of the line can be described either as a decay rate in decibels per second or as reverberation time in seconds. The decay of the 250-Hz octave band of noise pictured in Fig. 13-2 has two exponential slopes. The initial slope gives a reverberation time of 0.35 second and the final slope a reverberation time of 1.22 seconds. The slow decay that finally takes over once the level is low enough is probably a specific mode or group of modes encountering low absorption either by striking the absorbent at grazing angles or striking where there is little absorption. This is typical of one type of nonexponential decay, or stated more precisely, of a dual exponential decay.

Another type of nonexponential decay is illustrated in Fig. 13-3. The deviations from the straight line connecting the beginning and end of the decay are considerable. This is a decay of an octave band of noise centered on 250 Hz in a 400-seat chapel, poorly isolated from an adjoining room. Decays taken in the presence of acoustically coupled spaces are characteristically concave upward, such as in Fig. 13-3, and often the deviations from the straight line are even greater. When the decay traces are nonexponential, i.e., they depart from a straight line in a level vs. time plot, we must conclude that true diffuse conditions do not prevail.

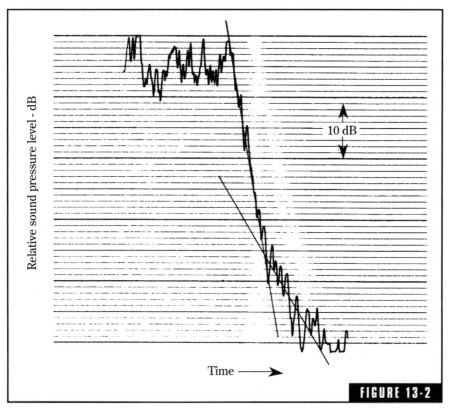

FIGURE 13-2

Typical double slope-decay, evidence of a lack of diffuse sound conditions. The slower decaying final slope is probably due to modes that encounter lower absorption.

Spatial Uniformity of Reverberation Time

When reverberation time for a given frequency is reported, it is usually the average of multiple observations at each of several positions in the room. This is the pragmatic way of admitting that reverberatory conditions differ from place to place in the room. Figure 13-4 shows the results of actual measurements in a small (22,000 cu ft) video studio. The multiple uses of the space required variable reverberation time, which was accomplished by hinged wall panels that can be closed, revealing absorbent sides, or opened, revealing reflecting sides. Multiple reverberation decays were recorded at the same three microphone positions for both "panels-reflective" and "panels-absorptive" conditions. The circles are the average values, and the light lines represent average reverberation time at each of the three positions. It is evident

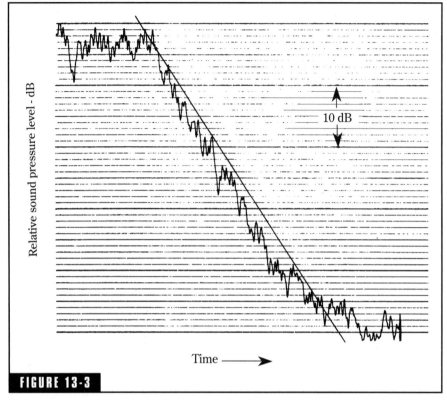

Relative sound pressure level - dB

10 dB

Time ⟶

FIGURE 13-3

The nonexponential form of this decay, taken in a 400-seat chapel, is attributed to acoustically coupled spaces. The absence of a diffuse sound field is indicated.

that there is considerable variation, which means that the sound field of the room is not completely homogeneous during this transient decay period. Inhomogeneities of the sound field are one reason that reverberation times vary from point to point in the room, but there are other factors as well. Uncertainties in fitting a straight line to the decay also contribute to the spread of the data, but this effect should be relatively constant from one position to another. It seems reasonable to conclude that spatial variations in reverberation time are related, at least partially, to the degree of diffusion in the space.

Standard deviations of the reverberation times give us a measure of the spread of the data as measured at different positions in a room. When we calculate an average value, all evidence of the spread of the data going into the average is lost. The standard deviation is the statistician's way of keeping an eye on the data spread. The method of calculating the standard deviation is described in the manuals of most scientific calculators. Plus or minus one standard deviation from the mean value embraces 68% of the data points if the distribution is normal (Gaussian), and reverberation data should qualify reasonably well. In Table 13-1, for 500 Hz, panels reflective, the mean RT60 is 0.56 seconds with a standard deviation of 0.06 seconds. For a normal distribution, 68% of the data points would fall between 0.50 and 0.62 second. That 0.06 standard deviation is 11% of the 0.56 mean. The percentages listed in Table 13-1 give us a rough appraisal of the precision of the mean.

In order to view the columns of percentage in Table 13-1 graphically, they are plotted in Fig. 13-5. Variability of reverberation time values at the

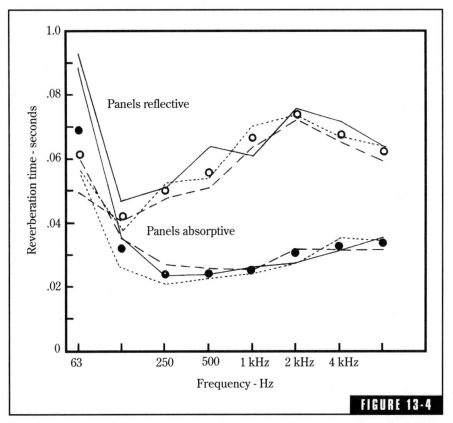

FIGURE 13-4

Reverberation time characteristics of a 22,000-cu ft studio with acoustics adjustable by hinged panels, absorbent on one side and reflective on the other. At each frequency, the variation of the average reverberation time at each of the three positions indicates non-diffuse conditions especially at low frequencies.

Table 13-1 Reverberation time of small video studio.

Octave band center frequency	Panels reflective			Panels absorptive		
	RT60	Std. dev.	Std. dev. % of mean	RT60	Std. dev.	Std. dev. % of mean
63	0.61	0.19	31.	0.69	0.18	26.
125	0.42	0.05	12.	0.32	0.06	19.
250	0.50	0.05	10.	0.24	0.02	8.
500	0.56	0.06	11.	0.24	0.01	4.
1 kHz	0.67	0.03	5.	0.26	0.01	4.
2 kHz	0.75	0.04	5.	0.31	0.02	7.
4 kHz	0.68	0.03	4.	0.33	0.02	6.
8 kHz	0.63	0.02	3.	0.34	0.02	6.

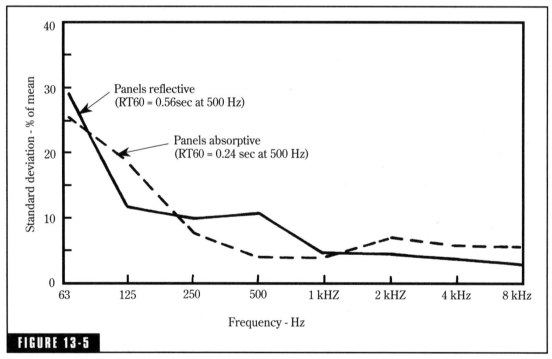

FIGURE 13-5

Closer examination of the reverberation time variations of the studio of Fig. 13-4. The standard deviation, expressed as a percentage of the mean value, shows lack of diffusion, especially below 250 Hz.

higher frequencies settles down to reasonably constant values in the neighborhood of 3% to 6%. Because we know that each octave at high frequencies contains an extremely large number of modes that results in smooth decays, we can conclude confidently that at the higher audible frequencies essentially diffuse conditions exist, and that the 3% to 6% variability is normal experimental measuring variation. At the low frequencies, however, the high percentages (high variabilities) are the result of greater mode spacing producing considerable variation in reverberation time from one position to another. We must also admit that these high percentages include the uncertainty in fitting a straight line to the wiggly decay characteristic of low frequencies. However, a glance at Fig. 13-4 shows that there are great differences in reverberation time between the three measuring positions. For this 22,000 cu ft studio for two different conditions of absorbance (panels open/closed), diffusion is poor at 63 Hz, somewhat better at 125 Hz, and reasonably good at 250 Hz and above.

Decay Shapes

If all decays have the same character at all frequencies and that character is smooth decay, complete diffusion prevails. In the real world, the decays of Fig. 7-10 with significant changes in character are more common, especially for the 63-Hz and 125-Hz decays.

Microphone Directivity

One method of appraising room diffusion is to rotate a highly directional microphone in various planes and record its output to the constant excitation of the room. This method has been applied with some success to large spaces, but the method is ill adapted to smaller recording studios, control rooms, and listening rooms, in which diffusion problems are greatest. In principle, however, in a totally homogeneous sound field, a highly directional microphone pointed in any direction should pick up a constant signal.

Room Shape

How can a room be built to achieve maximum diffusion of sound? This opens up a field in which there are strong opinions—some of them supported by quite convincing experiments—and some just strong without such support.

There are many possible shapes of rooms. Aside from the general desirability of a flat floor in this gravity-stricken world, walls can be splayed, ceilings inclined, cylindrical or polygonal shapes employed. Some shapes can be eliminated because they focus sound, and focusing is the opposite of diffusing. For example, parabolic shapes yield beautifully sharp focal points and cylindrical concavities less sharp but nonetheless concentrated. Even polygonal concave walls of 4, 5, 6, or 8 sides approach a circle and result in concentrations of sound in some areas at the expense of others.

The popularity of rectangular rooms is due in part to economy of construction, but it has its acoustical advantages. The axial, tangential, and oblique modes can be calculated with reasonable effort and their distribution studied. For a first approximation, a good approach is to consider only the more dominant axial modes, which is a very simple calculation. Degeneracies (mode pile-ups) can be spotted and other room faults revealed.

The relative proportioning of length, width, and height of a sound sensitive room is most important. If plans are being made for constructing such a room, there are usually ideas on floor-space requirements, but where should one start in regard to room proportions? Cubical rooms are anathema. The literature is full of early quasi-scientific guesses, and later statistical analyses of room proportions that give good mode distribution. None of them come right out and say, "This is the absolute optimum." Bolt[2] gives a range of room proportions producing the smoothest room characteristics at low frequencies in small rectangular rooms (Fig. 13-6). Volkmann's 2 : 3 : 5 proportion,[3] was in favor 50 years ago. Boner suggested the 1 : 1.26 : 1.59 ratio as optimum.[4] Sepmeyer[5]

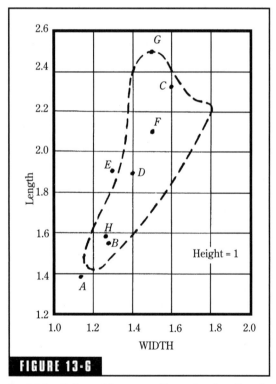

FIGURE 13-6

A chart of favorable room dimensional ratios to achieve uniform distribution of modal frequencies of a room. The broken line encloses the so-called "Bolt-Area."[2] The letters refer to Table 13-2.

published a computer statistical study in 1965 that yields several favorable ratios. An even later paper by Louden[6] lists 125 dimension ratios arranged in descending order of room acoustical quality.

Table 13-2 lists the best proportions suggested by all of these papers. To compare these with the favorable area suggested by Bolt, they are plotted in Fig. 13-6. Most of the ratios fall on or very close to the *Bolt area*. This gives confidence that any ratio falling in the Bolt area will yield reasonable low-frequency room quality as far as distribution of axial modal frequencies is concerned.

One cannot tell by looking at a room's dimensional ratio whether it is desirable or not, and it is preferable to make the evaluation, rather than just take someone's word for it. Assuming a room height of 10 ft, and the other two dimensions, an axial mode analysis such as Fig. 13-7 can be made for each. This has been done and these modes are plotted in Fig. 13-8. Each is keyed into Table 13-2 for source identification. All of these are relatively small rooms and therefore suffer the same fate of having axial-mode spacings in frequency greater than desired. The more uniform the spacing, the better. Degeneracies, or mode coin-

Table 13-2. Rectangular room dimension ratios for favorable mode distribution.

Author		Height	Width	Length	In Bolt's range?
1. Sepmeyer[5]	A	1.00	1.14	1.39	No
	B	1.00	1.28	1.54	Yes
	C	1.00	1.60	2.33	Yes
2. Louden[6]	D	1.00	1.4	1.9	Yes
3 best ratios	E	1.00	1.3	1.9	No
	F	1.00	1.5	2.5	Yes
3. Volkmann[3] 2 : 3 : 5	G	1.00	1.5	2.5	Yes
4. Boner[4] 1: $\sqrt[3]{2}$: $\sqrt[3]{4}$	H	1.00	1.26	1.59	Yes

	Length $L = 19$ ft 5 in $L = 19.417$ ft $f_1 = 565/L$	Width $W = 14$ ft 2 in $W = 14.17$ ft $f_1 = 565/W$	Height $H = 8$ ft $H = 8.92$ ft $f_1 = 565/H$	Arranged in ascending order	Diff
				29.1	
					10.8
f_1	29.1	39.9	63.3	39.9	
					18.3
f_2	58.1	79.7	126.7	58.2	
					5.1
f_3	87.3	119.6	190.0	63.3	
					16.4
f_4	116.4	159.5	253.4	79.7	
					7.6
f_5	145.5	199.4	316.7	87.3	
					29.1
f_6	174.6	239.2		116.4	
					3.2
f_7	203.7	279.1		119.6	
					7.1
f_8	232.8	319.0		126.7	
					18.8
f_9	261.9			145.5	
					14.0
f_{10}	291.0			159.5	
					15.1
f_{11}	320.1			174.6	
					15.4
f_{12}				190.0	
					9.4
f_{13}				199.4	
					4.3
f_{14}				203.7	
					29.1
f_{15}				232.8	
					6.4
				239.2	
					14.2
				253.4	
					8.5
				261.9	
					17.2
				279.1	
					11.9
				291.0	
					25.7
				316.7	

FIGURE 13-7

A convenient data form for studying the effects of room proportions on the distribution of axial modes.

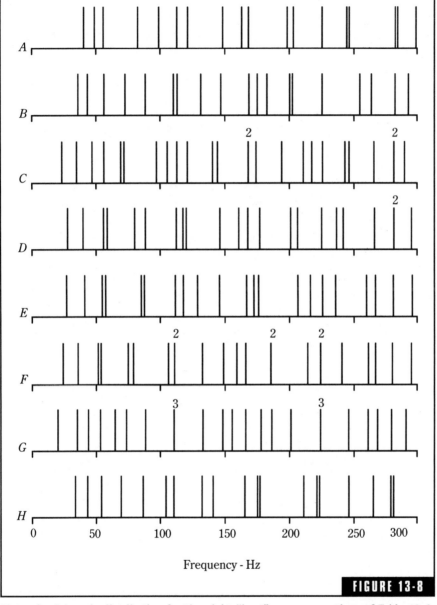

FIGURE 13-8

Plots of axial mode distribution for the eight "best" room proportions of Table 13-2. The small numbers indicate the number of modes coincident at those particular frequencies. A room height of 10 ft is assumed.

cidences, are a potential problem, and they are identified by the 2 or 3 above them to indicate the number of resonances piled up. Modes very close together, even though not actually coincident, can also present problems. With these rules to follow, which of the 8 "best" distributions of Fig. 13-8 are really the best and which the worst? First, we reject *G* with two triple coincidences greatly spaced from neighbors. Next, *F* is eliminated because of three double coincidences associated with some quite wide spacings. We can neglect the effect of the double coincidences near 280 Hz in *C* and *D* because colorations are rarely experienced above 200 Hz. Aside from the two rejected outright, there is little to choose between the remainder. All have flaws. All would probably serve quite well, alerted as we are to potential problems here and there. This simple approach of studying the axial-mode distribution has the advantage of paying attention to the dominant axial modes knowing that the weaker tangential and oblique modes can only help by filling in between the more widely spaced axial modes.

Figure 13-7 illustrates a data form that makes it easy to study the axial modes of a room. Analyzing the results requires some experience and a few rules of thumb are suggested. A primary goal is to avoid coincidences (pile-ups) of axial modes. For example, if a cubical space were analyzed, all three columns would be identical; the three fundamentals and all harmonics would coincide. This produces a triple coincidence at each modal frequency and great gaps between. Unquestionably, sound in such a cubical space would be highly colored and acoustically very poor. The room of Fig. 13-7 has 22 axial modes between 29.1 and 316.7 Hz. If evenly spaced, the spacing would be about 13 Hz, but spacings vary from 3.2 to 29.1 Hz. However, there are no coincidences—the closest pair are 3.2 Hz apart. If a new room is to be constructed, you have the freedom on paper to move a wall this way or that or to raise or lower the ceiling a bit to improve distribution. The particular room proportions of Fig. 13-7 represent the end product of many hours of cut and try. While this cannot be represented as the best proportioning possible, this room, properly treated, will yield good, uncolored sound. The proper starting point is proper room proportions.

In adapting an existing space, you lack the freedom to shift walls as on paper. A study of the axial modes as per Fig. 13-7, however, can still be very helpful. For example, if such a study reveals problems and space permits, a new wall might improve the modal situation markedly. By splaying this wall, other advantages discussed later may accrue. If the study points to a coincidence at 158 Hz, well separated from neighbors, one is alerted to potential future problems with an understanding of the cause. There is always the possibility of introducing a Helmholtz resonator tuned to the offending coincidence to control its effect (see pp. 226–229). All these things are related to sound diffusion.

Splaying Room Surfaces

Splaying one or two walls of a sound-sensitive room does not eliminate modal problems, although it might shift them slightly and produce somewhat better diffusion.[7] In new construction, splayed walls cost no more, but may be quite expensive in adapting an existing space. Wall splaying is one way to improve general room diffusion, although its effect is nominal. Flutter echoes definitely can be controlled by canting one of two opposing walls. The amount of splaying is usually between 1 foot in 20 feet and 1 foot in 10 feet.

Nonrectangular Rooms

The acoustical benefit to be derived from the use of nonrectangular shapes in audio rooms is rather controversial. Gilford[10] states, "...slanting the walls to avoid parallel surfaces.... does not remove colorations; it only makes them more difficult to predict." Massive trapezoidal shaped spaces, commonly used as the outer shell of recording studio control rooms, guarantee asymmetrical low-frequency sound fields even though it is generally conceded that symmetry with the control position is desirable.

Computer studies based on the finite element approach have revealed in minute detail what happens to a low-frequency sound field in nonrectangular rooms. The results of a study using this method conducted by van Nieuwland and Weber at Philips Research Laboratories, The Netherlands, are given in Figs. 13-10 through 13-13.[8] Highly contorted sound fields are shown, as expected, for the nonrectangular case, for modes 1,0, 1,3, 0,4, and 3,0. A shift in frequency of the standing wave from that of the rectangular room of the same area is indicated: −8.6%, −5.4%, −2.8%, and +1% in the four cases illustrated. This would tend to support the common statement that splay-

The use of distributed sound-absorbing modules is an economical way to achieve maximum absorption as well as to enhance sound diffusion in the room. *(World Vision International)*

ing of walls helps slightly in breaking up degeneracies, but shifts of 5% or more are needed to avoid the effects of degeneracies. The proportions of a rectangular room can be selected to eliminate, or at least greatly reduce, degeneracies, while in the case of the nonrectangular room, such a prior examination of degeneracies is completely impractical. Making the sound field asymmetrical by splaying walls only introduces unpredictability in listening room and studio situations.

If the decision is made to splay walls in an audio room, say 5%, a reasonable approximation would be to analyze the equivalent rectangular room having the same volume.

Geometrical Irregularities

Many studies have been made on what type of wall protuberances provide the best diffusing effect. Somerville and Ward[9] reported years ago that geometrical diffusing elements reduced fluctuations in a swept-sine steady-state transmission test. The depth of such geometrical diffusors must be at least ½ of a wavelength before their effect is felt. They studied cylindrical, triangular, and rectangular elements and found that the straight sides of the rectangular-shaped diffusor provided the greatest effect for both steady-state and transient phenomena. BBC experience indicates superior subjective acoustical properties in studios and concert halls in which rectangular ornamentation in the form of coffering is used extensively.

Absorbent in Patches

Applying all the absorbent in a room on one or two surfaces does not result in a diffuse condition, nor is the absorbent used most effectively. Let us consider the results of an experiment that shows the effect of

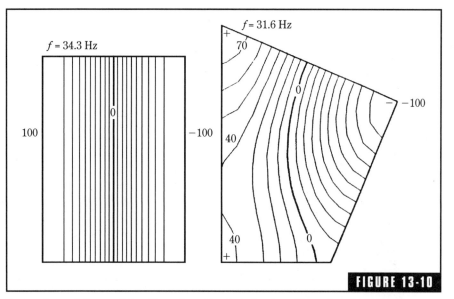

Comparison of the modal pattern for a 5 x 7 meter two-dimensional room and a non-rectangular room of the same area. This sound field of the 1,0 mode is distorted in the nonrectangular room and the frequency of the standing wave is shifted slightly.[8]

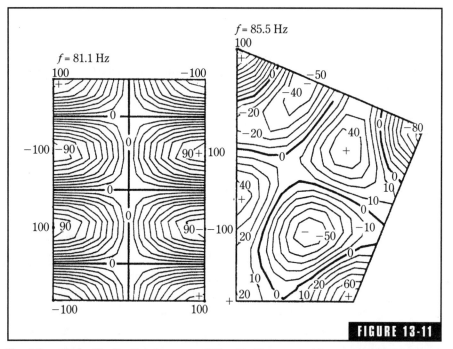

The 1,3 mode for the 5 x 7 meter room of Fig. 13-10 compared to a nonrectangular room of the same area. The sound field is distorted and the frequency is shifted.[8]

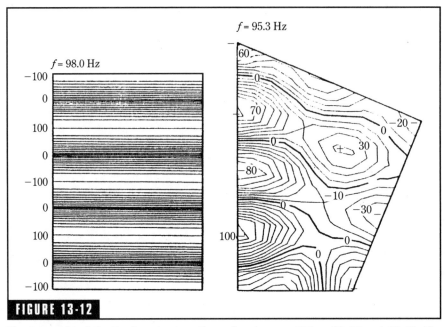

FIGURE 13-12

The 0,4 mode of the 5 x 7 meter two-dimensional room of Figs. 13-10 and 13-11. The high distortion in the rectangular room is accompanied by a shift in standing wave frequency.[8]

distributing the absorbent.[1] The experimental room is approximately a 10-ft cube and it was tiled (not an ideal recording or listening room, but acceptable for this experiment). For test 1, reverberation time for the bare room was measured and found to be 1.65 seconds at 2 kHz. For test 2, a common commercial absorber was applied to 65% of one wall (65 sq ft), and the reverberation time at the same frequency was found to be about 1.02 seconds. For test 3, the same area of absorber was divided into four sections, one piece mounted on each of four of the room's six surfaces. This brought the reverberation time down to about 0.55 seconds.

The startling revelation here is that the area of the absorber was identical between tests 2 and 3; the only difference was that in test 3 it was in four pieces, one on each of 3 walls and one piece on the floor. By the simple expedient of dividing the absorbent and distributing it, the reverberation time was cut almost in half. Inserting the values of reverberation time of 1.02 and 0.55 seconds and the volume and area of the room into the Sabine equation, we find that the average absorption coefficient of the room increased from 0.08 to 0.15 and the number of absorption units from 48 to 89 sabins. Where did all this extra absorption come from? Laboratory testing personnel measuring absorption coefficients in reverberation chambers have agonized over the problem for years. Their conclusion is that there is an edge effect related to diffraction of sound that makes a given sample appear to be much larger acoustically. Stated another way, the sound-absorbing efficiency of 65 sq ft of absorbing material is only about half that of four 16-sq ft pieces distributed about the room, and the edges of the four pieces total about twice that of the single 65-sq ft piece. So, one advantage of distributing the absorbent in a room is that its sound-absorbing efficiency is greatly increased, at least at certain frequencies. But be warned: The above statements are true for 2 kHz, but at 700 Hz and 8 kHz, the difference between one large piece and four distributed pieces is small.

Another significant result of distributing the absorbent is that it contributes to diffusion of sound. Patches of absorbent with reflective

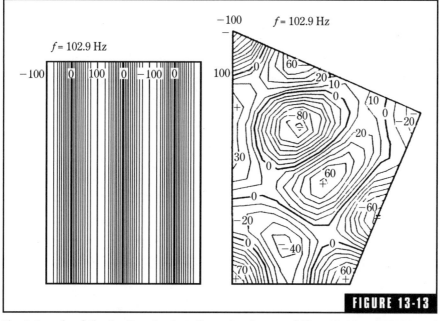

FIGURE 13-13

The 3,0 mode of the 5 x 7 meter two-dimensional room of Figs. 13-10 and 13-11, and 13-12 and resulting distortion of the modal pattern when changed to a nonrectangular room of the same area.[8]

walls showing between the patches have the effect of altering wave-fronts, which improves diffusion. Sound-absorbing modules in a recording studio such as in Fig. 13-9 distribute the absorbing material and simultaneously contribute to the diffusion of sound.

Concave Surfaces

A concave surface such as that in Fig. 13-14A tends to focus sound energy and consequently should be avoided because focusing is just the opposite of the diffusion we are seeking. The radius of curvature determines the focal distance; the flatter the concave surface, the greater the distance at which sound is concentrated. Such surfaces often cause problems in microphone placement. Concave surfaces might produce some awe-inspiring effects in a whispering gallery where you can hear a pin drop 100 ft away, but they are to be avoided in listening rooms and small studios.

Convex Surfaces: The Poly

One of the most effective diffusing elements, and one relatively easy to construct, is the polycylindrical diffusor (poly), which presents a convex section of a cylinder. Three things can happen to sound falling on

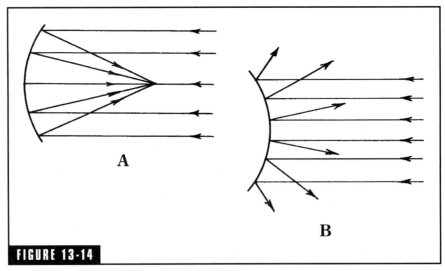

FIGURE 13-14

Concave surfaces (A) tend to focus sound, convex surfaces (B) tend to diffuse it. Concave surfaces should be avoided if the goal is to achieve well-diffused sound.

such a cylindrical surface made of plywood or hardboard: The sound can be
reflected and thereby dispersed as in Fig.
13-14B; the sound can be absorbed; or the
sound can be reradiated. Such cylindrical
elements lend themselves to serving as
absorbers in the low frequency range
where absorption and diffusion are so
badly needed in small rooms. The reradiated portion, because of the diaphragm
action, is radiated almost equally throughout an angle of roughly 120° as shown in
Fig. 13-15A. A similar flat element reradiates sound in a much narrower angle,
about 20°. Therefore, favorable reflection,
absorption, and reradiation characteristics
favor the use of the cylindrical surface.
Some very practical polys and their
absorption characteristics are presented in

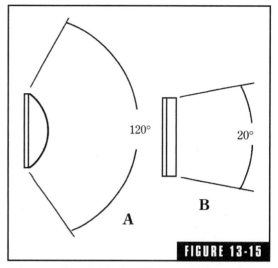

FIGURE 13-15

(A) A polycylindrical diffusor reradiates sound
energy not absorbed through an angle of about 120°.
(B) A similar flat element reradiates sound in a
much smaller angle.

Chap. 9. The dimensions of such diffusors are not critical, although to
be effective their size must be comparable to the wavelength of the
sound being considered. The wavelength of sound at 1,000 Hz is a bit
over 1 ft, at 100 Hz about 11 ft. A poly element 3 or 4 ft across would be
effective at 1000 Hz, much less so at 100 Hz. In general, poly base or
chord length of 2 to 6 ft with depths of 6 to 18 inches meet most needs.

It is important that diffusing elements be characterized by randomness. A wall full of polys, all of 2-ft chord and of the same depth, might
be beautiful to behold, like some giant washboard, but not very effective as diffusors. The regularity of the structure would cause it to act as
a diffraction grating, affecting one particular frequency in a much different way than other frequencies, which is opposite to what the ideal
diffusor should do.

Axes of symmetry of the polys on different room surfaces should be
mutually perpendicular.

Plane Surfaces

Geometrical sound diffusing elements made up of two flat surfaces to
give a triangular cross section or of three or four flat surfaces to give a

polygonal cross section may also be used. In general, their diffusing qualities are inferior to the cylindrical section.

Endnotes

[1]Randall, K.E. and F.L. Ward, *Diffusion of Sound in Small Rooms*, Proc. Inst. Elect. Engs., Vol 107B (Sept. 1960), p. 439-450.

[2]Bolt, R.H., *Note on Normal Frequency Statistics for Rectangular Rooms*, J. Acous. Soc. Am., 18, 1 (July 1946) p. 130-133.

[3]Volkmann, J.E., *Polycylindrical Diffusers in Room Acoustical Design*, J. Acous. Soc. Am., 13 (1942), p. 234-243.

[4]Boner, C.P., *Performance of Broadcast Studios Designed with Convex Surfaces of Plywood*, J. Acous. Soc. Am., 13 (1942) p. 244-247.

[5]Sepmeyer, L.W., *Computed Frequency and Angular Distribution of the Normal Modes of Vibration in Rectangular Rooms*, J. Acous Soc. Am., 37, 3 (March 1965), p. 413-423.

[6]Louden, M.M., *Dimension-Ratios of Rectangular Rooms with Good Distribution of Eigentones*, Acustica, 24 (1971), p. 101-103.

[7]Nimura, Tadamoto and Kimio Shibayama, *Effect of Splayed Walls of a Room on Steady-State Sound Transmission Characteristics*, J. Acous. Soc. Am., 29, 1 (January 1957), p. 85-93.

[8]van Nieuwland, J.M. and C. Weber, *Eigenmodes in Non-Rectangular Reverberation Rooms*, Noise Control Eng., 13, 3 (Nov/Dec 1979), 112-121.

[9]Somerville, T. and F.L. Ward, *Investigation of Sound Diffusion in Rooms by Means of a Model*, Acustica, 1, 1 (1951), p. 40-48.

[10]Gilford, Christopher, *Acoustics for Radio and Television Studios*, (1972), London, Peter Peregrinus, Ltd.

The Schroeder Diffusor

In a strange turn of events, diffraction gratings have become important in acoustics. The subject of diffraction gratings brings to mind several memories, widely separated in time and space.

There are three large Morpho butterflies mounted in a frame in our home, a sweet reminder of first seeing their gorgeous flashes of color as they flitted about in a jungle clearing in Panama. Their color changes in a most dazzling fashion. They were even more fascinating to me when I learned that this color is not pigmentation—it is *structural color*. The butterfly's wing is basically a tawdry tan—the vivid coloring is the result of breaking down the light that falls upon it into rainbow colors by diffraction. Viewed through a microscope, the wing surface is made up of a myriad of very small bumps and grooves.

At Mt. Wilson Observatory is the diffraction grating used by Edwin Hubble to measure the light of distant stars. Using this diffraction grating, he formulated his theory of the expanding universe based on the "red shifts" of starlight he observed. It is a glass plate with precise, parallel lines cut on it, many thousands to the inch. Sunlight falling on this grating is broken down into its component colors, just as the light from the stars.

In early days, the diffraction of X-rays by crystals was studied in the Physics Department at Stanford University in a safety cage made of chicken wire screen in a remote corner of the basement. Crystals are basically naturally occurring three-dimensional diffraction gratings. The X-rays, having a wavelength of the same order as the spacings of the atoms in the crystal lattice, were scattered by the crystal in an orderly, predictable fashion.

Applying the principle of the diffraction grating to acoustics is further evidence of the ever-growing cutting edge of science.

Schroeder's First Acoustic Diffusor

In a remarkable outpouring of fresh, new ideas, Manfred R. Schroeder of the University of Göttingen, Germany, and AT&T Bell Laboratories at Murray Hill, New Jersey, has opened new vistas on the diffusion of sound. Schroeder has lifted from computer and number theory the idea that a wall with grooves arranged in a certain way will diffuse sound to a degree unattainable in the past. Maximum-length codes can be used to create pseudo-random noise by application of certain sequences of +1 and −1. The power spectrum (from the Fourier transform) of such noise is essentially flat. A wide and flat power spectrum is related to reflection coefficients and angles, and this gave rise to the idea that by applying the +1 and −1 in a maximum-length sequence something acoustically interesting might result. The −1 suggested a reflection from the bottom of a groove in a wall with a depth of a quarter wavelength. The +1 reflection is a reflection from the wall itself without any groove.

Professor Schroeder's next move was to test the idea.[1] He ordered a piece of sheet metal to be bent into the shape of Fig. 14-1, planning to test the idea with 3-cm microwaves. This shape followed the binary maximum-length sequence with period length 15:

$$- + + - + - + + + + - - - + -$$

The resulting reflection pattern (Fig. 14-2(B)) indicated that this piece of sheet metal was much like any other piece of sheet metal; it gave a strong specular reflection, but little diffusion of energy. He asked how deep the steps were and learned that they were half-wavelength, not the quarter wavelength ordered. No wonder it reflected the microwaves like a flat sheet of metal! The grooves one-half wavelength deep gave a reflection coefficient of +1, just like a flat sheet. Back to the shop for a new metal sheet with grooves a quarter wavelength deep.

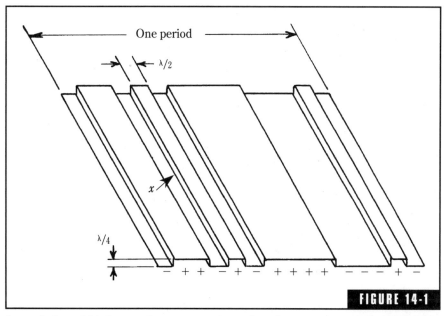

One period

λ/2

x

λ/4

− + + − + − + + + + − − − + −

FIGURE 14-1

A metal sheet folded to conform to a maximum-length sequence used by Schroeder to check the diffusion of 3-cm radio waves. The wells are all 1/4 wavelength in depth.

This time the reflection pattern of Fig. 14-2A resulted, a very encouraging development.

Professor Schroeder's associate, who shaped the metal grating, was one who always liked to check Schroeder's theories. He quietly ran another test with a narrow strip of metal covering just one of the grooves, the one marked *x* in Fig. 14-1. The reflection pattern of Fig. 14-2B resulted, which shows essentially specular reflection of most of the energy back toward the source. In other words, covering only one of the slots almost completely destroyed the favorable reflection pattern of Fig.14-2A. This success encouraged further development of the basic idea for acoustical applications.

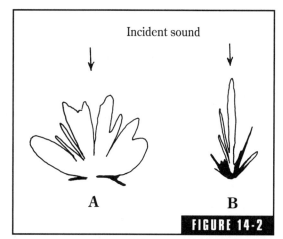

Incident sound

A   B

FIGURE 14-2

When the wells of Fig. 14-1 were accidentally made 1/2 wavelength deep, the almost pure specular reflection of B resulted, the same as for a flat sheet of metal. When another diffuser was made with the correct 1/4 wavelength depths, the very favorable diffusion pattern of A was obtained. *With permission of M.R. Schroeder and Applied Science Publishers, Ltd., London.*

Maximum-Length Sequences

This experiment was a turning point in acoustics. The reflection pattern of Fig.14-2A is far superior to the best sound diffusor previously available. For over a half century, diffusion was sought through adjustment of room proportions, splaying of walls, the use of semispherical, polycylindrical, triangular, cubical, and rectangular geometrical protrusions, and the distribution of absorbing material as shown in Chap. 13. The degree of diffusion achieved by such means was far from sufficient to meet the need. Considered in this light, the importance of Professor Schroeder's simple experiment with the piece of bent sheet metal becomes apparent.

Because of the one-quarter wavelength groove depth requirement of the binary maximum-length diffusor, the sound-diffusing properties of the surface depend on the wavelength of the incident sound. Experience has indicated that reasonable diffusion results over a band of plus or minus one-half octave of the frequency around which the diffusor is designed. For example, consider a maximum-length sequence diffusor with a sequence length of 15.[2] A design frequency of 1,000 Hz gives a one-half wavelength of 7.8 in, and a one-quarter wavelength groove depth of 3.9 in. A single period of this diffusor would be about 5 ft in width and effective from about 700 to about 1,400 Hz. Many such units would be required to provide diffusion over a reasonable portion of the audible band.

Reflection Phase-Grating Diffusors

The diffraction grating, so long the sole province of optics, is now being applied to acoustics. Commercial development of the idea is being aggressively pursued by RPG Diffusor Systems, Inc., of Upper Marlboro, Maryland.[8] In the past, the acoustical designer basically had only absorption and reflection to use. Diffusion was something earnestly sought, but only partially achieved. With the large assortment of grating-type diffusor units commercially available, this situation is rapidly changing.

The acoustical treatment of large spaces is well served by such diffusors. A *large space* is defined as one whose normal mode frequencies are so closely spaced as to avoid low-frequency resonance problems. This includes music halls, auditoriums, and churches. The sound quality of a music hall is influenced greatly by the reflections from side walls. No longer do we completely depend on side walls for necessary lateral reflections. A row of diffraction-grating diffusors down the center of the hall at ceiling level can diffuse the sound from the stage laterally to the people in the seats. Any troublesome specular reflection can be controlled by strategic placement of diffusors.

In churches, there is always conflict between the intelligibility of the spoken word and conditions for full enjoyment of the music. The rear wall is often the source of reflections that create disturbing echoes. To make this wall absorbent is often detrimental to music conditions. Making the rear wall diffusive, however, minimizes the echo problem while at the same time conserving precious music and speech energy. Music directors are often faced with the problem of singers or instrument players not hearing each other well. How can they play or sing together properly if they cannot hear each other? Surrounding the music group with an array of reflection phase grating diffusors both conserves music energy and spreads it around to achieve ensemble between musicians.

Difficult small-room acoustics are helped especially by diffraction-grating-type diffusing elements. The need to cant walls and distribute absorption material to achieve some semblance of sound diffusion is relaxed by the proper use of efficient grating diffusors. For the first time it is possible, by proper design, to get acceptable voice recordings from small "announce booths" because diffusing elements create a larger room sound. It is safe to say that future audio-room design will never be the same because of these diffusors.

Quadratic-Residue Diffusors

The principle of the reflection phase grating is now applied to sound, again through Schroeder's intuitive understanding of the limitations of the maximum-length diffusor. He reasoned that an incident sound wave falling on what physicists call a reflection phase grating would diffuse sound almost uniformly in all directions. The phase shifts (or time shifts) can be obtained by an array of wells of depths determined by a quadratic residue sequence. The maximum well depth is determined by the longest wavelength to be diffused. The well width is about a half wavelength at the shortest wavelength to be scattered.[3–5] The depths of the sequence of wells are determined by the statement:

$$\text{Well Depth Proportionality Factor} = n^2 \bmod p \qquad \textbf{(14-1)}$$

in which,

> p = a prime number.
> n = a whole number between zero and infinity.

Webster's Ninth New Collegiate Dictionary defines a prime number as *Any number (other than 0 or ±1) that is not divisible without remainder by any other integer (except ±1 and ± the integer itself).* Examples of prime numbers are 5, 7, 11, 13, etc. The *modulo* refers simply to residue. For example, inserting $p = 11$ and $n = 5$ into Eq. 14-1 gives 25 modulo 11. The modulo 11 means that 11 is subtracted from 25 until the significant residue is left. In other words, 11 is subtracted from 25 twice and the residue 3 is our answer. (For a review of number theory, see Endnotes.[6,7])

In Fig. 14-3 quadratic residue sequences are listed for the prime numbers 5, 7, 11, 13, 17, 19, and 23, a separate column for each. To check the above example for $n = 5$ and $p = 11$, enter the column marked 11, run down to $n = 5$ and find 3, which checks the previous computation. Figure 14-4 is a model for a quadratic residue reflection phase grating diffusor for $p = 17$. The numbers in each column of Fig. 14-3 are proportional to well depths of different quadratic residue diffusors. The design of Fig. 14-4 is based on the prime 17 sequence of numbers. At the bottom of each column of Fig. 14-3 is a sketch of a quadratic residue diffusor profile with well depths proportional to the numbers in the sequence. The broken lines indicate thin dividers between the wells.

The Hewlett-Packard HP-41C hand-held calculator has the modulo function in memory. As an example, for $n = 12$, $p = 19$. Equation 14-1 becomes 144 modulo 19, which becomes a bit messy to do by hand. The HP-41C solves this easily. In an appendix of Schroeder's excellent book[5] a modulo program for calculators is included.

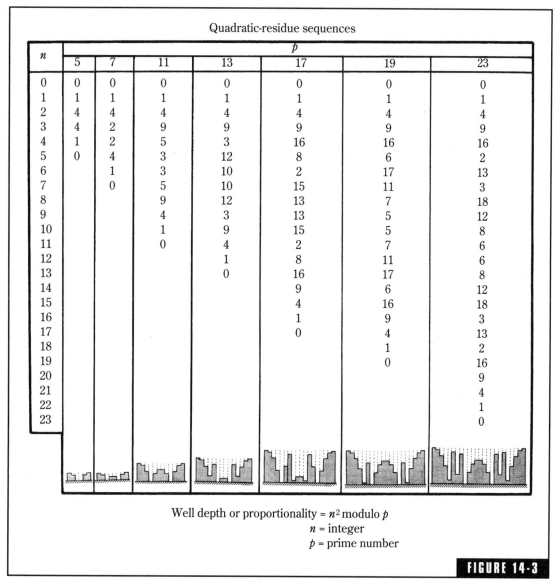

Quadratic-residue sequences

| n | \multicolumn{7}{c}{p} |
	5	7	11	13	17	19	23
0	0	0	0	0	0	0	0
1	1	1	1	1	1	1	1
2	4	4	4	4	4	4	4
3	4	2	9	9	9	9	9
4	1	2	5	3	16	16	16
5	0	4	3	12	8	6	2
6		1	3	10	2	17	13
7		0	5	10	15	11	3
8			9	12	13	7	18
9			4	3	13	5	12
10			1	9	15	5	8
11			0	4	2	7	6
12				1	8	11	6
13				0	16	17	8
14					9	6	12
15					4	16	18
16					1	9	3
17					0	4	13
18						1	2
19						0	16
20							9
21							4
22							1
23							0

Well depth or proportionality $= n^2$ modulo p

n = integer

p = prime number

FIGURE 14-3

Quadratic-residue sequences for prime numbers from 5–23. In the diffusor profile at the foot of each column, the depths of the wells are proportional to the sequence of numbers above.

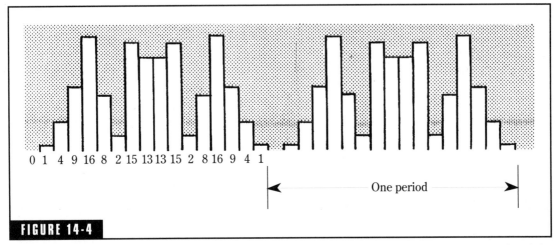

0 1 4 9 16 8 2 15 13 13 15 2 8 16 9 4 1

One period

FIGURE 14-4

A typical quadratic-residue diffusor based upon the prime number 17 column of Fig 14-3. The depths of the wells are proportional to the sequence of numbers in the prime-17 column. Two periods are shown illustrating how adjacent periods are fitted together.

Primitive-Root Diffusors

Primitive-root diffusors use a different number theory sequence, as follows:

$$\text{Well Depth Proportionality Factor} = g^n \text{ modulo } p \qquad \textbf{(14-2)}$$

in which:

p is a prime number, and
g is the least primitive root of p.

Figure 14-5 includes primitive-root sequences for six different combinations of p and g. The sketches at the bottom of each column are not symmetrical like those of the quadratic residue diffusors. In most cases this is a disadvantage but in some cases it is an advantage. There is an acoustical problem with the primitive root diffusors in that the specular mode is not suppressed as well as it is in the quadratic-residue diffusor. Commercial development has largely utilized quadratic residue sequences.

Separators between the wells, usually metallic, are commonly used to maintain the acoustical integrity of each well. Without separators the effectiveness of the diffusor is decreased. The stepped phase shifts for sound arriving at angles other than the perpendicular tend to be confused in the absence of dividers.

Primitive-root sequences

n	$p = 5$ $g = 2$	$p = 7$ $g = 3$	$p = 11$ $g = 2$	$p = 13$ $g = 2$	$p = 17$ $g = 3$	$p = 19$ $g = 2$
1	2	3	2	2	3	2
2	4	2	4	4	9	4
3	3	6	8	3	10	8
4	1	4	5	3	13	16
5		5	10	6	5	13
6		1	9	12	15	7
7			7	10	11	14
8			3	9	16	9
9			6	5	14	18
10			1	10	8	17
11				7	7	15
12				1	4	11
13					12	3
14					2	6
15					6	12
16					1	5
17						10
18						1

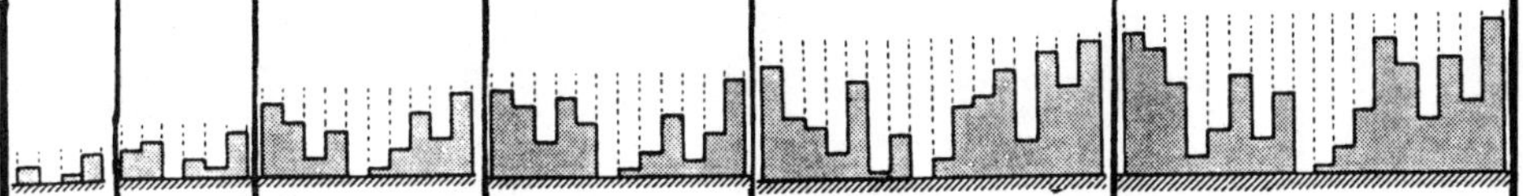

Well depth or proportionality = g^n modulo p
p = prime number
g = least primitive root of p

FIGURE 14-5

Primitive-root sequences for six combinations of prime number and least-primitive roots. Sound diffusor profiles at the foot of each column have depths proportional to the sequence of numbers above. Note that these diffusors are not symmetrical like the quadratic-residue diffusors.

Quadratic-Residue Applications

Numerous applications of the quadratic residue theory have been made by RPG Diffusor Systems, Inc.[8] For example, Fig. 14-6 shows two of their model QRD-4311™ below and a QRD Model 1911™ above. The 19 indicates that it is built on prime 19 and the 11 specifies well widths of 1.1 inch. The sequence of numbers in the prime 19 column of Fig. 14-3 specify the proportionality factors for well depths of the Model QRD 1911™ diffusor of Fig. 14-6.

Following the same pattern, the QRD™ Model 4311 in the lower portion of Fig. 14-6 is based on prime 43 with well widths also of 1.1 inch. For practical reasons, the columns of Fig. 14-3 stopped at prime 23; primes between 23 and 43 are 29, 31, 37, and 41. A good practical exercise would be to solve n^2 modulo p for $n = 0$ to 43 and $p = 43$ and compare the well depths of the Model 4311 in Fig. 14-6 knowing that deep wells are dark and dense, shallow wells are light.

This particular cluster of QRD diffusors offers excellent diffusion in the horizontal hemidisc of Fig. 14-7A. In Fig. 14-6 the specular reflection from the face of the diffusor is shown in Fig. 14-7B. The vertical wells of the QRD 4311™ scatter horizontally and the horizontal wells of the QRD 1911™ scatter vertically. Together they produce a virtual hemisphere of diffusion.

Performance of Diffraction-Grating Diffusors

In designing an audio space of any kind, the acoustician has three building blocks at his disposal: absorption, reflection, and diffusion. In the past, it was a common experience to find too much reflection but too little diffusion in the room to be treated. Only absorption and a limited number of geometrical tricks were available for the designer to carry out his assignment. Absorption, reflection, and diffusion are now in better balance, at least potentially.

The effects on the incident sound of the three physical principles of absorption, reflection, and diffusion are compared in Fig. 14-8. Sound impinging on the surface of a sound absorber is largely absorbed, but a tiny fraction is reflected. The temporal response shows a greatly attenuated reflection from the surface of the absorber.

FIGURE 14-6

A cluster of commercial quadratic-residue diffusors. Below are two QRD-4311™ diffusing modules with a single QRD-1911™ mounted above. The hemidisc of diffusion for the
lower unit is horizontal, that of the upper unit is vertical. Peter D'Antonio, RPG Diffusor System,
Inc. and the Audio Engineering Society, Inc.

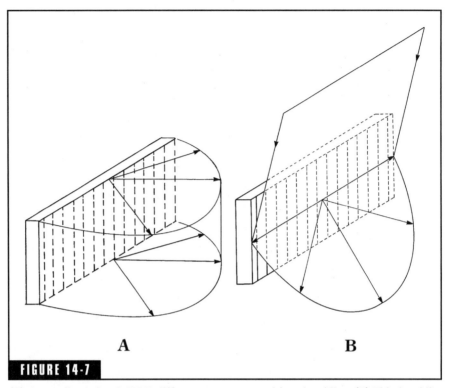

FIGURE 14-7

(A) A one-dimensional QRD™ diffusor scatters around in a hemidisc. (B) This hemidisc may be one specularly directed by orientation of the source with respect to the diffusor.

The same sound wave falling on a hard, reflective surface yields a reflection of almost the same intensity as the sound falling on the surface itself, just reduced slightly by losses at the reflective surface. The polar plot shows the energy concentrated about the angle of reflection. The width of the polar plot is a function of wavelength and the size of the reflecting surface.

A sound wave falling upon a diffusor, such as a quadratic residue type, is diffracted throughout the hemidisc of Fig. 14-7A. The diffused energy falls off exponentially. The polar diagram shows energy spread more or less equally throughout 180 degrees, but somewhat reduced at grazing angles.

The uniformity of the angular distribution of scattered energy through a wide range of frequency is shown dramatically in Fig. 14-9. These experimental polar plots, obtained for a commercial quadratic-residue diffusor, are smoothed by averaging over octave bands. Comparable computer-generated polar plots, for single frequencies based on a far-field diffraction theory, show a host of tightly packed lobes that have little practical significance. Near-field Kirchhoff diffraction theory shows less lobing.

The left column of Fig. 14-9 shows polar distribution of octave bands of energy centered from 250 Hz to 8,000 Hz, a span of five octaves. The right column shows the effect for sound incident at 45 degrees for the same frequencies. The results for all angles of incident sound for the lowest frequency are dependent on the well depth. The upper frequency is directly proportional to the number of wells per period and inversely proportional to the well width. In Fig. 14-9, the diffusion of a flat panel is shown by a light line for comparison.

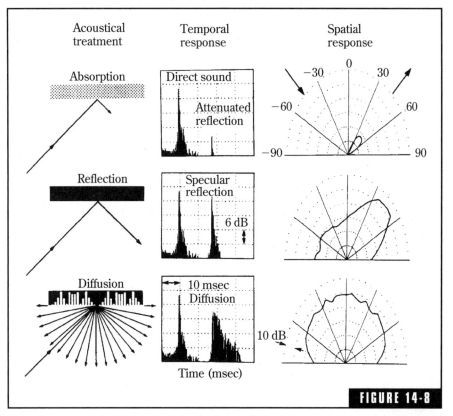

FIGURE 14-8

A comparision of the three physical principles of absorption, reflection, and diffusion.
Peter D'Antonio, RPG Diffusor Systems, Inc. and the Audio Engineering Society.

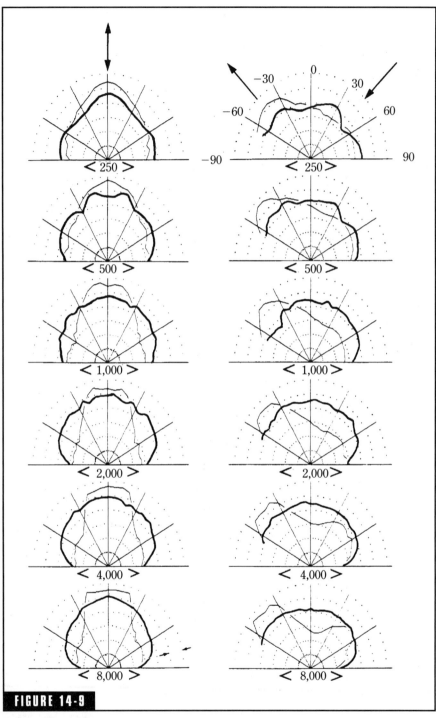

FIGURE 14-9

Experimental polar plots of a commercial quadratic-residue diffusor, smoothed by averaging over octave bands. The angular distribution of the energy is excellent over a wide frequency range and angles of incidence. Peter D'Antonio, RPG Diffusor Systems, Inc.

The uniformity of spatial diffusion is determined by the length of the period. Good broad-bandwidth, wide-angle diffusion, then, requires a large period with a large number of deep, narrow wells. This is why the QRD®-4311 has 43 wells of only 1.1 inches width and a maximum well depth of 16 inches.

A good question about this time would be, "What does a diffusor do that a hard, flat wall would not do?" The answer is to be found in Fig. 14-10. This figure compares the return from a flat panel with that

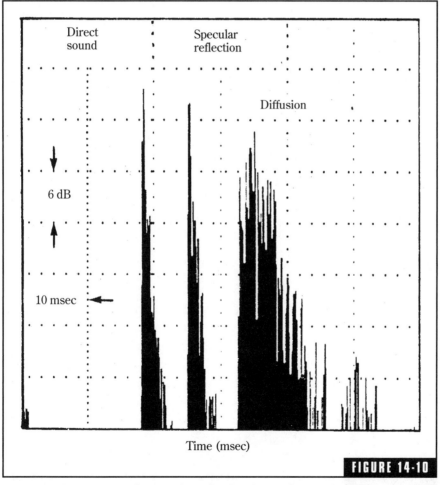

FIGURE 14-10

An energy-time contrast of a specular reflection from a flat panel and the energy diffused from a quadratic residue diffusor. The peak energy from a diffusing surface is somewhat lower than that of the flat panel but it is spread out in time. Peter D'Antonio, RPG Diffusor Systems, Inc. and the Audio Engineering Society.

from a quadratic-residue diffusor. The large peak to the left is the direct sound. The second large peak is the specular reflection from a flat panel. Note carefully that the energy of this sharp specular panel reflection is only a few dB below that of the incident sound. The diffused energy from the quadratic residue diffusor is spread out significantly in time. Most importantly, and revealed in the polar diagrams of Fig. 14-9, the gratings diffract sound throughout 180 degrees, not just in the specular direction like the flat panel.

There is a rich and growing literature on the development of the Schroeder diffraction grating sound diffusor. Keep abreast of the rapid changes in the field.[5,9–25]

Expansion of the QRD Line

The appearance of reflection phase grating diffusors Models QRD-4311™ and QRD-1711™ was followed by numerous other models designed to meet specific needs. For example, Fig. 14-11 shows Model QRD-734, which is a 2 ft × 4 ft model suitable for use in suspended ceiling T-frames as well as other ways. Also shown in Fig. 14-11 is the Abffusor™, which combines absorption and diffusion in the same unit. The Triffusor™, also shown in Fig. 14-11, has reflective, absorptive, and diffusive faces. A group of these, set into a wall, offers wide options in the acoustics of a space by rotating the individual units.

Solving Flutter Problems

If two opposing reflective surfaces of a room are parallel there is always the possibility of *flutter echoes*. This applies to either horizontal or vertical modes. Successive, repetitive reflections, equally spaced in time, can even produce a perception of a pitch or timbre coloration of music and a degradation of intelligibility of speech. The lack of ornamentation in modern architecture results in a greater flutter possibility.

Flutter can be reduced by careful placement of sound absorbing material, or by splaying walls. Splaying is impractical in most cases and increasing absorption often degrades the acoustical quality of a space. What is needed is a wall treatment that reduces reflections by scattering rather than absorption. The RPG Flutterfree™ does exactly this. It is an architectural hardwood molding 4 in wide by 4 or 8 ft length, shown in

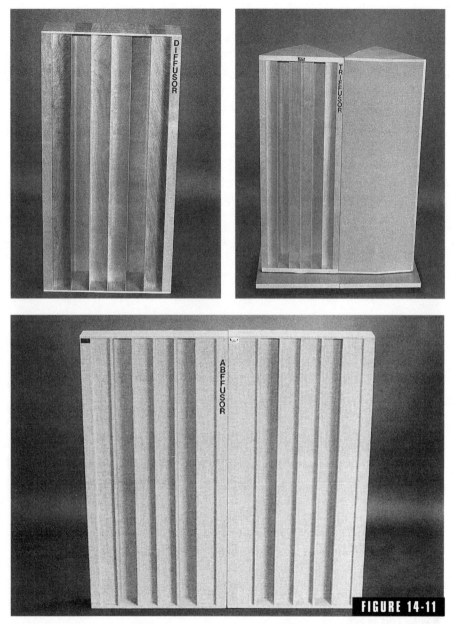

FIGURE 14-11

Three proprietary sound diffusing systems: (Upper Left) a broad bandwidth and wide-angle QRD 734 Diffusor™, (Below) the Abffusor™ broad bandwidth diffusor/absorber, and (Upper Right) the Triffusor™ having one side absorptive, one diffusive, and one reflective for acoustical variability. Peter D'Antonio, RPG Diffusor Systems, Inc. and the Audio Engineering Society.

Fig. 14-12, which reduces specular reflection as well as contributing to the diffusion of sound. The molding works as a one-dimensional reflection phase grating diffusor because of the wells routed into its surface. The depths of the wells follow a prime-7 quadratic residue sequence. These moldings may be affixed to a wall butted together or spaced, horizontally or vertically. If they are vertical, spectral reflections are controlled in the horizontal plane and *vice versa*.

The Flutterfree™ moldings can be employed as slats for a Helmholtz slat-type low-frequency absorber (Chap. 9). All the while low-frequency sound is being absorbed by the Helmholtz absorber, the surface of each slat performs as a mid-high-frequency sound diffusor.

Application of Fractals

Certain production limitations have been encountered by RPG Diffusor Systems, Inc. in the development of reflection phase grating diffusors. For example, the low-frequency limit is determined principally

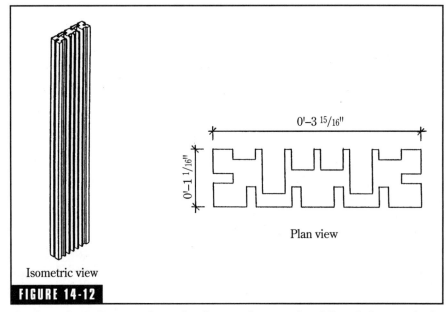

0'-3 15/16"

0'-1 1/16"

Plan view

Isometric view

FIGURE 14-12

The Flutterfree™ is a non-absorptive flutter echo control molding. It is a quadratic-residue diffusor based on the prime-7. It can also serve as slats on a slat-type Helmholtz low-frequency absorber. Peter D'Antonio, RPG Diffusor Systems, Inc. and the Audio Engineering Society.

by well depth, and the high-frequency limit is determined principally by well depth, and the high-frequency limit is determined principally by well width. Manufacturing constraints place a limit of 1 in on well width and 16 in on well depth, beyond which the units become diaphragmatic.

To increase the effective bandwidth, the self-similarity principle has been applied in the form of fractals producing units called Diffractals™.[26] These are really diffusors within diffusors within diffusors as shown in the progressive illustration of Fig. 14-13. Three sizes of quadratic residue diffusors are required to make up the complete Diffractal™. The operation is much like that of the multi-unit loudspeaker. The various diffusors making up a Diffractal™ operate independently even as the woofer, mid-range, and tweeter loudspeaker units operate independently to create the wide-band system.

Figure 14-14 shows the DFR-82LM Diffractal™ that is 7 ft 10 in high and 11 ft wide with a depth of 3 ft. This is a second generation, low-mid-frequency unit that covers the range 100 Hz to 5 kHz. The low-frequency portion is based on a prime-7 quadratic residue sequence. A mid-range Diffractal™ is embedded at the bottom of each well of the larger unit. The frequency range of each section as well as the crossover points of these composite units are completely calculable.

Figure 14-15 shows the larger DFR-83LMH™ unit that is 16 ft wide, 6 ft 8 in high and 3 ft deep. This is a three-way unit covering a frequency range of 100 Hz to 17 kHz. The depths of the wells of the

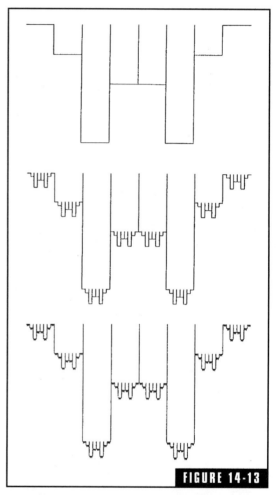

FIGURE 14-13

The Diffractal™ wide-band diffusor is a two- or three-way crossed over system composed of high-frequency fractal sound diffusors nested within a low-frequency diffusor. The three-way system illustrated is a diffusor within a diffusor within a diffusor, analogous to a three-way loudspeaker system composed of woofer, mid-range unit, and tweeter. Peter D'Antonio, RPG Diffusor Systems, Inc. and the Audio Engineering Society.

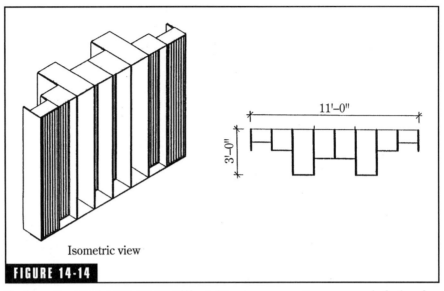

Isometric view

FIGURE 14-14

The large sized Model DFR-83LMH Diffractal™, a two-way system composed of a low-fre-qency unit (prime-7) with a mid-range Diffractal embedded at the bottom of each well. *Peter D'Antonio, RPG Difusser Systems, Inc.*

low-frequency unit follow a prime-7 quadratic residue sequence. Fractals are set in the wells of fractals that are set in the low-frequency wells.

Diffusion in Three Dimensions

All of the reflection phase grating diffusors discussed previously have rows of parallel wells. These can be called one dimensional units because the sound is scattered in a hemidisc as shown in Fig. 14-16-A. There are occasions in which hemispherical coverage is desired, as shown in Fig. 14-16B. RPG Diffusor Systems™ have met this need by offering the Omniffusor™. The Omniffusor™ consists of a symmetrical array of 64 square cells as shown in Fig. 14-17. The depth of these cells is based on the phase-shifted prime-7 quadratic residue number theory sequence.

An off-shoot of the Omniffusor™ (which is made of wood) is the FRG Omniffusor™, which consists of an array of 49 square cells based on the two-dimensional phase-shifted quadratic residue number theory. This unit is made of fiberglass-reinforced gypsum. It is lighter in weight and one-third the cost of the Omniffusor™, is therefore better adapted for application to large surface areas.

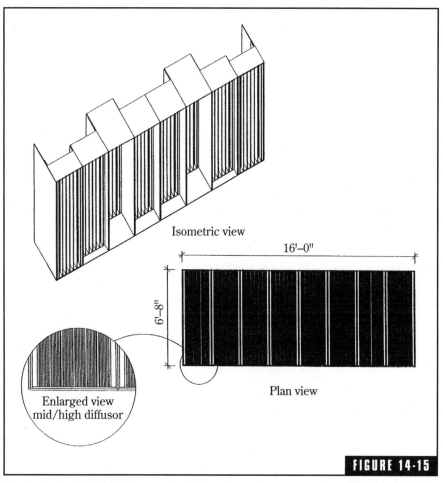

Isometric view

16'–0"

6'–8"

Enlarged view
mid/high diffusor

Plan view

FIGURE 14-15

The large-sized Model DFR-83LMH Diffractal™, a three-way unit covering a wide frequency range. Fractals are set in the wells of diffractals in this model. Peter D'Antonio, RPG
Diffusor Systems, Inc. and the Audio Engineering Society.

Acoustic Concrete Blocks

The Cinderblox and its many derivatives have been in use since
1917. In 1965 The Proudfoot Company came out with their Sound-
Blox®, which supplies not only load-bearing ability and the mass
required for sound isolation, but also enhanced low-frequency
absorption through Helmholtz resonators formed by slots and cavities in the blocks. RPG Diffusor System introduced their Diffusor-
Blox® in 1990 that goes one step further: load-bearing ability,

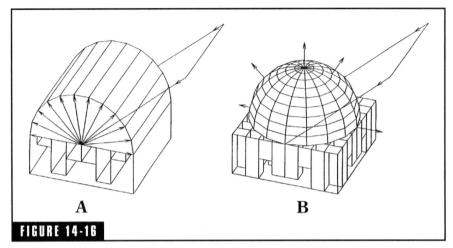

FIGURE 14-16

Comparison of the diffraction patterns of (A) the hemicylindrical form of the one-dimensional quadratic-residue diffusor and (B) the hemispherical form of the two-dimensional diffusor. Peter D'Antonio, RPG Diffusor Systems, Inc.

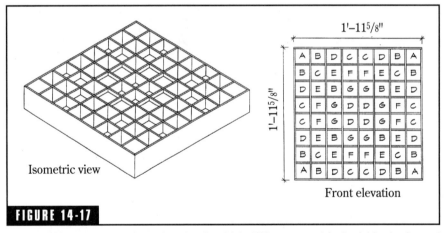

FIGURE 14-17

The Omniffusor™ a two-dimensional unit which diffuses sound in both the horizontal and vertical planes for all angles of incidence. Peter D'Antonio, RPG Diffusor Systems, Inc.

transmission loss, low-frequency absorption, and sound diffusion. The DiffusorBlox® system is made up of three distinct blocks, all of which are nominally $8 \times 16 \times 12$ inches. A typical block is shown in Fig. 14-18. These concrete blocks are characterized by a surface containing a partial sequence of varying well depths, separated by dividers; an internal 5-sided cavity that can accept a fiberglass insert; an optional rear half-flange for reinforced construction; and an optional low-frequency absorbing slot. Diffusorblox® are fabricated on standard automatic block machines using molds licensed from RPG Diffusor Systems, Inc. Typical walls constructed of DiffusorBlox® are illustrated in Fig. 14-19.

Measuring Diffusion Efficiency

A measure of the effectiveness of a diffusor can be obtained by comparing the intensity in the specular direction with the intensity at ±45° of that direction. This can be expressed as:[27]

$$\text{Diffusion Coefficient} = \frac{I\,(\pm 45°)}{I\,(\text{specular})}$$

The diffusion coefficient is 1.0 for the perfect diffusor. This coefficient varies with frequency and is commonly expressed in graphical form. The variation of diffusion coefficient with frequency for several typical units is shown in Fig. 14-20. For comparison, diffusion from a flat panel is included as a broken line. These measurements were all made under reflection-free condition on sample areas of 64 sq ft and using the time-delay-spectrometry technique.

The number of wells and well widths affect the performance of the units. The QRD Model 4311® (Fig. 14-6) having the deepest well

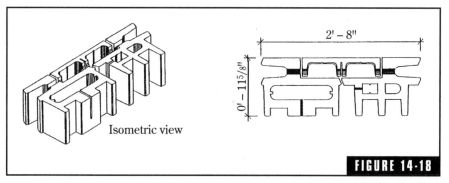

Isometric view

2′ – 8″

0′ – 11 5/8″

FIGURE 14-18

The DiffusorBlox™ concrete block, which offers the good transmission loss of a heavy wall, absorption via Helmholtz resonator action, and diffusion through quadratic residue action. The blocks are formed on standard block machines using licensed molds. Peter D'Antonio, RPG Diffusor Systems, Inc.

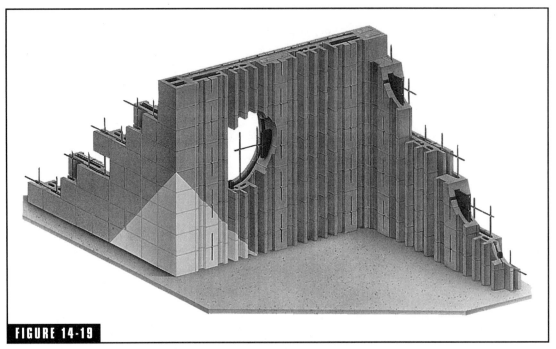

Typical wall configurations using Diffusorblox™. Peter D'Antonio, RPG Diffusor Systems, Inc.

depths and the narrowest well widths (feasible from a manufacturing standpoint) gives the highest diffusion coefficient over the greatest frequency range. For comparison, two other units are shown in Fig. 14-20, the 1925 and the 734, built with the primes 19 and 7 and well widths of 2.5 and 3.4 inches. The performance of these, while good, is somewhat inferior to the QRD Model 4311®.

Comparison of Gratings with Conventional Approaches

Figure 14-21 compares the diffusing properties of the flat panel (a and b), flat panel with distributed absorption on it (c and d), the monocylinder (e and f), the bicylinder (g and h), and quadratic residue diffusors (i and j). The left column is for sound at 0° incidence and the right column is for sound at 45° incidence. The "fore-and-aft" scale is diffraction from 90° through 0° to −90°. The horizontal frequency scale is basically from 1 through 10 kHz. These three-dimensional read-outs

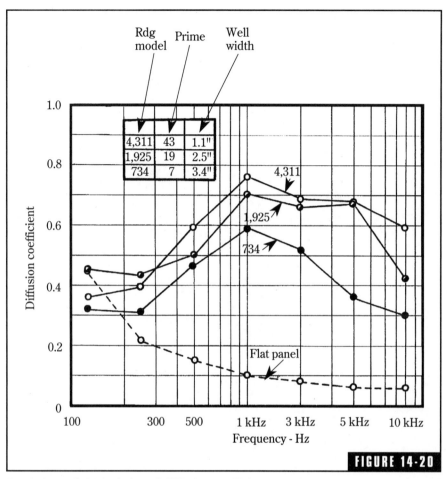

Comparison of the variation of diffusion coefficient with frequency of three RPG™ diffusors and a flat panel. Peter D'Antonio, RPG Diffusor Systems, Inc.

cover all pertinent variables and provide a wealth of information, which requires some skill in interpretation. Specific comments Dr. D'Antonio has made include the following:

■ The first six energy-frequency-curves contain artifacts of the measurement process which should be disregarded because they are not in the anechoic condition.

■ For 0° incidence, the specular properties of the flat panel with distributed absorption are quite evident by the pronounced peak at 0° the specular angle.

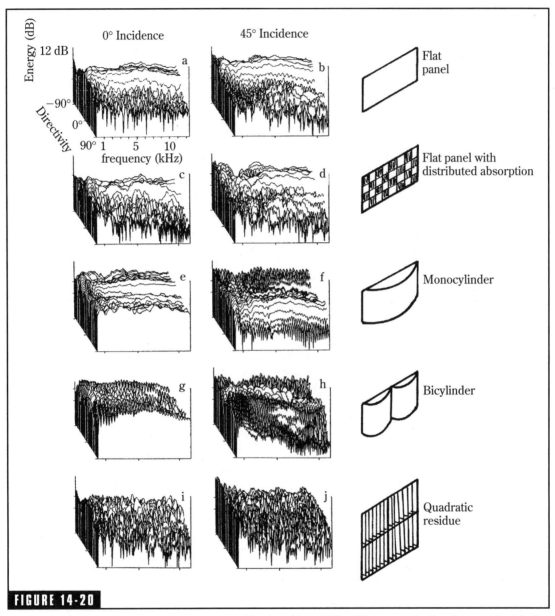

FIGURE 14-20

Energy-frequency-directivity plots comparing diffusion from (a,b) a flat panel, (c,d) a flat panel with distributed absorption, (e,f) monocylinder, (g,h) bicylinder, and (i,j) quadratic-residue diffusor. These plots compare many of the early attempts at diffusion with modern quadratic-residue diffusors. *Peter D'Antonio, RPG Diffusor Systems, Inc. and Acoustical Society of America, Inc.*

■ The good spatial diffusion of the monocylinder (e) is illustrated by the relatively constant energy response from 90° to −90°. The bicylinder (g) shows two closely spaced peaks in the time response. Although the spatial diffusion looks good, there is appreciable equal-spaced comb filtering and broadband high-frequency attenuation. This accounts, in part, for the poor performance of cylindrical diffusors.

■ The quadratic residue diffusors (i and j) maintain good spatial diffusion even at 45° incidence. The dense notching is uniformly distributed across the frequency spectrum and the energy is relatively constant with scattering angle.

Endnotes

[1]Mackenzie, Robin, *Auditorium Acoustics* (1975) London, Applied Science Publishers, Ltd., see Chapter 17 especially, "New Results and Ideas for Architectural Acoustics" by M. R. Schroeder, p. 206-209.

[2]Schroeder, M.R., *Diffuse Sound Reflection by Maximum-Length Sequences*, J. Acous. Soc. Am., 57, 1 (Jan 1975) 149-150.

[3]Schroeder, M.R., and R.E. Gerlach, *Diffuse Sound Reflection Surfaces*, Proc. Ninth International Congress on Acoustics, Madrid, 1977, paper D-8.

[4]Schroeder, M.R., *Binaural Dissimilarity and Optimum Ceilings for Concert Halls: More Lateral Sound Diffusion*, J. Acous. Soc. Am., 65, 4 (Apr 1979) 958-963.

[5]Schroeder, M.R., *Number Theory in Science and Communication*, 2nd enlarged ed., Berlin, Springer (1986).

[6]Beiler, Albert H., *Recreations in the Theory of Numbers*, 2nd. Ed., (1966) New York, Dover Publications, Inc.

[7]Davenport, H., *The Higher Arithmetic*, (1983) New York, Dover Publications, Inc.

[8]RPG Diffuser Systems, Inc., 651-C Commerce Drive, Upper Marlboro, MD 20772. Telephone: (301) 249-0044, FAX: (301) 249-3912.

[9]Strube, Hans Werner, *Scattering of a Plane Wave by a Schroeder Diffusor: A Mode-Matching Approach*, J. Acous. Soc, Am., 67, 2 (Feb 1980) 453-459.

[10]deJong, B.A. and P.M. van den Berg, *Theoretical Design of Optimum Planar Sound Diffusors*, J. Acous. Soc. Am., 68, 4 (Oct 1980) 1,154-1,159.

[11]Berkout, D.W., van Wulfften Palthe, and D. deVries, *Theory of Optimal Plane Diffusors*, J. Acous. Soc. Am. 65, 5 (May 1979) 1334-1336.

[12]D'Antonio, Peter and John H. Konnert, *The Reflection Phase Grating: Design Theory and Application*, J. Audio Eng. Soc., 32, 4 (Apr 1984) 228-238.

[13]D'Antonio, Peter and John H. Konnert, *The Schroeder Quadratic-Residue Diffusor: Design Theory and Application*, 74th Audio Eng. Soc. Convention, New York, October 1983, preprint #1999.

[14]D'Antonio, Peter and John H. Konnert, *The RFZ/RPG Approach to Control Room Monitoring*, 76th Audio Eng. Soc. Convention, New York, October, 1984, preprint #2157.

[15]D'Antonio, Peter and John H. Konnert, *The RPG Reflection-Phase-Grating Acoustical Diffusor: Applications*, 76th Audio Eng. Soc. Convention, New York, October, 1984, preprint #2156.

[16]D'Antonio, Peter and John H. Konnert, *The RPG Reflection-Phase-Grating Acoustical Diffusor: Experimental Measurements*. 76th Audio Eng. Soc. Convention, New York, October, 1984, preprint 2158.

[17]D'Antonio, Peter and John H. Konnert, *The Role of Reflection-Phase-Grating Diffusors in Critical Listening and Performing Environments*, 78th Audio Eng. Soc. Convention, Anaheim, CA, May 1985, preprint #2255.

[18]D'Antonio, Peter and John H. Konnert, *The Acoustical Properties of Sound Diffusing Surfaces: The Time, Frequency, and Directivity Energy Response*, 79th Audio Eng. Soc. Convention, New York, October, 1985, preprint #2295.

[19]D'Antonio, Peter and John H. Konnert, *Incorporating Reflection-Phase-Grating Diffusors in Worship Spaces*, 81st Audio Eng. Soc. Convention, Los Angeles, November, 1986, preprint #2364.

[20]D'Antonio, Peter and John H. Konnert, *New Acoustical Materials and Designs Improve Room Acoustics*, 81st Audio Eng. Soc. Convention, Los Angeles, November 1986, preprint #2365.

[21]D'Antonio and John H. Konnert, *The Reflection-Phase-Grating Acoustical Diffusor: Application in Critical Listening and Performing Environments*, Proc. 12th International Congress on Acoustics.

[22]D'Antonio, Peter, *The Reflection-Phase-Grating Acoustical Diffusor: Diffuse It or Lose It*, dB The Sound Engineering Magazine, 19, 5 (Sept/Oct 1985) 46-49.

[23]D'Antonio, Peter and John H. Konnert, *Advanced Acoustic Design of Stereo Broadcast and Recording Facilities*, 1986 NAB Eng. Conference Proc., p. 215-223.

[24]D'Antonio, Peter and John H. Konnert, *New Acoustical Materials Improve Broadcast Facility Design*, 1987 NAB Eng. Conference Proc., p. 399-406.

[25]D'Antonio, Peter, *Control-Room Design Incorporating RFZ, LFD, and RPG Diffusors*, dB The Sound Engineering Magazine, 20, 5 (Sept/Oct 1986) 47-55. Contains bibliography with 40 entries.

[26]D'Antonio, Peter and John Konnert, *The QRD Diffractal: A New One- or Two-Dimensional Fractal Sound Diffusor*, J. Audio Eng. Soc., 40, 3 (Mar 1992), 117-129.

[27]D'Antonio, Peter and John Konnert, *The Directional Scattering Coefficient: Experimental Determination*, J. Audio Eng. Soc., 40, 12 (Dec 1992), 997-1017.

Modal Resonances in Enclosed Spaces

Hermann Von Helmholtz (1821-1894) performed some interesting acoustical experiments with resonators. His resonators were a series of metal spheres of graded sizes, each fitted with a neck, appearing somewhat like the round-bottom flask found in the chemistry laboratory. In addition to the neck there was another small opening to which he applied his ear. The resonators of different sizes resonated at different frequencies, and by pointing the neck toward the sound under investigation he could estimate the energy at each frequency by the loudness of the sound of the different resonators.

There were numerous applications of this principle long before the time of Helmholtz. There is evidence that bronze jars were used by the Greeks in their open-air theaters, possibly to provide some artificial reverberation. A thousand years ago Helmholtz-type resonators were embedded in church walls in Sweden and Denmark with the mouths flush with the wall surface, apparently for sound absorption.[1] The walls of the modern sanctuary of Tapiola Church in Helsinki, Finland, are dotted with slits in the concrete blocks[2] (Fig. 15-1). These are resonator "necks" that open into cavities behind, together forming resonating structures. Energy absorbed from sound in the room causes each resonator to vibrate at its own characteristic frequency. Part of the energy is absorbed, part reradiated. The energy reradiated is sent in every direction, contributing to the diffusion of sound in the room.

FIGURE 15-1

Helmholtz-type resonators built into the wall of the Tapiola Church in Helsinki, Finland. The slots and cavities behind them act as both absorbers and diffusors of sound.

The resonator principle, old as it is, continually appears in modern, up-to-the-minute applications.

Resonance in a Pipe

The two ends of the pipe of Fig. 15-2 can be likened to two opposing walls of a listening room or recording studio. The pipe gives us a simple, one-dimensional example to work with. That is, what happens between opposite walls of a rectangular room can be examined without being bothered by the reflections from the other four surfaces. The pipe, closed at both ends and filled with air, is a resonator capable of vibrating at its characteristic frequencies when excited in some way. Air inside an organ pipe can be set to vibrating by blowing a stream of air across a lip at the edge of the pipe. It is simpler to place a small loudspeaker inside the pipe. A sine wave signal is fed to the loudspeaker and varied in frequency. A small hole drilled in the pipe in the end opposite the loudspeaker makes possible hearing the low-level tones radiated by the loudspeaker. As the frequency is increased, nothing unusual is noted until the frequency radiated from the loudspeaker coincides with the natural frequency of the pipe. At this frequency, f_1, modest energy from the loudspeaker is strongly reinforced and a relatively loud sound is heard at the ear hole. As the frequency is increased, the loudness is again low until a frequency of $2f_1$ is reached, at which point another strong reinforcement is noted. Such resonant peaks can also be detected at $3f_1$, $4f_1$ . . . etc.

Now let us assume that means of measuring and recording the sound pressure all along the pipe are available. In Fig. 15-2 the graphs below the sketch of the pipe show how the sound pressure varies along the length of the pipe for different excitation frequencies. A sound wave traveling to the right is reflected from the right plug and a sound wave traveling to the left from the left plug. The left-going waves react with the right-going waves to create, by superposition, a *standing* wave at the natural frequency of the pipe or one of its multiples. Measuring probes inserted through tiny holes along the pipe could actually measure the high pressure near the closed ends and zero at the center, etc. Similar nodes (zero points) and antinodes (maxima) can be observed at $2f_1$, $3f_1$, $4f_1$ . . . etc., as shown in Fig. 15-2. The dimensions of a studio or listening room determine its characteristic frequencies much as though there were a north-south pipe, an east-west pipe, and a vertical pipe, the pipes corresponding to the length, width, and height of the room, respectively.

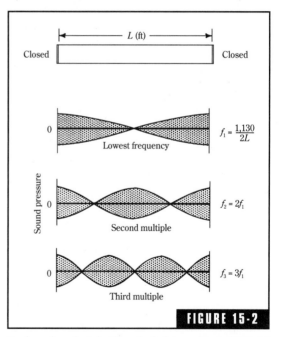

Bathroom Acoustics

Why is it that singing in the shower or tub is such a satisfying experience (to the singer, at least)? Because here one's voice sounds

FIGURE 15-2

A pipe closed at both ends helps us to understand how resonance occurs between two opposing walls of a listening room or studio. The distance between the walls determines the characteristic frequency of resonance.

richer, fuller, and more powerful than anywhere else! The case of the bathroom baritone clearly illustrates the effect of resonance in a small room and the resulting reinforcement of sound at certain frequencies related to the dimensions of the room. Exciting the air in the bathroom at frequencies far removed from these characteristic frequencies results in weaker sounds, except at multiples of these frequencies, where the effect can be very much like that at the lowest natural frequencies.

The person singing in the bathroom is, in a sense, inside a Helmholtz resonator or an immense organ pipe, but with one important difference; it is now a three rather than an essentially one-dimensional system like the pipe. The hard walls of the bathroom are highly reflective. There is a characteristic modal frequency of resonance associated with the length, another with the width, and still another with the height of the bathroom. In the case of the cubical bathroom, all three modal frequencies coincide to give a mighty reinforcement to the baritone's voice at the basic characteristic modal frequency and multiples of it.

Reflections Indoors

Anyone can appreciate the difference between sound conditions indoors and sound outdoors. Outdoors the only reflecting plane may be the earth's surface. If that surface happens to be covered with a foot of snow, which is an excellent absorber of sound, it may be difficult to carry on a conversation with someone 20 feet away. Indoors the sound energy is contained, resulting in a louder sound with a given effort. A speaker can be heard and understood by hundreds of people with no reinforcement but that of reflecting surfaces.

Consider sound reflections from a single wall. In Fig. 15-3, a point source of sound, S, is a given distance from a massive wall. The spherical wave fronts (solid lines, traveling to the right) are reflected from this surface (broken lines). Physicists working in various forms of radiation (light, radio waves, sound) resort to the concept of images because it makes their mathematical studies much easier. In Fig. 15-3, the reflections from the surface traveling to the left act exactly as though they were radiated from another identical point source, S_1, an equal distance from the reflecting surface but on the opposite side. This is the simplest image case of one source, one image, and a reflecting surface, all in free space.

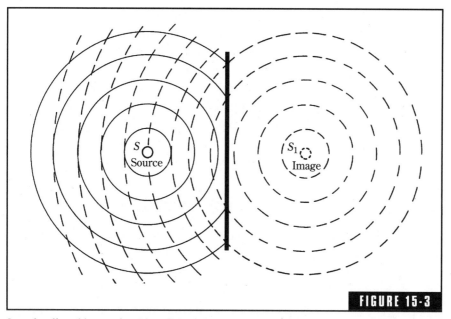

Sound radiated by a point *S* is reflected by the rigid wall. The reflected wave can be considered as coming from *S*₁, an image of *S*.

The isolated reflecting surface of Fig. 15-3 is now brought down to earth and made into the north wall of a rectangular room as in Fig. 15-4. Source S still has its image S_1 in what is now the north wall of the room. The source also has other images. S_2 is the image in the east wall reflecting surface, S_3 is the image of S in the west wall, and S_4 is the image of S in the south wall. Use your imagination to visualize S_5, the image in the floor, and S_6, the image in the ceiling. All of these six images are assumed to be pulsating just like S and sending sound energy back into the room. The farther the images are from the reflecting plane, the weaker will be their contribution at a given point, P, in the room, but they all make their contribution.

There are images of the images as well. The S_1 image has its image in the south wall at S_1', the image of the S_2 image in the west wall at S_2' and similarly, images S_3, S_4, S_5, and S_6 appear at S_3', S_4', S_5', and S_6' (some off the page). And, then, there are the images of the images of the images, and so on *ad infinitum*. The more remote images are so weak that they can be neglected for the sake of simplicity. Going further in the discussion of the image is beyond the scope of this book. We have discussed

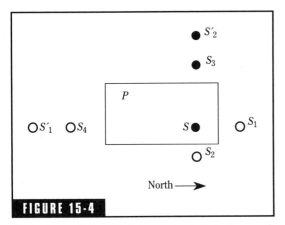

FIGURE 15-4

The surface of Fig. 15-3 is now made into the north wall of an enclosure. The source S now has six primary images, one in each of the six surfaces of the enclosure. Images of the images result in a theoretically infinite number of images of S. Sound intensity at point P is made up of the direct sound from S plus the contributions from all images.

them to this extent only to show how to visualize how the sound field at some point P in a room is built up from the direct sound from the source S plus the vector sum of the contributions of all the images of S. This is just another way of saying that the sound at P is built up from the direct sound from S plus single or multiple reflections from all six surfaces.

Two-Wall Resonance

Figure 15-5 shows two parallel, reflective walls of infinite extent. When a loudspeaker radiating pink noise excites the space between the walls, the wall-air-wall system exhibits a resonance at a frequency of $f_o =$ 1,130/2L or 565/L, when L = the distance in feet between the two walls and 1,130 the speed of sound in feet per second. A similar resonance occurs at $2f_o$, $3f_o$, $4f_o$ . . . , etc. up through the spectrum. The fundamental frequency f_o is considered a natural frequency of the space between the reflective walls, and it is accompanied by a train of modes each of which also exhibits resonance. Other names that have been applied to such resonances are *eigentones* (obsolete), *room resonances*, *permissible frequencies*, *natural frequencies*, or just plain *modes*, which is preferred to make a studio or listening room. In adding two more mutually perpendicular pairs of walls, to form a rectangular enclosure, we also add two more resonance systems, each with its own fundamental and modal series.

Actually, the situation is far more complicated. So far only *axial modes* have been discussed, of which each rectangular room has three, plus modal trains for each. Each axial mode involves only two opposite and parallel surfaces. *Tangential modes*, on the other hand, involve four surfaces. *Oblique modes* involve all six surfaces of the room.

Waves vs. Rays

What the diagrams in Fig. 15-6 offer in terms of clarity, they lack in rigor. In these diagrams, rays of sound are pictured as obeying the law:

The angle of incidence equals the angle of reflection. For higher audio frequencies, the ray concept is quite fruitful. When the size of the enclosure becomes comparable to the wavelength of the sound in it, however, special problems arise and the ray approach collapses. For example, a studio 30 ft long is only 1.3 wavelengths long at 50 Hz. Rays lose all meaning in such a case. Physicists employ the wave approach to study the behavior of sound of longer wavelengths.

Frequency Regions

The audible spectrum is very wide when viewed in terms of wavelength. At 16 Hz, considered the low-frequency limit of the average human ear, the wavelength is 1130/16 = 70.6 ft. At the upper extreme of hearing, say 20,000 Hz, the wavelength is only 1130/20,000 = 0.0565 ft or about $1\frac{1}{16}$ of an inch. The behavior of sound is affected greatly by the wavelength of the sound in comparison to the size of objects encountered. In a room, sound of $1\frac{1}{16}$-in wavelength is scattered (diffused) significantly by a wall irregularity of a few inches. The effect of the same irregularity on sound of 70 ft wavelength would be

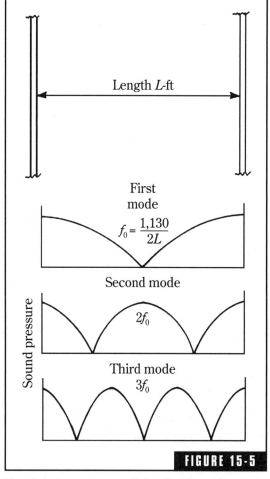

FIGURE 15-5

The air between two parallel, reflective walls can be considered a resonant system with a frequency of resonance of f_0 = 1,130/2L. This system is also resonant at integral multiples of f_0.

insignificantly small. The heart of the acoustical problem in the usual audio room is that no single analytical approach can cover sound of such a wide range of wavelengths.

In considering the acoustics of small rooms, the audible spectrum is divided arbitrarily into four regions, A, B, C, and D (Fig. 15-7). Region A is the very low frequency region below a frequency of 1130/2L or 565/L where L is the longest dimension of the room. Below the frequency of this lowest axial mode there is no resonant support

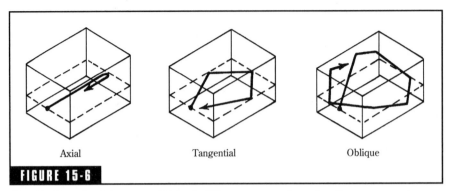

Visualization of axial, tangential, and oblique room modes using the ray concept.

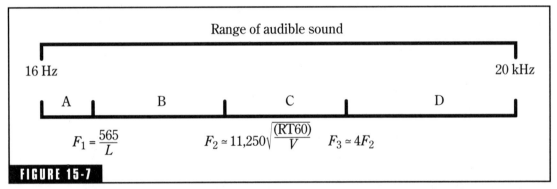

When dealing with the acoustics of enclosed spaces, it is convenient to consider the audible frequency range as composed of four regions: A, B, C, and D demarked by the frequencies F_1, F_2, and F_3. In region D, specular reflections and ray acoustics prevail. In region B, room modes dominate. Region C is a transition zone in which diffraction and diffusion dominate. There is no modal boost for sound in region A.

for sound in the room. This does not mean that such very low frequency sound cannot exist in the room, only that it is not boosted by room resonances because there are none in that region.

Region B is that region we have studied in detail in which the dimensions of the room are comparable to the wavelength of sound being considered. It is bounded on the low-frequency end by the lowest axial mode, 565/L. The upper boundary is not definite but an approximation is given by what has been called the *cutoff* or *crossover* frequency given by the equation:

$$F_2 \simeq 11{,}250 \sqrt{\frac{(\text{RT60})}{V}} \qquad (15\text{-}1)$$

where

F_2 = cutoff or crossover frequency, Hz.

RT60 = reverberation time of the room, seconds.

V = volume of the room, cu ft.

A $10 \times 16 \times 23.3$ ft room has a volume of 3,728 cu ft. For a reverberation time of 0.5 second the F_2 crossover frequency is 130 Hz. In Fig. 15-7, $F_1 = 565/23.3 = 24.2$ Hz, and $F_3 \simeq 4F_2 \simeq (4)(130) \simeq 520$ Hz.

Region D covers the higher audible frequencies for which the wave lengths are short enough for geometric acoustics to apply. Specular reflections (angle of incidence equals angle of reflection) and the sound ray approach to acoustics prevail. In this region statistical approaches are generally possible.

Retracing our steps, region C is a transition region between region B, in which wave acoustics must be used, and region D in which ray acoustics are valid. It is bounded on the low-frequency end approximately by the cutoff frequency F_2 of Eq. 15-1 and on the high end approximately by $4F_2$. It is a difficult region dominated by wavelengths often too long for ray acoustics and too short for wave acoustics.

For the $10 \times 16 \times 23.3$ ft room, below $565/23.3 = 24.2$ Hz is region A in which there is no resonant boost for sound. Between 24.2 Hz and 130 Hz (Eq. 15-1) the wave acoustical approach of modal resonances is essential. Between 130 Hz and (4) (130) = 520 Hz is the transition region C. Above about 520 Hz the modal density is very high, statistical conditions generally prevail, and the simpler geometrical acoustics can be used. Room size determines how the audible spectrum must be divided for acoustical analysis. Very small rooms, with too few modal resonances spaced too far apart, are characterized by domination of a great stretch of the audible spectrum by modal resonances. This is the "small studio problem" in a nutshell.

Dividing the Audio Spectrum

The very wide frequency range of the ear forces us to apply different approaches for different frequency ranges in studying the small-room sound field. An appreciation of this fact is necessary before these different approaches are presented. The audible spectrum in Fig. 15-7 is divided into four parts. At the high-frequency extreme is the D region, in which the use of sound rays is applicable. In this region, rays of sound bounce around like billiard balls, following the general rule that the angle of incidence equals the angle of reflection. In this region, the wavelength of the sound is very small compared to the dimensions of the room.

Next we slide down the audible spectrum to the range of frequencies labeled B. In this region, the wavelength is comparable to the room dimensions. Here the ray concept is without meaning, and the concept of wave acoustics must be applied.

In between frequency regions D and B of Fig. 15-7 is region C, which is transitional in nature. The wavelength of sound in this range of frequencies does not fit either the wave or ray approach; it is too short for wave acoustics, too long for the application of the ray approach. It is a region in which diffusion and diffraction of sound dominate.

Region A of Fig. 15-7 is a sort of acoustical never-never land. We are still indoors, there are reflecting surfaces all about us, yet sound does not behave like it does outdoors. Nor does it behave like sound indoors at somewhat higher frequencies. When the wavelength of the sound is of the same order of magnitude as the dimensions of the room, as in region B, the modal resonances increase the loudness of the sound. The upper frequency of region A, however, is well marked by the axial mode of lowest frequency as determined by the largest dimension of the room. Below that frequency there is no resonant boost of modes. The sound of the very low frequencies of region A is not of outdoor character because it is contained by reflections from the walls, ceiling, and floor, but without resonance boost. The low-frequency response of this room is low in region A, is boosted by room resonances in region B, and is subject to the vagaries of diffraction and absorption in regions C and D.

Wave Acoustics

At times like this, it is very difficult to keep the treatments in this book nonmathematical. If we could dispense with such restraints, the first order of business would be to write down the wave equation. As a mere glance at this partial differential equation in three dimensions might strike terror into the hearts of many, such a temptation is resisted. Instead, attention is drawn to one of its solutions for sound in rectangular enclosures. The geometry used is that of Fig. 15-8, which judiciously fits the familiar, mutually perpendicular x, y, z coordinates of three-dimensional space to our studio or listening room. To satisfy the need for order, the longest dimension L is placed on the x axis, the next longest dimension W (for width) is placed on the y axis and the smallest dimension H (for height) on the z axis. The goal is to be able to calculate the permissible frequencies corresponding to the modes of a rectangular enclosure. Skipping the mathematics, we come directly to the answer from the equation stated by Rayleigh in 1869:

$$\text{Frequency} = \frac{c}{2}\sqrt{\frac{p^2}{L^2} + \frac{q^2}{W^2} + \frac{r^2}{H^2}} \qquad \text{(15-2)}$$

where

c = speed of sound, 1,130 ft/sec
L,W,H = room length, width, and height, ft.
p,q,r = integers 0, 1,2,3 . . ., etc.

You might flinch a little even at Eq. 15-2. Although it is not difficult, it can become a bit messy and tedious. The importance of this equation is that it gives the frequency of every axial, tangential, and oblique mode of a rectangular room.

The integers p, q, and r are the only variables once L, W, and H are set for a given room. These integers not only provide the key to the frequency of a given mode, but also serve to identify the mode as axial, tangential, or oblique. If $p = 1$, $q = 0$, and $r = 0$ (shorthand, the 1,0,0 mode), the width and height terms drop out and Eq. 15-1 becomes:

$$\text{Frequency} = \frac{c}{2}\sqrt{\frac{p^2}{L^2}} = \frac{c}{2L} = \frac{1,130}{2L} = \frac{565}{L}$$

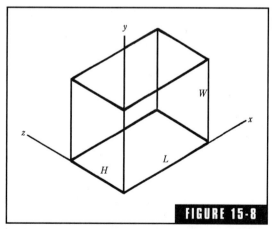

FIGURE 15-8

Orientation of rectangular room of length L, width W, and height H with respect to the x, y, and z coordinates for calculating room modal frequencies.

This is the axial mode corresponding to the length of the room. The width axial mode (0,1,0) and the height axial mode

(0,0,1) are calculated similarly by substituting the appropriate dimension. If two of the integers are zero, an axial mode frequency is identified because only one pair of surfaces is involved. In a similar way one zero identifies tangential modes, and no zero at all specifies oblique modes.

Mode Calculations—An Example

The utility of Eq. 15-2 is best demonstrated by an example. The room is small, but convenient for experimental verification. The dimensions of this room are: length L = 12.46 ft, width W = 11.42 ft, and average height H = 7.90 ft (the ceiling actually slopes along the length of the room with a height of 7.13 ft on one end and 8.67 ft on the other). These values of L, W, and H have been inserted in Eq. 15-2 along with an assortment of combinations of integers p, q, and r.

Room modes are possible only when p, q, and r are whole numbers (or zero) because this is the condition that creates a *standing* wave. There are many combinations of integers when fundamentals (associated with 1), second modes (associated with 2), and third modes (associated with 3), etc., are introduced. Table 15-1 lists some of the various combinations of p, q, and r and the resulting permissible modal frequency for each combination. Furthermore, each frequency is identified as axial, tangential, or oblique by the number of zeros in that particular p, q, r combination. The lowest natural room frequency is 45.3 Hz, which is the axial mode, associated with the longest dimension, the length L, of the room below which there is no modal boost of the sound. In this case, p = 1, q = 0, and r = 0. Mode 7, the 2,0,0 mode, yields a frequency of 90.7 Hz, which is the second axial mode associated with the length L. In the same manner, mode 18, with integers 3,0,0 is the third mode of the length axial mode, and mode 34 the fourth mode. However, there are many tangential and oblique modes between these, and Table 15-1 provides the means for carefully identifying all modes and raises the possibility of studying their relationships.

Axial modes have been emphasized in studio design, and the reason for this is given later. The message of Eq. 15-2 and Table 15-1 is that there is much more to room acoustics than axial modes and their spacing. In between the axial modal frequencies are many other modal frequencies that have an effect, even though weaker.

Table 15-1. Mode calculations. Room dimensions: $12.46 \times 11.42 \times 7.90$ ft.

Mode number	Integers p q r	Mode frequency, Hz	Axial	Tangential	Oblique
1	1 0 0	45.3	x		
2	0 1 0	49.5	x		
3	1 1 0	67.1		x	
4	0 0 1	71.5	x		
5	1 0 1	84.7		x	
6	0 1 1	87.0		x	
7	2 0 0	90.7	x		
8	2 0 1	90.7		x	
9	1 1 1	98.1			x
10	0 2 0	98.9	x		
11	2 1 0	103.3		x	
12	1 2 0	108.8		x	
13	0 2 1	122.1		x	
14	0 1 2	122.1		x	
15	2 1 1	125.6			x
16	1 2 1	130.2			x
17	2 2 0	134.2		x	
18	3 0 0	136.0	x		
19	0 0 2	143.0	x		
20	3 1 0	144.8		x	
21	0 3 0	148.4	x		
22	2 2 1	152.1			x
23	3 0 1	153.7		x	
24	1 1 2	158.0			x
25	3 1 1	161.5			x
26	0 3 1	164.8		x	
27	3 2 0	168.2		x	
28	2 0 2	169.4		x	
29	1 3 1	170.9			x
30	0 2 2	173.9		x	
31	2 3 0	173.9		x	
32	2 1 2	176.4			x
33	1 2 2	179.7			x
34	4 0 0	181.4	x		
35	3 2 1	182.8			x

Table 15-1. *Continued.*

Mode number	Integers p q r	Mode frequency, Hz	Axial	Tangential	Oblique
36	2 3 1	188.1			x
37	2 2 2	196.2			x
38	0 4 0	197.9	x		
39	3 0 2	197.9		x	
40	3 3 0	201.3		x	
41	3 1 2	203.5			x
42	0 3 2	206.1		x	
43	1 3 2	211.1			x
44	0 0 3	214.6	x		
45	1 0 3	219.3		x	
46	0 1 3	220.2		x	
47	3 2 2	220.8			x
48	1 1 3	224.8			x
49	2 3 2	225.2			x
50	2 0 3	232.9		x	
51	4 3 0	234.4		x	
52	0 2 3	236.3		x	
53	2 1 3	238.1		x	
54	3 4 0	240.2		x	
55	1 2 3	240.6			x
56	3 3 2	247.0			x
57	2 2 3	253.1			x
58	3 0 3	254.0		x	
59	0 3 3	260.9		x	
60	3 2 3	272.6			x
61	2 3 3	276.2			x
62	4 0 3	281.0		x	
63	0 0 4	286.1	x		
64	0 4 3	291.1		x	
65	3 0 4	316.8		x	
66	0 3 4	322.3		x	

Experimental Verification

All of the modal frequencies listed in Table 15-1 make up the acoustics of this particular room for the frequency range specified. To evaluate their relative effects, a swept-sine-wave transmission experiment was set up. In effect, this measures the frequency response of the room. Knowing that all room modes terminate in the corners of a room, a loudspeaker was placed in one low tricorner and a measuring microphone in the diagonal high tricorner of the room. The loudspeaker was then energized by a slowly swept sine-wave signal. The room response to this signal, picked up by the microphone, was recorded on a graphic level recorder having a paper speed of 3 mm/second. This resulted in a linear sweep from 50 to 250 Hz in 38 seconds. The resulting trace is shown in Fig. 15-9.

In the past, attempts have been made to identify the effects of each mode in reverberation rooms with all six surfaces hard and reflective. In such cases, the prominent modes stand out as sharp spikes on the recording. The test room in which the recording of Fig. 15-9 was made is a spare bedroom, not a reverberation chamber. Instead of concrete, the walls are of frame construction covered with gypsum board (drywall); carpet over plywood makes up the floor, closet doors almost cover one wall. There is a large window, pictures on the wall, and some furniture, including a couch. It is evident that this is a fairly absorbent room. The reverberation time at 125 Hz (as we will consider more fully later) was found to be 0.33 second. This room is much closer acoustically to studios and control rooms than to reverberation chambers, and this is why it was chosen.

Mode Identification

A careful study of Fig. 15-9 in an attempt to tie the peaks and valleys of the transmission run to specific axial, tangential, and oblique modes is rather disappointing. For one thing, the loudspeaker (JBL 2135) response is included, although it is quite smooth over this frequency region. The signal generator and power amplifier are very flat. Most of the ups and downs must be attributed to modes and the interaction of modes. Modes close together would be expected to boost room response if in phase, but cancel if out of phase. There are 11 axial modes, 26 tangential modes, and 21 oblique modes in this 45.3- to

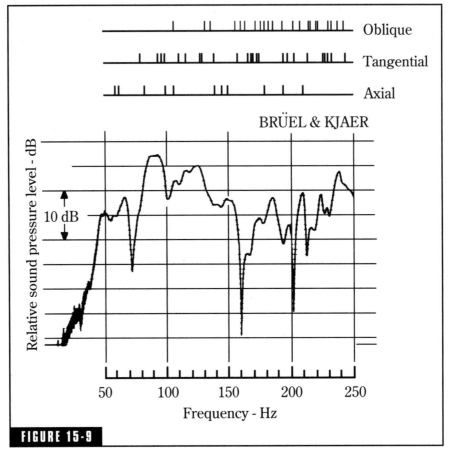

FIGURE 15-9

Swept sine-wave transmission run in the test room of Table 15-1. The location of each axial, tangential, and oblique modal frequency is indicated.

254.0-Hz record, and the best we can say is that the transmission trace shown in Fig. 15-9 is the composite effect of all 58 modes.

The three major dips are so narrow that they would take little energy from a distributed speech or music spectrum. If they are neglected, the remaining fluctuations are more modest. Fluctuations of this magnitude in such steady-state swept-sine transmission tests are characteristic of even the most carefully designed and most pleasing of studios, control rooms, and listening rooms. The ear commonly accepts such deviations from flatness of response. The modal structure of a space always gives rise to these fluctuations. They are normally neglected because attention is focused on the response of microphone, amplifier, loudspeaker, and other reproducing equipment. The response of the ear and of the room have yet to receive the attention they deserve.

Mode Decay

The steady-state response of Fig. 15-9 tells only part of the story. The ear is very sensitive to transient effects, and speech and music are made up almost entirely of transients. Reverberation decay is one transient phenomenon that is easily measured. When broadband sound such as speech or music excites the modes of a room, our interest naturally centers on the decay of the modes. The 58 modes in the 45.3- to 254.0-Hz band of Fig. 15-9 are the microstructure of the reverberation of the room. Reverberation is commonly measured in octave bands. Octave bandwidths of interest are shown in Table 15-2.

Each reverberatory decay by octave bands thus involves an average of the decay of many modes, but it takes an understanding of the decay of the individual modes to explain completely the decay of the octave band of sound. The higher the octave center frequency, the greater the number of modes included.

All modes do not decay at the same rate. Mode decay depends, among other things, on the way absorbing material is distributed in the room. Carpet on the floor of the test room has no effect on the 1,0,0 or 0,1,0 axial modes involving only walls. Tangential and oblique modes, which involve more surfaces, would be expected to die away faster than axial modes that involve only two surfaces. On the other hand, absorption is greater for the axial modes in which the sound impinges on the surfaces at right angles than for

Table 15-2. Modes in octave bands.

Octave	Limits (−3 dB points)	Modes		
		Axial	Tangential	Oblique
63 Hz	45– 89 Hz	3	3	0
125 Hz	89–177 Hz	5	13	7
250 Hz	177–254 Hz	3*	10*	14*

*Partial octave.

the low angle of incidence common for tangential and oblique modes.

The actual reverberation decays in the test room using sine-wave excitation are shown in Figs. 15-10 and 15-11. The measured reverberation times vary as shown in Table 15-3. The measured reverberation time varies almost two to one for the selected frequencies in the 180 to 240-Hz range. The decay of the 125-Hz octave band of noise (0.33 sec) and the 250-Hz octave band (0.37 sec) of Fig. 15-11 averages many modes and should be considered more or less the "true" values for this frequency region, although normally many decays for each band would be taken to provide statistical significance.

The dual slope decay at 240 Hz in Fig. 15-10 is especially interesting because the low value of the early slope (0.31 sec) is probably dominated by a single mode involving much absorption, later giving way to other modes that encounter much less absorption. Actually identifying the modes from Table 15-1 is difficult, although you might expect mode 44 to die away slowly and the group of three near 220 Hz (45, 46, 47) to be highly damped. It is common to force nearby modes into oscillation, which then decay at their natural frequency.

Mode Bandwidth

Normal modes are part and parcel of room resonances. Taken separately, each normal mode exhibits a resonance curve such as shown in

Table 15-3. Measured reverberation time of test room.

Frequency Hz	Average reverberation time, seconds
180	0.38
200	0.48
210	0.53
220	0.55
240	0.31 and 0.53 (double slope)
125 Hz octave noise	0.33
250 Hz octave noise	0.37

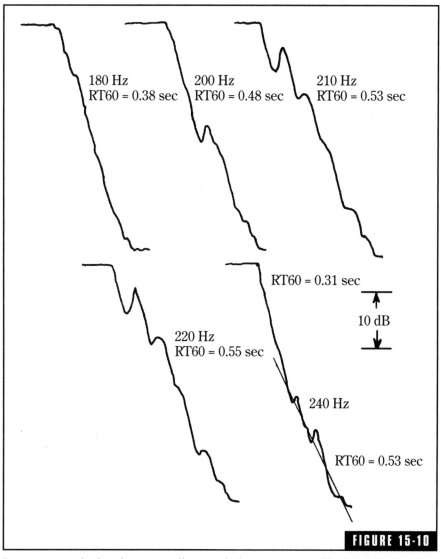

180 Hz
RT60 = 0.38 sec

200 Hz
RT60 = 0.48 sec

210 Hz
RT60 = 0.53 sec

RT60 = 0.31 sec

10 dB

220 Hz
RT60 = 0.55 sec

240 Hz

RT60 = 0.53 sec

FIGURE 15-10

Pure-tone reverbation-decay recordings made in a test room. Single modes decaying alone yield smooth, logarithmic traces. Beats between neighboring modes cause the irregular decay. The two-slope pattern bottom right reveals the smooth decay of a single prominent mode for the first 20 dB, after which one or more lightly absorbed modes dominate.

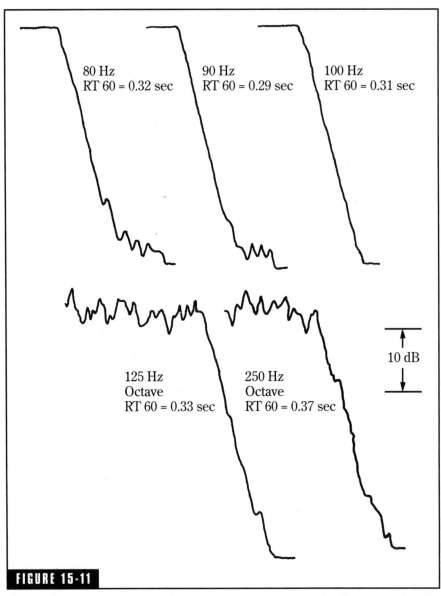

80 Hz
RT 60 = 0.32 sec

90 Hz
RT 60 = 0.29 sec

100 Hz
RT 60 = 0.31 sec

10 dB

125 Hz
Octave
RT 60 = 0.33 sec

250 Hz
Octave
RT 60 = 0.37 sec

FIGURE 15-11

The top three pure-tone reverbation decays are dominated by single prominent modes as indicated by the smooth adherence to logarithmic form. The two lower records are of octave bands of pink noise, which give the average decay of all modes in those octaves.

Fig. 15-12. Each mode, therefore, has a definite bandwidth determined by the simple expression:

$$\text{Bandwidth} = f_2 - f_1 = \frac{2.2}{\text{RT60}} \qquad \text{(15-3)}$$

RT60 = reverberation time, seconds

Bandwidth is inversely proportional to the reverberation time. In electrical circuits, the sharpness of the tuning curve depends on the amount of resistance in the circuit; the greater the resistance, the broader the tuning curve. In room acoustics, the reverberation time depends on absorption (resistance). The analogy is fitting (the more absorption, the shorter the reverberation time, and the wider the mode resonance). For convenient reference, Table 15-4 lists a few values of bandwidth in relation to reverberation time.

The mode bandwidth for a typical studio is in the general region of 5 Hz. This means that adjacent modes tend to overlap for rooms with short reverberation time, which is desirable. As the skirts of the resonance curves of adjacent modes A and B overlap, energizing mode A by driving the room at frequency A will also tend to force mode B into excitation. When the tone of frequency A is removed, the energy stored in B decays at its own frequency B. The two will beat with each other during the decay. The very uniform decays, such as the top three in Fig. 15-11, are probably single modes sufficiently removed from neighbors so as to act independently without irregularities caused by beats.

Figure 15-13 shows an expanded version of the 40- to 100-Hz portion of Fig. 15-9. The axial modes from Table 15-1 falling within this range are plotted with bandwidth of 6 Hz at the −3dB points. The axial modes (Fig. 15-13A) are peaked at zero reference level. The tangential modes have only one-half the energy of

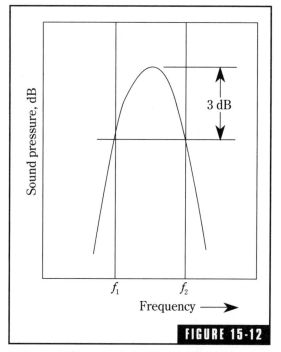

FIGURE 15-12

Each mode has a finite bandwidth. The more absorbent the room, the greater the bandwidth. As measured at the customary −3-dB points, the modal bandwidth of a recording studio is near 5 Hz.

Table 15-4. Mode bandwidth.

Reverberation time, seconds	Mode bandwidth Hz
0.2	11
0.3	7
0.4	5.5
0.5	4.4
0.8	2.7
1.0	2.2

the axial modes,[3] which justifies plotting their peaks 3 dB (10 log 0.5) below the axial modes in Fig. 15-13B. The oblique modes have only one-fourth the energy of the axial modes, hence the lone oblique mode at 98.1 Hz that falls within this range is plotted 6 dB (10 log 0.25) below the axial mode peak in Fig. 15-13C.

The response of the test room, Fig. 15-13D is most certainly made up of the collective contributions of the various modes tabulated in Table 15-1. Can Fig. 15-13D be accounted for by the collective contributions of axial modes A, tangential modes B, and the single oblique mode C? It seems reasonable to account for the 12-dB peak in the room response between 80 and 100 Hz by the combined effect of the two axial, three tangential, and one oblique mode in that frequency range. The fall off below 50 Hz is undoubtedly due to

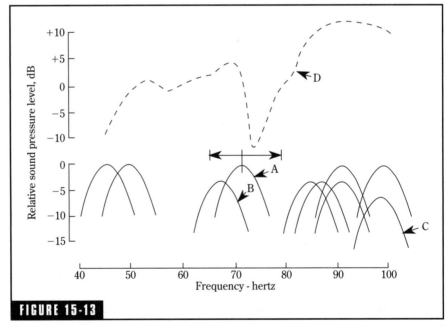

FIGURE 15-13

Attempted correlation of calculated modes and the measured swept-sine response of the test room over the frequency range 40 Hz to 100 Hz. (A) Axial modes. (B) Tangential modes. (C) Oblique mode. (D) Reproduction of the 40- to 100-Hz portion of room response of Fig. 15-9.

loudspeaker response. This leaves the 12-dB dip at 74 Hz yet to be accounted for.

Examining the axial mode at 71.5 Hz, it can be seen that this dip is the vertical mode of the test room working against a sloping ceiling. The frequency corresponding to the average height is 71.5 Hz, but the one corresponding to the height at the low end of the ceiling is 79.3 Hz, and for the high end it is 65.2 Hz. The uncertainty of the frequency of this mode is indicated by arrows in Fig. 15-13. If this uncertain axial mode were shifted to a slightly lower frequency, the 12-dB dip in response could be better explained. It would seem that a dip in response should appear near 60 Hz, but none was found.

Rather than being an absolute experimental verification of theory, this test-room experiment was conducted only to explain the basic features of the theory. Conditions and techniques lacked the necessary rigor to produce exact results. The test room is not a rectangular parallelepiped, which is the basis of Eq. 15-1. Neither is the loudspeaker response known accurately. In addition, there is the overriding fact that in combining the effects of the modes of Fig. 15-13 A, B, and C to get Fig. 15-13 D, phase must be taken into account. These components must be combined vectorially with both magnitude and phase fully considered. The main purpose of Fig. 15-13 is to emphasize that *room response is made up of combined modal responses.*

Mode Pressure Plots

It is easy to say that the modal pattern of a given room creates a very complex sound field, but to really drive this point home several sketches of sound pressure distributions are included. The one dimension organ pipe of Fig. 15-2 is a starting point that can be compared to the 1,0,0 mode of Fig. 15-14 for a three-dimensional room. The pressure is higher near the ends (1.0) and zero along the center of the room. Figure 15-15 shows sound pressure distribution when only the 3,0,0 axial mode is energized. The sound pressure nodes and anti-nodes in this case are straight lines as shown in Fig. 15-16.

Three-dimensional sketches of sound pressure distribution throughout a room become difficult, but Fig. 15-17 is an attempt for the 2,1,0 tangential mode. We see sound pressure "piled up" in each corner of the room with two more "piles" at the center edges. This is

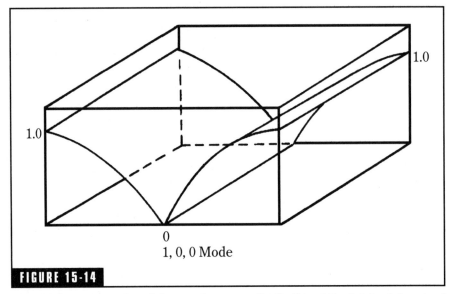

1.0

1.0

0

1, 0, 0 Mode

FIGURE 15-14

A graphical representation of the sound pressure distribution of the 1,0,0 axial mode of a room. The sound pressure is zero in the vertical plane at the center of the room and maximum at the ends of the room. This is comparable to f_1 of the organ pipe of Fig. 15-2.

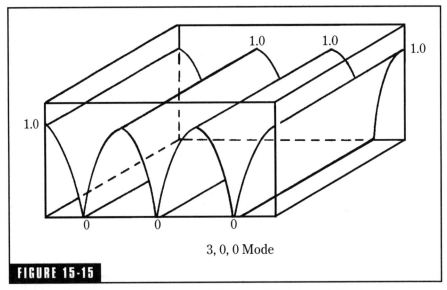

1.0   1.0   1.0

1.0

0   0   0

3, 0, 0 Mode

FIGURE 15-15

Representation of the sound pressure distribution of the 3,0,0 axial mode of a room.

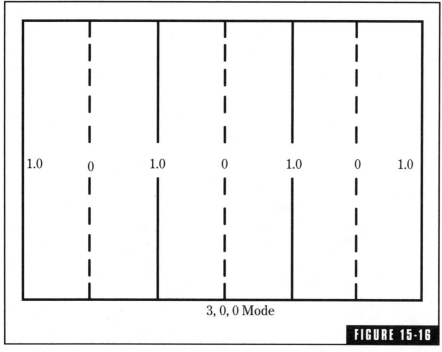

3, 0, 0 Mode

FIGURE 15-16

Sound pressure contours on a section through a rectangular room for the 3,0,0 axial mode.

more graphically portrayed in Fig. 15-18 in which the pressure contour lines are drawn. The broken lines crisscrossing the room between the "piles" of sound mark the zero pressure regions.

Imagine how complicated the sound pressure pattern would be if all the modes were concurrently or sequentially excited by voice or music energy chasing up and down the spectrum while constantly shifting in intensity. The plot of Fig. 15-18 shows pressure maxima in the corners of the room. These maxima always appear in room corners for all modes. To excite all modes, place the sound source in a corner. Conversely, if you wish to measure all modes, a corner is the place to locate the microphone.

Modal Density

Modal density increases with frequency. In Table 15-1 the one-octave spread between 45 and 90 Hz, only 8 modal frequencies are counted. In the next highest octave there are 25 modes. Even in this very limited

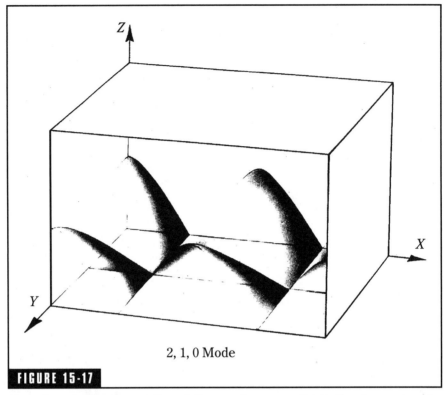

2, 1, 0 Mode

FIGURE 15-17

Three-dimensional representation of the sound pressure distribution in a rectangular room for the tangential mode 2,1,0. Bruël and Kjaer Instruments, Inc.

low-frequency range below 200 Hz, you can see modal density increasing with frequency. Figure 15-19 shows that at somewhat higher frequencies the rate of increase dramatically rises. Above about 300 Hz or so, the mode spacing is so small that the room response smooths markedly with frequency.

Mode Spacing and Coloration

Colorations largely determine the quality of sound for a small studio or listening room. The big task then is to determine which, if any, of the hundreds of modal frequencies in a room are likely to create colorations.

The spacing of the modal frequencies is the critically important factor. In the D-region of Fig. 15-7, the modal frequencies of a small

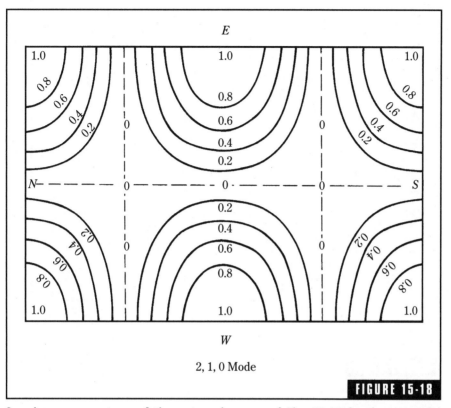

E

N— — — — — 0 — — — — 0 — — — — 0 — — — — S

W

2, 1, 0 Mode

FIGURE 15-18

Sound-pressure contours of the rectangular room of Fig. 15-16 for the tangential mode 2,1,0.

room are so close together that they tend to merge helpfully and harmlessly. In the B- and C-regions, below about 300 Hz, their separation is greater and it is in this region that problems can arise.

How close together must adjacent modal frequencies be to avoid problems of coloration? Gilford[4] states his opinion that an axial mode separated more than 20 Hz from the next axial mode will tend to be isolated acoustically. It will tend not to be excited through coupling due to overlapping skirts but will tend to act independently. In this isolated state it can respond to a component of the signal near its own frequency and give this component its proportional resonant boost.

Another criterion for mode spacing has been suggested by Bonello[5,6] who considers all three types of modes, not axial modes alone. He states that it is desirable to have all modal frequencies in a critical band at least 5% of their frequency apart. For example, one modal frequency

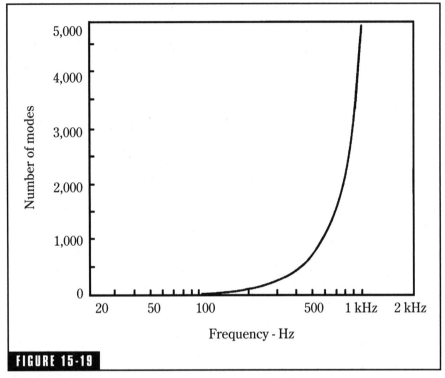

FIGURE 15-19

The number of modal frequencies increases dramatically with frequency.

at 20 Hz and another at 21 Hz would be barely acceptable. However, a similar 1-Hz spacing would not be acceptable at 40 Hz (5% of 40 Hz is 2 Hz). Thus we see that Gilford's concern was primarily how far apart axial modes must be spaced to avoid problems resulting from independent and uncoupled modal action. Bonello's concern has to do with separations to avoid degeneracy (coincident) effects.

Zero spacings between modal frequencies are a common source of coloration. Zero spacing means that two modal frequencies are coincident (called a degeneracy by acousticians), and such degeneracies tend to overemphasize signal components at that frequency.

Experiments with Colorations

Any ear can be offended by colorations caused by isolated modes, but even a critical and trained ear needs some instrumental assistance in identifying and evaluating such colorations. The BBC Research Department made an interesting study.[4] Observers listened to persons speaking at a microphone in the studio under investigation, the voices being reproduced in another room over a high-quality system. Observers' judgments were aided by a selective amplifier that amplified a narrow frequency band (10 Hz) to a level about 25 dB above the rest of the spectrum. The output was mixed in small proportions with the original signal to the loudspeaker, the proportions being adjusted until it is barely perceptible as a contribution to the whole output. Any colorations were then made clearly audible when the selective amplifier was tuned to the appropriate frequency.

In most studios tested this way, and we can assume that they were well designed, only one or two obvious colorations were found in each. Figure 15-20 is a plot of 61 male voice colorations observed over a period of 2 years. Most fall between 100 and 175 Hz. Female voice colorations occur most frequently between 200 and 300 Hz.

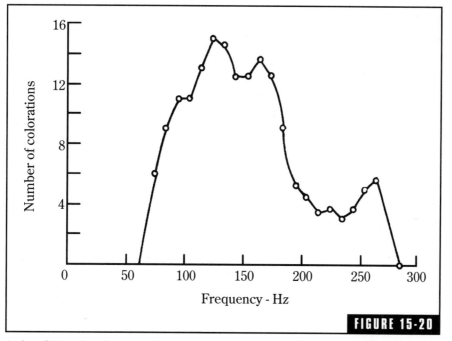

FIGURE 15-20

A plot of 61 male voice colorations observed over a period of 2 years in BBC studios. Most fall in the 100-175 Hz region. Female voice colorations occurr between 200 and 300 Hz. After Gilford.[4]

Simplified Axial Mode Analysis

Apply what you learned about axial modes to a specific rectangular listening room or studio. The dimensions of our specimen room are 28 × 16 × 10 ft. The 28-ft length resonates at 565/28 = 20.2 Hz, the two side walls 16 ft apart resonate at 565/16 = 35.3 Hz, and the floor-ceiling combination resonates at 565/10 = 56.5 Hz. These three axial resonances and the train of multiples for each are plotted in Fig. 15-21. There are 27 axial resonance frequencies below 300 Hz, and for this exercise, the horde of weaker tangential and oblique modes are neglected.

Because most signal colorations are traceable to axial modes, their spacings will be examined in detail. Table 15-5 illustrates a convenient form for this simplified analysis of axial modes. The resonance frequencies from the L, W, and H columns of Table 15-5 are arranged in ascending order in the fourth column. This makes it easy to examine that critical factor, axial mode spacing.

The L-f_7 resonance at 141.3 Hz coincides with the W-f_4 resonance. This means that these two potent axial modes act together to create a potential coloration of sound at that frequency. This coincidence is also separated 20 Hz from its neighbors. It is also noted from Fig. 15-20 that 141.3 Hz is in a frequency range that is especially troublesome. Here, then, is a warning of a potential problem.

At 282.5 Hz we see a "pile-up" of L-f_{14}, W-f_8, and H-f_5 modes which, together, would seem to be an especially troublesome source of col-

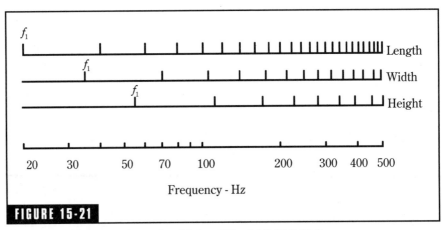

FIGURE 15-21

The axial modal frequencies and multiples of the 16 X 28 X 10 ft room.

Table 15-5. Axial mode analysis form.

Room dimensions: Length 28.0 ft.
Width 16.0 ft.
Height 10.0 ft.

	Axial mode resonances Hz			Arranged in ascending order	Axial mode spacing Hz
	L	W	H	20.2	15.1
f_1	20.2	35.3	56.5	35.3	5.1
f_2	40.4	70.6	113.0	40.4	16.1
f_3	60.5	105.9	169.5	56.5	4.0
f_4	80.7	141.3	226.0	60.5	10.1
f_5	100.9	176.6	282.5	70.6	10.1
f_6	121.1	211.9	339.0	80.7	20.2
f_7	141.3	247.2		100.9	5.0
f_8	161.4	282.5		105.9	7.1
f_9	181.6	317.8		113.0	8.1
f_{10}	201.8			121.1	20.2
f_{11}	222.0			141.3 ⎫	0
f_{12}	242.1			141.3 ⎭	20.1
f_{13}	262.3			161.4	8.1
f_{14}	282.5			169.5	7.1
f_{15}	302.7			176.6	5.0
				181.6	20.2
				201.8	10.1
				211.9	10.1
				222.0	4.0
				226.0	16.1
				242.1	5.1
				247.2	15.1
				262.3	20.2
				282.5 ⎫	0
				282.5 ⎬	0
				282.5 ⎭	20.2
				302.7	20.2

orations. They are also separated from neighbors by 20 Hz. However, looking at the experimentally derived plot of Fig. 15-20, practically no problem with voice colorations was found at 282 Hz, at least for male voices. The reason for this is probably the salutary presence of tangential and oblique modes neglected in this study.

With the threat of coloration at 141 Hz, adjustment of dimensions of a proposed room would be a logical attack. If it is an existing room, a Helmholtz resonator, sharply tuned and properly located, is a difficult but possible solution.

The Bonello Criterion

Proof that modal resonances in rooms are an international problem, this method of evaluating their effect comes from Buenos Aires. Bonello[5,6] suggests a novel method of determining, by computer, the acoustical desirability of the proportions of rectangular rooms. He divides the low end of the audible spectrum into bands ⅓ octave wide and considers the number of modes in each band below 200 Hz. The ⅓-octave bands are chosen because they approximate the critical bands of the human ear.

To meet Bonello's criterion, each ⅓ octave should have more modes than the preceding one, or at least the same number. Modal coincidences are not tolerated unless at least 5 modes are in that band.

How does a 15.4 × 12.8 × 10-ft room qualify by this criterion? Figure 15-22 shows that it passes this test with flying colors. The graph climbs steadily upward with no downward anomalies. The horizontal section at 40 Hz is allowed. The advantage of Bonello's plan is that it is well adapted to computer calculation and print-out.

Controlling Problem Modes

The general construction of Helmholtz resonators for normal room treatment is detailed in chapter 9, but building one with very sharp tuning (high Q) is more demanding.[7] The flexing of wooden boxes introduces losses that lower the Q. To attain a truly high-Q resonator with sharp tuning, the cavity must be made of concrete, ceramic, or other hard, nonyielding material, but fitted with some means of vary-

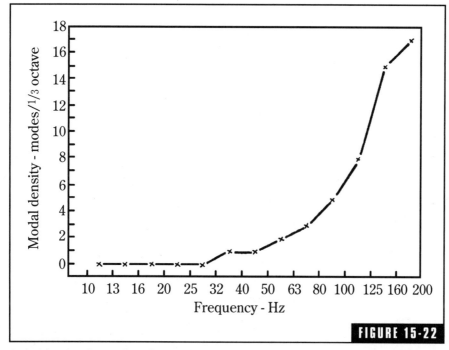

FIGURE 15-22

A plot showing the number of modes in 1/3 octave bands for the 15,4 X 12.8 X 10 ft room. The graph climbs steadily upward with no downward anomalies; hence the room meets the Bonello criterion.

ing the resonance frequency. The resonance frequency of a Coca Cola bottle was measured[8] at 185 Hz and was found to have a bandwidth (−3 dB points) of 0.67 Hz. This yields a $Q = 185/0.67 = 276$, a very high value. So, if you are fortunate enough to have your mode problem at 185 Hz ...!

It is also important where the Coke bottle (or other Helmholtz absorber) is placed, if the goal is to bring a mode or closely spaced group of modes under control. Let us say that the 2,1,0 mode of Fig. 15-18 is causing a voice coloration and that it is necessary to introduce a narrow sliver of absorption at the 2,1,0 frequency. If the Helmholtz absorber were placed at a pressure node (zero pressure) it would have, obviously, no effect. Placed at one of the antinodes (pressure peaks) it would have tight interaction with the 2,1,0 mode. Therefore, any corner would be acceptable, as would the pressure peaks on the E or W walls.

Mode Summary

- There are three types of acoustical resonances (natural frequencies, standing waves, normal modes) in a rectangular enclosure. These are the axial, tangential, and oblique modes.

- The axial modes are made up of two waves going in opposite directions, traveling parallel to one axis, and striking only two walls. Axial modes make the most prominent contribution to the acoustical characteristics of a space. Because there are three axes to a rectangular room, there are three fundamental axial frequencies, each with its own train of modes.

- The tangential modes are formed by four traveling waves that reflect from four walls and move parallel to two walls. Tangential modes have only half the energy of axial modes, yet their effect on room acoustics can be significant. Each tangential mode has its train of modes.

- The oblique modes involve eight traveling waves reflecting from all six walls of an enclosure. Oblique modes, having only one-fourth the energy of the axial modes, are less prominent than the other two.

- The number of normal modes increases with frequency. Small rooms whose dimensions are comparable to the wavelength of audible sound have the problem of excessive separation between modes, which can contribute to poor characteristics for recording or other critical work. A musical note falling "in the cracks" between widely separated modes will be abnormally weak and will die away faster than other notes. It is almost as though that particular note were sounded outdoors while the other notes were simultaneously sounded indoors.

- Axial, tangential, and oblique modes decay at different rates. Absorbing material must be located on surfaces near which a given modal pressure is high if it is to be effective in absorbing that mode. For example, carpet on the floor has no effect on the horizontal axial modes. Tangential modes are associated with more surface reflections than axial modes and oblique modes with even more than tangential ones.

- As frequency is increased, the number of modes greatly increases. Above 300 Hz, average mode spacing becomes so small that room response tends to become smoother.

- Colorations caused by acoustical anomalies of studios, monitoring rooms, listening rooms, and other small rooms are particularly devastating to speech quality. Gilford states[4] that axial modes spaced approximately 20 Hz or more, or a pair of modes coincident or very close, are frequent sources of colorations. He also states that colorations are likely to be audible when an axial mode coincides with a fundamental or first format of at least one vowel sound of speech, and are in the region of high-speech energy. Speech colorations below 80 Hz are rare because there is so little energy in speech in that part of the spectrum. There are essentially no speech colorations above 300 Hz for either male or female voices. Modal colorations are more noticeable in speech than in music.

- The bandwidth of room modes, measured at the −3 dB points, increases as reverberation time is shortened. Modes in ordinary studios have bandwidths on the order of 5 Hz. Harmonics of modes have the same bandwidths as their fundamentals.

The serious student wishing to pursue this subject further is urged to consult references 9 through 17. These papers, spanning a half century, trace the growth in our understanding of room modes.

Endnotes

[1]Brüel, Per V., *Sound Insulation and Room Acoustics*, London, Chapman and Hall (1951),

[2]Ruusuvuori, Aarno, Architect; Mauri Parjo, Acoustical Consultant.

[3]Morse, Philip M. and Richard H. Bolt, *Sound Waves in Rooms*, Reviews of Modern Physics, 16, 2 (Apr 1944) 69-150.

[4]Gilford, C.L.S., *The Acoustic Design of Talks Studios and Listening Rooms*, Proc. Inst. Elect. Engs., 106, Part B, 27 (May 1959) 245-258. Reprinted in J. Audio Eng. Soc., 27, 1/2 (1979) 17-31.

[5]Bonello, O.J., *A New Computer-Aided Method for the Complete Acoustical Design of Broadcasting and Recording Studios*, IEEE International Conference on Acoustics and Signal Processing, ICASSP 79, Washington 1979, p. 326-329.

[6]Bonello, O.J., *A New Criterion for the Distribution of Normal Room Modes*, J. Audio Eng. Soc., 29, 9 (Sept 1981) 597-606. Correction in 29, 12 (1981) 905.

[7]van Leeuwen, F.J., *The Damping of Eigentones in Small Rooms by Helmholtz Resonators*, European Broadcast Union Review, A, 62 (1960) 155-161.

[8]Siekman, William, Private communication. Measurements were made at Riverbank Acoustical Laboratories when Siekman was manager. He presented the results of these measurements at the April 1969 convention of the Acoustical Society of America.

[9]Hunt, Frederick V., *Investigation of Room Acoustics by Steady-State Transmission Measurements*, J. Acous. Soc. Am., 10 (Jan 1939) 216-227.

[10]Hunt, F.V., L.L. Beranek, and D.Y. Maa, *Analysis of Sound Decay in Rectangular Rooms*, J. Acous. Soc. Am., 11 (July 1939) 80-94.

[11]Bolt, R. H., *Perturbation of Sound Waves in Irregular Rooms*, J. Acous. Soc. Am., 13 (July 1942) 65-73.

[12]Bolt, R.H., *Note on Normal Frequency Statistics in Rectangular Rooms*, 18, 1 (July 1946) 130-133.

[13]Knudsen, Vern O., *Resonances in Small Rooms*, J. Acous. Soc. Am. (July 1932) 20-37.

[14]Mayo, C. G., *Standing-Wave Patterns in Studio Acoustics*, Acustica, 2, 2 (1952) 49-64.

[15]Meyer, Erwin, *Physical Measurements in Rooms and Their Meaning in Terms of Hearing Conditions*, Proc. 2nd International Congress on Acoustics, (1956) 59-68.

[16]Louden, M.M., *Dimension Ratios of Rectangular Rooms with Good Distribution of Eigentones*, Acustica, 24 (1971) 101-104.

[17]Sepmeyer, L.W., *Computed Frequency and Angular Distribution of the Normal Modes of Vibration in Rectangular Rooms*, J. Acous. Soc. Am., 37, 3 (March 1965) 413-423.

Sound Reflections in Enclosed Spaces

The perceptual effect of sound reflections depends on the size of the room. The situation in a recording studio, control room, or listening room is quite different from that in a music hall or large auditorium. The case of sound reflections in small rooms is considered first in this discussion.

Law of the First Wavefront

Imagine two people in a small room as illustrated in Fig. 16-1. The first sound of the person speaking to reach the listener is that traveling a direct path because it travels the shortest distance. This direct sound, which arrives first at the ears of the listener, establishes the perception of the direction from which the sound came. Even though it is immediately inundated by a stream of reflections from the various surfaces of the room, this directional perception persists, tending to lock out the effects of all later reflections insofar as direction is concerned. Cremer has called this *the law of the first wavefront*. This fixation of the direction to the source of sound is accomplished within a small fraction of a millisecond and, as already mentioned, is unaffected by the avalanche of reflections following the arrival of the direct sound.

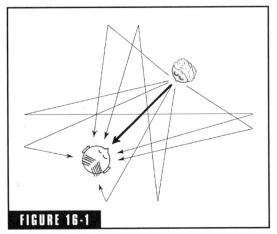

FIGURE 16-1

Of the many reflections of sound in a small room, the direct wave arrives first and establishes the receiver's sense of the direction of arrival of the sound. Later reflections do not affect this directional perception.

Mean Free Path

The average (mean) distance sound travels between successive reflections is called *the mean free path*. This average distance is given by the expression $4V/S$, in which V is the volume of the space and S is the surface area of the space. In a room $25 \times 20 \times 10$ ft, the sound travels a distance of 10.5 ft between reflections on the average. Sound travels 1.13 ft per millisecond. At that speed it takes only 9.3 ms to traverse the mean-free distance of 10.5 ft. Viewed another way, about 107 reflections take place in the space of a second.

Figure 16-2 shows cathode-ray oscillograms of the reflections occurring during the first 0.18 second in a recording studio having a volume of 16,000 cu ft. The microphone was placed, successively, in four different locations in the room. The impulsive sound source was in a fixed position.

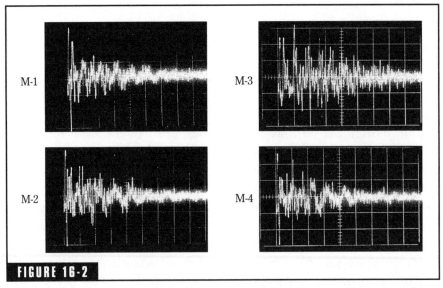

FIGURE 16-2

Individual reflections are resolved in these echograms taken at four different positions in a studio of 16,000 cu ft volume and having a reverberation time of 0.51 second. The horizontal time scale is 20 msec/div.

The sound source was a pistol that punctured a paper with a blast of air, giving an intense pulse of sound of less than a millisecond duration. The reflection patterns at the four positions show differences but, in each, scores of individual reflections are clearly resolved. These "echograms" define the transient sound field of the room during the first 0.18 second as contrasted to the steady-state condition. The question now reduces to the perceptual effects of all these reflections on the human auditory system.

The Effect of Single Reflections

Numerous research studies on the audibility of simulated reflections have been conducted over a period of many years. Most of these have used an arrangement of loudspeakers very much like the traditional high-fidelity, stereophonic arrangement as shown in Fig. 16-3. The observer is seated at the apex so that lines drawn to the two loudspeakers are approximately 60 degrees apart (this angle varies with the investigator). The mono signal (commonly speech) is fed to one of the loudspeakers, which represents the direct signal. The signal to the other loudspeaker can be delayed any amount: This represents a lateral reflection. The two variables under study are: *the level of the reflection* compared to that of the direct, and *the delay of the reflection* with respect to the direct signal.

Perception of Sound Reflections

The earlier researchers in this field were interested primarily in the effect of reflections on the perception of music in music halls. Recent work by Olive and Toole[1,2] was aimed more specifically at listening conditions in small rooms, such as recording studio control rooms and home listening rooms. The work of Olive and Toole is summarized in Fig. 16-4. This graph plots reflection level against reflection delay, the two variables mentioned above. A reflection

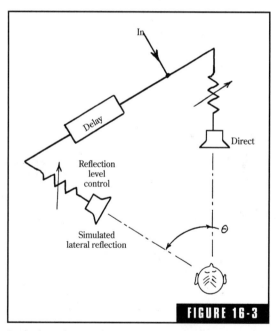

FIGURE 16-3

Typical equipment arrangement used by many investigators of the audible effect of a simulated lateral reflection on the direct signal. Reflection level (with respect to the direct) and reflection delay are the variables under control.

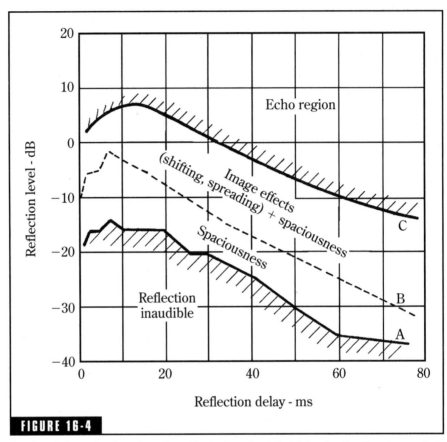

FIGURE 16-4

Results of investigations into the effect of a simulated lateral reflection in anechoic environment and with speech as the test signal. Curve A is the absolute threshold of detection of the reflection (Olive and Toole[1]). Curve B is the image-shift threshold (Olive and Toole[1]). Curve C indicates the points at which the reflection is heard as a discreet echo. This is a composite of the results of Refs. 3 and 5. Journal of the Audio Engineering Society.

level of 0 dB means that the reflection is the same level as the direct signal. A reflection level of −10 dB means that the reflection level is 10 dB below the direct. In all cases reflection delay is in milliseconds later than the direct signal.

Figure 16-4, curve A is the absolute threshold of audibility of the echo. This means that at any particular delay, the reflection is not heard for reflection levels below this line. Note that for the first 20 milliseconds, this threshold is essentially constant. At greater delays, progressively lower reflection levels are required for a just-audible reflection. It will soon become apparent that for a home listening room or other small room, delays in the 0–20 ms range are of greatest significance. In this range the reflection audibility threshold varies little with delay.

Perception of Spaciousness

Assume a reflection delay of 10 ms with the reflection coming from the side. As the level of the reflection is increased from a very low level, the reflection is completely inaudible. As the level of the reflection is increased, it finally becomes audible as its level reaches about 15 dB below the direct signal. As the reflection level is increased beyond this point, the room takes on a sense of spaciousness; the anechoic room in which the tests were made sounded more like a normal room. The listener is not aware of the reflection as a discrete event, nor of any directional effect, only this sense of spaciousness.

Image Changes

As the level of the reflection is further increased other effects become audible. At about 10 dB above the threshold of audibility of the reflection, changes in the apparent size and location of the front auditory image become apparent. At greater delays, the image tends to become smeared toward the reflection.

Reviewing what happens in the 10–20 ms delay range, as the reflection level is increased above the threshold of audibility, spatial effects dominate. As the reflection level is increased roughly 10 dB above the audibility threshold, image effects begin to enter, including image size and shifting of position of the image.

Discrete Echoes

Reflections having a level another 10 dB above the image shift threshold introduce another perceptual threshold. The reflections now are discrete echoes superimposed on the central image.[3] Such discrete echoes are very damaging to sound quality in any practical situation. For this reason, reflection level/delay combinations that result in such echoes are to be shunned in practical installations.

Effect of Angle of Incidence on Audibility of Reflection

German researchers have shown that the direction from which the reflection arrives has practically no effect on the perception of the reflection with one important exception. When the reflection arrives from the *same direction* as the direct signal, it can be up to 5 to 10 dB louder than the direct before it is detected. This is due to masking of the reflection by the direct signal. If the reflection is recorded along with the direct signal and reproduced over a loudspeaker, it will be masked by this 5 to 10 dB amount.

Effect of Signal Type on Audibility of Reflection

The type of signal has a major effect on the audibility of reflections. Consider the difference between continuous and noncontinuous sounds. Impulses, in the form of 2 clicks per second, are of the noncontinuous type. Pink noise illustrates the continuous type. Speech and music lie in between the two types. In Fig. 16-5 the differences in audibility thresholds of continuous vs. noncontinuous sounds is illustrated. Anechoic speech comes closer to being a noncontinuous sound than either music or pink noise. At delays less than 10 ms, the level of impulses for threshold detection must be much higher than continuous sounds. The threshold curves for Mozart music and pink noise are very close together. This confirms the belief that pink noise is a reasonable surrogate for music for measurements.

Effect of Spectrum on Audibility of Reflection

Most of the researchers used sounds having the same spectrum for both the direct and reflected simulations. In real life, reflections depart from the original spectrum because sound-absorbing materials invariably absorb high frequencies more than low frequencies. In addition to this, off-axis loudspeaker response lowers the high-frequency content even more. Threshold audibility experiments have shown that rather radical low-pass filtering of the reflection signal produced only minor differences in thresholds. The conclusion is that alteration of reflection spectrum does not change audibility thresholds appreciably.

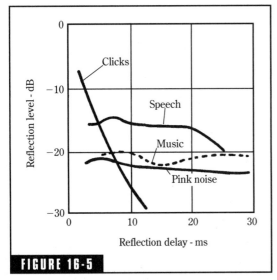

FIGURE 16-5

Absolute thresholds of perception of reflections of different types of signals, ranging from 2/second clicks (noncontinuous), to pink noise (continuous). The closeness of pink noise to Mozart music gives assurance that pink noise is a reasonable surrogate for music in measurements. After Olive and Toole.[1] Journal of the Audio Engineering Society.

Using Reflection Data

Figure 16-4 is a remarkable concentration of the perceptual effects of lateral reflections. A single lateral reflection affects the size and position of the auditory image and controls spaciousness. How about the two-loudspeaker stereo installation? Olive and Toole investigated this and found that the effects obtained from a single loudspeaker are directly applicable to the stereo case. This means that the information presented earlier is immediately available for application to stereo listening situations such as the home music-reproduction room.

Those interested in the reproduction of high-fidelity sound will see the practicality of the results of these reflection studies. The possibility of adjusting the spaciousness of the listening room as well as the stereo image sharpness is intriguing. All this is possible by careful and knowledgeable manipulation of lateral reflections. Of course, lateral reflections can come into their own only after interfering early reflections are reduced. These possibilities are explored further in later chapters.

Large Spaces

Echoes

Larger enclosed spaces, such as performance theaters, music halls, churches, and auditoriums, are potential producers of discrete echoes. Architects must be continually alert to surfaces that might produce reflections of sufficient level and delay to be perceived as discrete echoes. This is a gross defect for which there is little tolerance, audible to everyone with normal hearing.

Figure 16-6 is plotted on the same general coordinates as Fig. 16-4.[4,5] The main difference is that the delay scale is carried out to 600 ms, in line with larger spaces.

Reverberation time affects the audibility of echoes. The reverberation time of the graph of Fig. 16-6 is 1.1 seconds. The heavy broken line shows the decay rate representing this reverberation time (60 dB in 1.1 seconds or 30 dB in 550 ms). The shaded area represents combinations of echo level and echo delay, experimentally determined, which result in echoes disturbing to people. The upper edge is for 50%, the lower edge is for 20% of the people disturbed by the echo. Concert hall reverberation time usually hovers around 1.5 seconds; many churches are closer to 1 second to favor speech. Measurements of this type, made in spaces having other reverbera-tion times, show that the shaded area indicating the level/delay region causing troublesome echoes is close to being tangent to the reverberation decay line. This presents the possibility of estimating the echo threat of any large room by drawing the reverberation line. For example, the lighter bro-ken line in Fig. 16-6 is drawn for a reverberation time of 0.5 second (60 dB decay in 500 ms or 30 dB decay in 250 ms). An echo-interference area just above this decay line can be very roughly inferred in this way.

Spaciousness

Acoustical consultants and architects routinely design music halls to give lateral reflections of appropriate levels and delays to add a sense of

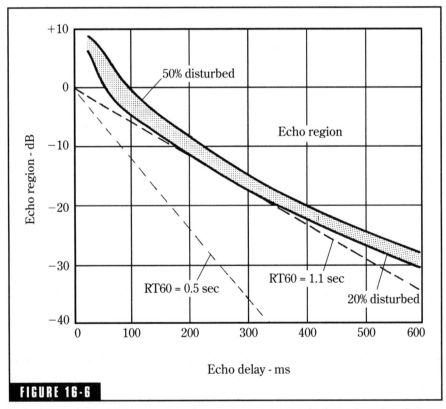

FIGURE 16-6

Acceptable echo levels for speech under reverberant conditions (Reverberation Time 1.1 seconds). After Nickson, Muncey, and Dubout.[4]

spaciousness to the music for those in the audience. This is a highly specialized area of limited interest to readers of this book. This application does emphasize the importance of informed manipulation of reflections in large spaces to achieve desirable results, even as in smaller rooms.

Endnotes

[1]Olive, Sean E. and Floyd E. Toole, *The Detection of Reflections in Typical Rooms*, J. Audio Eng. Soc., 37, 7/8, (July/Aug 1989) 539-553.

[2]Toole, Floyd E., *Loudspeakers and Rooms for Stereophonic Sound Reproduction*, Proc. Audio Eng. Soc. 8th International Conference, Washington, D.C., (1990) 71-91.

[3]Meyer, E. and G.R. Schodder, *On The Influence of Reflected Sound on Directional Localization and Loudness of Speech*, Nachr. Akad. Wiss., Göttingen; Math. Phys. Klasse IIa, 6, (1952) 31-42.

[4]Nickson, A.F.B., R.W. Muncey, and P. Dubout, *The Acceptability of Artificial Echoes with Reverberant Speech and Music*, Acustica, 4 (1954) 515-518.

[5]Lochner, J.P.A. and J.F. Burger, *The Subjective Masking of Short Time Delayed Echoes by Their Primary Sounds and Their Contribution to the Intelligibility of Speech*, Acustica, 8 (1958) 1-10.

Comb-Filter Effects

The term *comb filter* has been widely used in the popular audio press as an explanation of delayed reflection effects. Comb filtering is a steady-state phenomenon. It has limited application to music and speech, which are highly transient phenomena. With transient sounds, the audibility of a delayed replica is more the result of successive sound events. A case might be made for proper application of combing to brief snatches of speech and music that approach steady state, but already there is an etymological impasse. The study of the audible effects of delayed reflections is better handled with the generalized threshold approach of Chap. 16.

What Is a Comb Filter?

A filter changes the shape of the frequency response or transfer function of a system. An electronic circuit used to shape the frequency response of a system to achieve a certain desired end could be a filter. A filter could also be a system of pipe and cavities used to change an acoustical system, such as is used in some microphones to adjust the pattern.

In the early days of multitrack recording, experimenters were constantly developing new, different, and distinctive sounds. *Phasing* and *flanging* were popular words among these experimenters.[1] At first multiple-head tape recorders were used to provide delayed replicas of sounds that were then mixed with the original sound to produce some unusual and eerie effects. Currently special electronic circuits are used to generate these delays. Whatever the means, these audible colorations of sound are the result of comb filters.[2]

Superposition of Sound

A sound contractor was concerned about the aiming of his horns in a certain auditorium. The simplest mechanical mounting would cause the beam of one horn to cut across the beam of the other horn. What happens in that bit of space where the two beams intersect? Would the beams tend to spread out? Would sound energy be lost from the beams as one beam interacts with the other? Relax—nothing happens.

Imagine a physics lab with a large, but shallow, ripple tank of water on the lecture table. The instructor positions three students around the tank, directing them to drop stones in the tank simultaneously. Each stone causes circular ripples to flow out from the splash point. Each set of ripples expands as though the other two ripple patterns were not there.

The principle of superposition states that every infinitesimal volume of the medium is capable of transmitting many discrete disturbances in many different directions, all simultaneously and with no detrimental effect of any one on the others. If you were able to observe and analyze the motion of a single air particle at a given instant under the influence of several disturbances, you would find that its motion is the vector sum of the various particle motions required by all the disturbances passing by. At that instant, the air particle moves with amplitude and direction of vibration to satisfy the requirements of each disturbance just as a water particle responds to several disturbances in the ripple tank.

At a given point in space, assume an air particle responds to a passing disturbance with amplitude A and 0° direction. At the same instant another disturbance requires the same amplitude A, but with a 180° direction. This air particle satisfies both disturbances at that instant by not moving at all.

A microphone is a rather passive instrument. Its diaphragm responds to whatever fluctuations in air pressure occur at its surface. If the rate of such fluctuations (frequency) falls within its operating range, it obliges with an output voltage proportional to the magnitude of the pressure involved. In Fig. 17-1, a 100-Hz tone from loudspeaker A actuates the diaphragm of a microphone in free space, and a 100-Hz voltage appears at the microphone terminals. If a second loudspeaker B lays down a second 100-Hz signal at the diaphragm of the microphone identical in pressure but 180 degrees out of phase with the first signal, one acoustically cancels the other, and the microphone voltage falls to zero. If an adjustment is made so that the two 100-Hz acoustical signals of identical amplitude are in phase, the microphone delivers twice the output voltage, an increase of 6 dB. The microphone slavishly responds to the pressures acting on its diaphragm. In short, the microphone responds to the vector sum of air pressure fluctuations impinging upon it. This characteristic of the microphone is intimately involved in acoustical comb-filter effects.

Tonal Signals and Comb Filters

A 500-Hz tone is shown as a line on a frequency scale in Fig. 17-2A. All of the energy concentrated in this pure tone is located at this frequency. Figure 17-2B shows an identical signal except it is delayed by 0.5 ms in respect to the signal of A. The signal has the same frequency and amplitude, but the timing is different. Consider both A and B as acoustical signals combining at the diaphragm of a microphone. Signal A could be a direct signal and B a reflection of A off a nearby sidewall. What is the nature of the combined signal the microphone puts out?

Because signals A and B are pure tones, simple sine waves, both vary from a positive peak to a negative peak 500 times per second. Because of the 0.5 ms delay, these two tonal signals will not reach their positive or negative peaks at the same instant. Often along the time axis both are positive, or both are negative, and at times one is positive while the other is negative. When the sine wave of sound pressure representing signal A and the sine wave of sound pressure representing signal B combine (with due respect to positive and negative

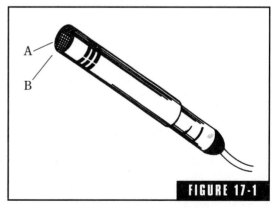

FIGURE 17-1

The microphone diaphragm responds to the vector sum of sound pressures from multiple sources.

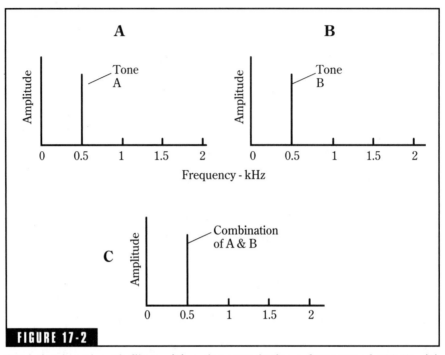

FIGURE 17-2

Tonal signals and comb filters; (A) a sine wave having a frequency of 500 Hz, (B) another sine wave of 500 Hz that is delayed 0.5 ms from A, and C is a summation of A and B. The 500-Hz signal and its delayed counterpart reach their peaks at slightly different times, but adding them together simply yields another sine wave; there is no comb filtering (see Figs. 1-11, 1-13).

signs) they produce another sine wave of the same frequency, but of different amplitude.

Figure 17-2 treats the two 500-Hz tones as lines on a frequency scale. Figure 17-3 treats the same 500-Hz direct tone and the delayed tone in the more familiar sine-wave form. The delay is accomplished by feeding the 500-Hz tone into a digital delay device and combining the original and the delayed tones in a common three-resistor summing circuit.

In Fig. 17-3A the direct 500-Hz tone is shown originating at zero time. It takes 2 ms for one cycle of a 500-Hz tone ($1/500 = 0.002$ sec). One cycle is also equivalent to 360 degrees. The 500-Hz signal, e, is plotted according to the time and degree scales at the bottom of the figure.

A delay of 0.1 ms is equivalent to 18 degrees; a delay of 0.5 ms is equivalent to 90 degrees; a delay of 1 ms is equivalent to 180 degrees. The effect of these three delays on the tonal signals is shown in Fig. 17-3B (later the same delays will be compared with music and speech signals). The combination of e and e_1 reaches a peak of approximately twice that of e (+6 dB). A shift of 18 degrees is a very small shift, and e and e_1 are practically in phase. The curve $e + e_2$, at 90 degree phase difference has a lower amplitude, but still a sine form. Adding e to e_3 (delay 1 ms, shift of 180 degrees) gives zero as adding two waves of identical amplitude and frequency but with a phase shift of 180 degrees results in cancellation of one by the other.

Adding direct and delayed sine waves of the same frequency results in other sine waves of the same frequency. Adding direct and delayed sine waves of different frequencies gives periodic waves of irregular wave shape. *Conclusion:* Adding direct and delayed periodic waves does not create comb filters. Comb filters require signals having distributed energy such as speech, music, and pink noise.

Combing of Music and Speech Signals

The spectrum of Fig. 17-4A can be considered an instantaneous slice of music, speech, or any other signal having a distributed spectrum. Figure 17-4B is essentially the same spectrum but delayed 0.1 ms from Fig. 17-4A. Figure 17-4C is the acoustical combination of the A and B sound pressure spectra at the diaphragm

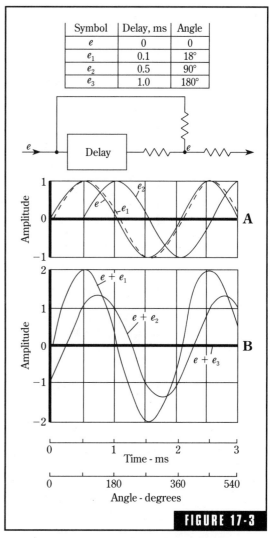

Symbol	Delay, ms	Angle
e	0	0
e_1	0.1	18°
e_2	0.5	90°
e_3	1.0	180°

FIGURE 17-3

An exercise to demonstrate that combining sine waves yields not comb filters, but simply other sine waves. A distributed spectrum is required for the formation of comb filters. 500 Hz sine waves are displayed with delays of 0.1, 0.5, and 1.0 ms to conform to the distributed spectrum cases in Fig. 17-5.

of the microphone. The resulting overall response of Fig. 17-4C appears like a sine wave, but combining spectra is different from combining tonal signals. This sine-wave appearance is natural and is actually a sine-wave shape with the negative loops made positive.

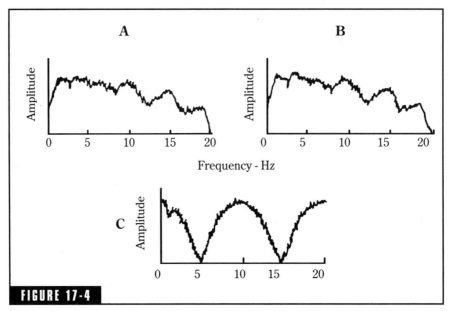

FIGURE 17-4

Combing of signals having distributed spectra; (A) instantaneous spectrum of music signal. (B) A replica of A, which is delayed 0.1 ms from A. (C) A summation of A and C showing typical comb filtering.

Combing of Direct and Reflected Sound

The 0.1 ms delay in Fig. 17-4 could have been from a digital-delay device, or it could have been a reflection from a wall or other object. The spectral shape of a signal will be changed somewhat upon reflection, depending on the angle of incidence, the acoustical characteristics of the surface, etc.

A reflection delayed 0.1 ms will have traveled (1,130 ft/sec) (0.001 sec) = 1.13 ft further than the direct signal. This difference in path length, only about 1¹¹⁄₃₂ inch, could result from a grazing angle with both source and listener, or microphone, close to the reflecting surface. Greater delays are expected in more normal situations such as those of Fig. 17-5. The spectrum of Fig. 17-5A is from a noise generator. This is a "shhh" sound similar to the between-station noise of an FM radio receiver. Random noise of this type is used widely in acoustic measurements because it is a continuous signal, its energy is distributed throughout the audible frequency range, and it is closer to speech and music signals than sine or other periodic waves. In Fig. 17-5B, this random noise signal drives a loudspeaker, which faces a reflective surface; a

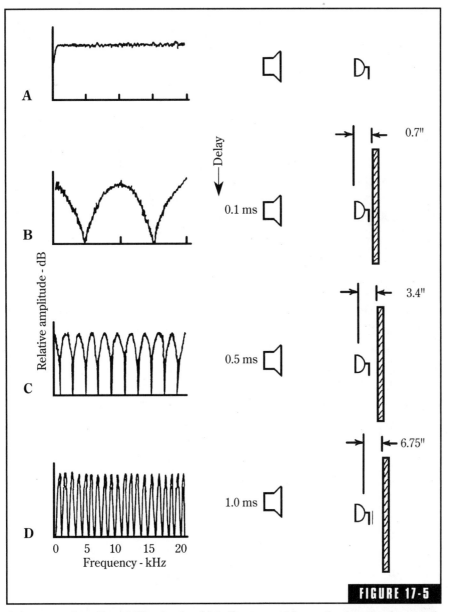

A demonstration of comb filtering in which direct sound from a loudspeaker is acoustically combined with a reflection from a surface at the diaphragm of a microphone. (A) No surface, no reflection. (B) Placing the microphone 0.7 in from the surface creates a delay of 0.1 ms and the combination of the direct and the reflected rays shows cancellations at 5 and 15 kHz and every 10 kHz. (C) A delay of 0.5 ms creates cancellations much closer together. (D) A delay of 1 ms results in cancellations even more closely together. If t is taken as the delay in seconds, the first null is $1/2t$ and spacing between nulls or between peaks is $1/t$.

nondirectional microphone is then placed at varying distances from the reflective surface.

In Fig. 17-5B, the microphone diaphragm is placed about 0.7 inches from the reflective surface. Interference takes place between the direct sound the microphone picks up from the loudspeaker and the sound reflected from the surface. The output of the microphone shows the comb-filter pattern characteristic of a 0.1 ms delay.

Placing the microphone diaphragm about 3.4 inches from the reflective barrier, as in Fig. 17-5C, yields a 0.5 ms delay, which results in the comb-filter pattern shown. Plotted on a linear frequency scale, the pattern looks like a comb; hence, the name *combfilter*. Increasing the delay from 0.1 to 0.5 ms has increased the number of peaks and the number of nulls five-fold.

In Fig. 17-5D, the microphone is 6.75 inches from the reflective barrier giving a delay of 1.0 ms. Doubling the delay has doubled the number of peaks/nulls once again.

Increasing the delay between the direct and reflected components increases the number of constructive and destructive interference events proportionally. Starting with the flat spectrum of Fig. 17-5A, the far-from-flat spectrum of B is distorted by the presence of a reflection delayed only 0.1 ms. An audible response change would be expected. One might suspect that the distorted spectrum of D might be less noticeable because the multiple, closely spaced peaks and narrow notches tend to average out the overall response aberrations.

Reflections following closely after the arrival of the direct component are expected in small rooms because the dimensions of the room are limited. Conversely, reflections in large spaces would have greater delays, which generate more closely spaced comb-filter peaks and nulls. Thus, comb-filter effects resulting from reflections are more commonly associated with small room acoustics. The size of various music halls and auditoriums renders them relatively immune to audible comb-filter distortions, because the peaks and nulls are so numerous and packed so closely together that they merge into an essentially uniform response.

Figure 17-6 illustrates the effect of straining a music signal through a 2 ms comb filter. The relationship between the nulls and peaks of response is related to the piano keyboard as indicated. Middle C, (C4), has a frequency of 261.63 Hz, and is close to the first null of 250 Hz. The next higher C, (C5), has a frequency close to twice that of C4 and is treated favorably with a +6-dB peak. Other Cs up the keyboard will be either discriminated against with a null, or especially favored with a peak in response—or something in between. Whether viewed as fundamental frequencies or a series of harmonics, the timbre of the sound suffers.

The comb filters illustrated in Figs. 17-4, 17-5, and 17-6 are plotted to a linear frequency scale. In this form the *comb* appearance and visualization of the delayed effects are most graphic. A logarithmic-frequency scale, however, is more common in the electronics and audio industry. A comb filter resulting from a delay of 1 ms plotted to a logarithmic frequency scale is shown in Fig. 17-7.

Comb Filters and Critical Bands

Is the human auditory system capable of resolving the perturbations of Fig. 17-5D? The resolution of human hearing is circumscribed by the

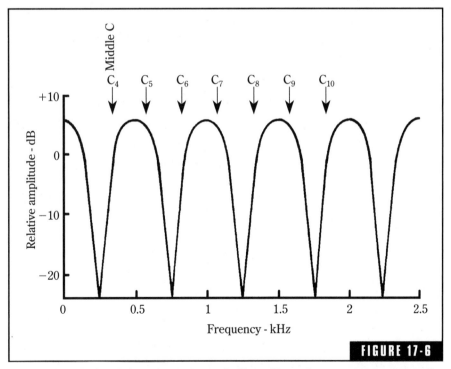

FIGURE 17-6

Passing a music signal through a 2 ms comb filter affects the components of that signal as indicated. Components spaced one octave can be boosted 6 dB at a peak or essentially eliminated at a null, or can be given values between these extremes.

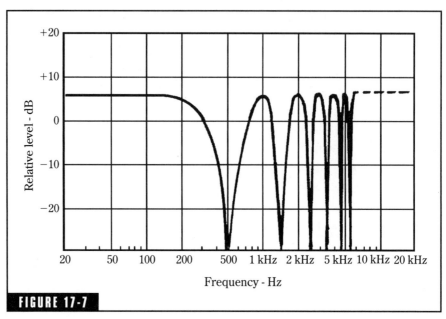

FIGURE 17-7

Up to this point all comb filters have been plotted to a linear scale to demonstrate the origin of the term *comb*. Plotting to the more convenient and familiar logarithmic scale aids in estimating the effects of a given comb on a given signal.

critical band tuning curves of the inner ear. The critical bandwidths at representative frequencies are recorded in Table 17-1. For example, the average critical bandwidth of the human auditory system at 1,000 Hz is about 128 Hz. A peak-to-peak comb-filter frequency of 125 Hz corresponds to a reflection delay of about 8 ms ($\frac{1}{0.008} = 125$ Hz), which corresponds to a difference in path length between the direct and reflected components of about 9 ft (1,130 ft/sec × 0.008 sec = 9.0 ft). This situation is plotted in Fig. 17-8B. To illustrate what happens for greater delays, Fig. 17-8C is sketched for a delay of 40 ms. Shorter delays are represented by Fig. 17-8A for a delay of 0.5 ms.

The relative coarseness of the critical band cannot analyze and delineate the

Table 17-1 Auditory critical bands.

Center frequency (Hz)	Width of critical band* (Hz)
100	38
200	47
500	77
1,000	128
2,000	240
5,000	650

*Calculated equivalent rectangular band as proposed by Moore and Glasberg.[3]

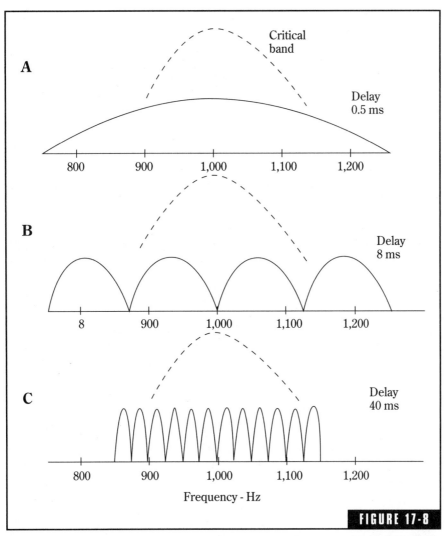

FIGURE 17-8

The audibility of combing is an important but not a well-understood factor. To assist in estimating the perceptual importance of comb filters, they are compared to the auditory critical band effective at a frequency of 1,000 Hz. (C) At a delay of 40 ms the width of the critical band is so coarse, relatively, that no analyzing of the comb filter is possible. (A) On the other hand the width of the auditory critical band is comparable to the comb peak at 0.5 ms delay. (B) is an example in between A and C. This would seem to confirm the observation that in large spaces (long delays) comb filters are inaudible, while they often are very troublesome in small spaces (short delays).

numerous peaks and nulls resulting from a 40-ms delay (Fig. 17-8C). Therefore, the human ear would not be expected to interpret response aberrations resulting from 40-ms combing as a coloration of the signal. On the other hand, the combing resulting from the 0.5 ms delay (Fig. 17-8A) could be delineated by the ear's critical band at 1,000 Hz resulting in a perceived coloration of the signal. Figure 17-8B illustrates an intermediate example in which the ear is marginally able to analyze the combed signal. The width of the critical bands of the auditory system increases rapidly with frequency. It is difficult to imagine the complexity of the interaction between a set of critical bands and a constantly changing music signal, with diverse combing patterns from a host of reflections. Only carefully controlled psychoacoustical experiments can determine whether the resulting colorations are audible (Chap. 16).

Comb Filters in Stereo Listening

In the standard stereo listening arrangement, the input signals to each ear come from two loudspeakers. These signals are displaced in time with respect to each other because of the loudspeaker spacing; the result is the generation of comb-filters. Blauert indicated that comb-filter distortion is not generally audible.[4] The auditory system has the ability of disregarding these distortions as the perception of timbre is formed. This is called *binaural suppression of colorations of timbre*; however, no generally accepted theory exists to explain how the auditory system accomplishes this.[4] Distortion can be heard by plugging one ear; however, this destroys the stereo effect. Comparing the timbre of signals from two loudspeakers producing comb-filter distortion and one loudspeaker (that does not) will demonstrate that stereo comb-filter distortion is barely audible. The timbre of the two is essentially the same. Furthermore, the timbre of the stereo signal changes little as the head is turned.

Coloration and Spaciousness

A reflected wave reaching the ear of a listener is always somewhat different from its direct wave. The characteristics of the reflecting wall vary with frequency. By traveling through the air, both the direct and reflected components of a sound wave are altered slightly, due to the air's absorption of sound, which varies with frequency. The amplitudes and timing of the direct and reflected components differ. The human auditory system responds to the frontal, direct component somewhat differently than to the reflection from the side. The perception of the reflected component is always different than the direct component. The amplitudes and timing will be related, but with an interaural correlation less than maximum.

Weakly correlated input signals to the ears contribute to the impression of spaciousness. If no reflections occur, such as when listening outdoors, there is no feeling of spaciousness (see Fig. 16-4). If the input signals to the ears are correct, the perception of the listener is that of being completely enveloped and immersed in the sound. The lack of strong correlation is a prerequisite for the impression of spaciousness.

Combing in Stereo Microphone Pickups

Because two microphones separated in space pick up a sound at slightly different times, their combined output will be similar to the single microphone with delayed reflections. Therefore, spaced microphone stereo-pickup arrangements are susceptible to comb-filter problems. Under certain conditions the combing is audible, imparting a *phasiness* to the overall sound reproduction, interpreted by some as room ambience. It is not ambience, however, but distortion of the time and intensity cues presented to the microphones. It is evident that some people find this distortion pleasing, so spaced microphone pickups are favored by many producers and listeners.

Audibility of Comb-Filter Effects

Chapter 16 showed that spaciousness is the result of reflections combining with the direct signal. This chapter demonstrated that combining a signal with a close replica of itself delayed a small amount creates comb filters. The audibility of comb filters is thus clearly stated in the Olive-Toole thresholds of Fig. 16-4. Only through psychoacoustical measurements of this type can the audibility of comb-filters be determined.

Comb Filters in Practice

Example 1: Figure 17-9 illustrates three microphone placements that produce comb filters of varying degree. A close source-to-microphone distance is shown in Fig. 17-9A. The direct component travels 1 ft and the floor-reflected component travels 10.1 ft (see Table 17-2). The difference between these (9.1 ft) means that the floor reflection is delayed 8.05 ms (9.1/1130 = 0.00805 sec.). The first null is therefore at 62 Hz with subsequent null and peak spacing of 124 Hz. The level of the

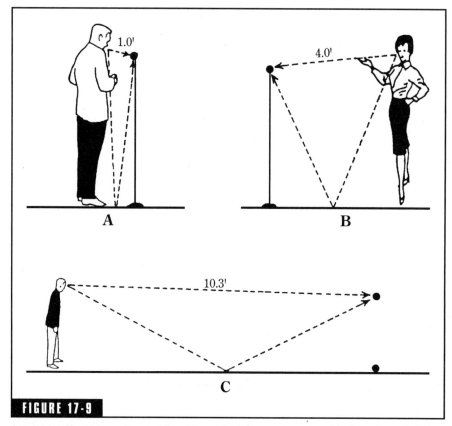

Common microphone placements compared with respect to production of comb filters (see Table 17-1). (A) Reflection 20 dB down, minimum comb-filter problems. (B) Reflection only 8 dB down, comb-filter problems expected. (C) Reflection almost same level as direct, comb-filter problems certain. A microphone on the floor of (C) would reduce the difference in path length between direct and reflected components (and the combing) to almost zero.

Table 17-2 Comb-filter situations (Refer to Fig. 17-9).

Fig 8-13	Path length, ft.		Difference		First null 1/2t	Pk/null spacing 1/t	Refl. level
	Direct	Refl.	Ft.	(t) ms.	Hz	Hz	dB
A	1.0	10.1	9.1	8.05	62	124	−20
B	4.0	10.0	6.0	5.31	94	189	−8
C	10.3	11.5	1.2	1.06	471	942	−1

reflection is -20 dB referred to the direct component (20 log 1.0/10.1 = 20 dB).

Similar calculations for Fig. 17-9B and C are included in Table 17-2. In A the direct component is 10 times stronger than the floor reflection. The effect of the comb filter would be negligible. Figure 17-9C has a reflection almost as strong as the direct, and the comb-filter effect would be maximum. Figure 17-9B is intermediate between A and C. A microphone is shown on the floor in Fig. 17-9C. A floor bounce would occur, but the difference between the direct and reflected path length would be very small, essentially eliminating the comb filter.

Example 2: Two microphones on a podium, Fig. 17-10, are very common. Are they used as stereo microphones? Stereo reproduction systems are quite rare in auditoriums. The chances are very good that the two microphones are fed into the same mono system and thus become an excellent producer of comb-filter effects. The common excuse for two microphones is "to give the speaker greater freedom of movement" or "to provide a spare microphone in case of failure of one." Assuming the microphones are properly polarized and the talker is dead center, there would

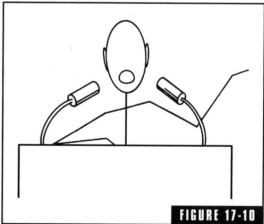

FIGURE 17-10

An infamous example of comb-filter production, two microphones feeding into the same mono amplifier with a sound source that moves about.

be a helpful 6-dB boost in level. Assume also that the microphones are 24 in apart and the talker's lips are 18 in from a line drawn through the two microphones and on a level with the microphones. If the talker moves laterally 3 in, a 0.2 ms delay is introduced, reducing important speech frequencies. If the talker does not move, the speech quality would probably not be good, but it would be stable. Normal talker movements shift nulls and peaks up and down the frequency scale with quite noticeable shifts in quality.

Example 3: A common situation with comb-filter possibilities is the singing group with each singer holding a microphone (Fig. 17-11), and each microphone fed to a separate channel but ultimately mixed together. The voice of A, picked up by both microphones, is mixed, producing comb filters resulting from the path difference. Each singer's voice is picked up by all microphones but only adjacent singers create noticeable comb filters. Experiments reported by Burroughs[5] indicate that if singer A's mouth is at least three times farther from singer B's microphone than from A's own microphone, the

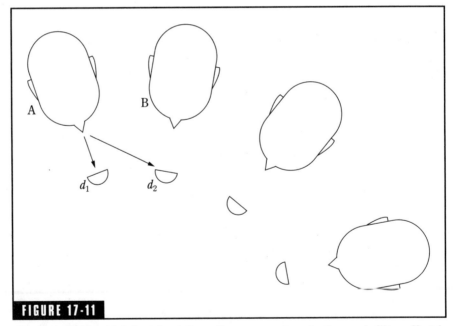

For group singing, if d_2 is at least three times as great as d_1, the comb-filter effect is minimized.

comb-filter effects are overshadowed by other problems. This 3:1 rule works because maintaining this distance means that delayed replicas are at least 9 dB below the main signal. This assures that comb-filter peaks and nulls are $+/-$ 1 dB or less in amplitude and thus essentially imperceptible.

Example 4: Dual mono loudspeakers, one on stage left and the other on stage right, or variations of this theme, are quite common (Fig. 17-12). Two sources radiating identical signals create comb filters over the audience area. On the line of symmetry (often down the center aisle) both signals arrive at the same time and no comb filters are produced. Equi-delay contours range out from stage center over the audience area, the 1-ms delay contour nearest the center line of symmetry, and greater delays as the sides of the auditorium are approached.

Example 5: Multi-element loudspeakers can have their own comb-filter sources. In Fig. 17-13 it is apparent that frequency f_1 is radiated by both bass and mid-range units, that both are essentially equal in magnitude, and that the two radiators are physically displaced. These are the ingredients for comb-filter production in the audience area.

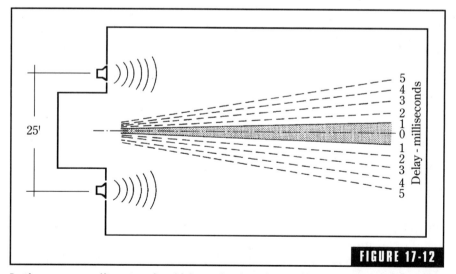

FIGURE 17-12

In the common split system in which two loudspeakers radiate identical signals, zones of constructive and destructive interference result which degrade sound quality in the audience area.

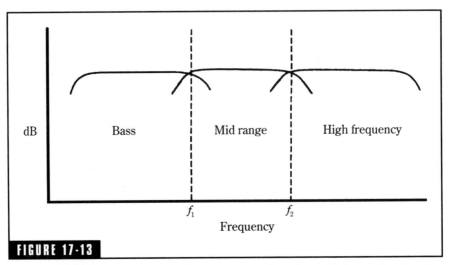

FIGURE 17-13

Comb-filter distortion can occur in the crossover region of a multielement loudspeaker because the same signal is radiated from two physically separated units.

The same process is at work between the mid-range and the tweeter units. Only a narrow band of frequencies is affected, the width of which is determined by the relative amplitudes of the two radiations. The steeper the crossover curves, the narrower the frequency range affected.

Example 6: Permanently mounted microphones may be flush-mounted with the advantage of an approximate +6-dB gain in sensitivity due to the pressure rise at the table surface. Another advantage is minimizing comb-filter distortions. In Fig. 17-14, a direct ray from the source activates the microphone diaphragm, which is shielded from reflections.

Estimating Comb-Filter Response

Remembering a few simple relationships enables you to estimate the effect of comb filters on the response of a system. If the delay is t seconds, the spacing between peaks and the spacing between nulls is $1/t$ Hz. For example, a delay of 0.001 second (1 ms) spaces the peaks $1/0.001 = 1,000$ Hz, and the nulls will also be spaced the same amount (Table 17-3).

The frequency at which the first null (i.e., the null of lowest frequency) will occur is $1/(2t)$ Hz. For the same delay of 0.001 s, or 1 ms,

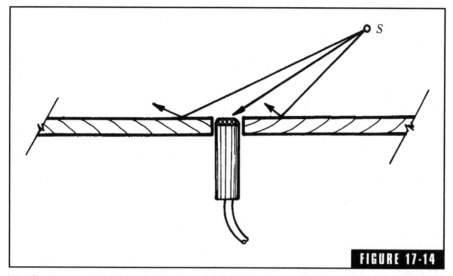

FIGURE 17-14

The flush-mounted microphone. Sounds from the source *S* that strike the surface do not reach the microphone, thus avoiding comb-filter effects. Another advantage of this mounting is an increase of level due to the pressure buildup near the reflecting surface.

Table 17-3 Comb-filter peaks and nulls.

Frequency of Delay (ms)	Spacing between nulls lowest null (Hz)	Spacing between peaks (Hz)
0.1	5,000	10,000
0.5	1,000	2,000
1.	500	1,000
5.	100	200
10.	50	100
50.	10	20

the first null will occur at $1/[2 \times 0.001] = 500$ Hz. For this 1-ms delay, you can almost figure out the system response in your head; the first null is at 500 Hz, nulls are spaced 1,000 Hz, and peaks are spaced 1,000 Hz apart. Of course, there is a peak between each adjacent pair of nulls at which the two signals are in phase. Adding two sine waves with the same frequency, the same amplitude, in phase, doubles the amplitude, yielding a peak 6 dB higher than either component by itself

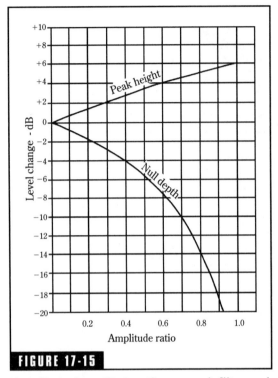

FIGURE 17-15

The effect of amplitude ratios on comb-filter peak height and null depth.

(20 log 2 = 6.02 dB). The nulls, of course, will be at a theoretical minimum of minus infinity as they cancel at phase opposition. In this way, the entire response curve can be sketched as the phase of the two waves alternates between the in-phase and the phase-opposition condition down through the spectrum.

An important point to observe is that the $1/(2t)$ expression above gives a null at 500 Hz, which robs energy from any distributed signal subject to that delay. A music or speech signal passing through a system having a 1-ms delay will have important components removed or reduced. This is nothing short of signal distortion, hence the common phrase *comb-filter distortion*.

If the mathematics of the $1/t$ and the $1/(2t)$ functions seem too laborious, Figs. 17-15 and 17-16 are included as graphical solutions.

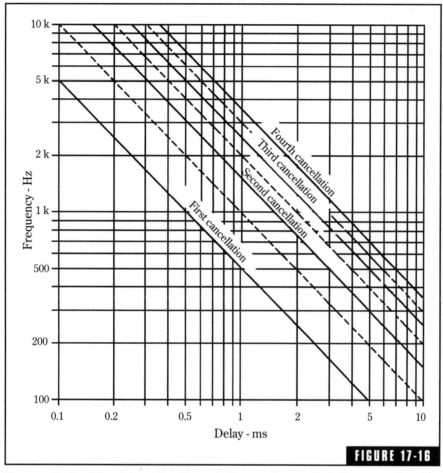

FIGURE 17-16

The magnitude of the delay determines the frequencies at which destructive interference (cancellations) and constructive interference (peaks) occur. The broken lines indicate the peaks between adjacent cancellations.

Endnotes

[1]Bartlett, Bruce, *A Scientific Explanation of Phasing (Flanging),* J. Audio Eng. Soc., 18, 6 (1970) 674–675.

[2]*The New Stereo Soundbook* by Ron Streicher and F. Alton Everest, Audio Engineering Associates, 1029 N. Allen Ave., Pasadena, CA 91104.

[3]Moore, Brian C.J. and Brian Glasberg, *Suggested Formulae for Calculating Auditory-Filter Bandwidths,* J. Acous. Soc. Am., 74, 3 (Sept 1983) 750–753.

[4]Blauert, Jens, *Spatial Hearing,* (1983), Cambridge, MA, MIT Press, 325–326.

[5]Burroughs, Lou, *Microphones: Design and Application,* Plainview, NY, Sagamore Publishing Co., (1974), Chapters 10 and 11.

Quiet Air for the Studio

The background noise levels in recording studios, control rooms, and listening rooms must be kept under control if these rooms are to be of maximum use in their intended way. Hums, buzzes, rumbles, aircraft noises, tooting auto horns, dogs barking, or typewriter sounds are most incongruous if audible during a lull in a program. Such sounds might not be noticed outside the studio when they are a natural part of the situation, but during a pause or a quiet musical or speech passage they stand out like the proverbial sore pollex.

In a studio, interfering sounds can come from control-room monitors operating at high level or from equipment in adjacent areas. Control rooms have their own noise problems, some intruding from the outside, some generated by recorders, equipment cooling fans, etc. There is one source of noise, however, that is common to all sound-sensitive rooms, and that is the noise coming from the air-conditioning diffusors or grilles, the subject of this chapter. A certain feeling of helplessness in approaching air-conditioning noise problems is widespread and quite understandable.

The control of air-conditioning noise can be expensive. A noise specification in an air-conditioning contract for a new structure can escalate the price. Alterations of an existing air-conditioning system to correct high noise levels can be even more expensive. It is important for studio designers to have a basic understanding of potential noise problems in air-conditioning systems so that adequate control and supervision can be exercised during planning stages and installation. This applies equally to the most ambitious studio and to the budget job.

Selection of Noise Criterion

The single most important decision having to do with background noise is the selection of a noise-level goal. The almost universal approach to this is embodied in the family of Balanced Noise Criterion (NCB) curves of Fig. 18-1.[1] The selection of one of these contours establishes the goal of maximum allowable noise-pressure level in each octave band. Putting the noise goal in this form makes it easily checkable by instruments. The downward slope of these contours reflects both the lower sensitivity of the human ear at low frequencies and the fact that most noises with distributed energy drop off with frequency. To determine whether the noise in a given room meets the contour goal selected, sound-pressure level readings are made in each octave and plotted on the graph of Fig. 18-1. The black dots represent such a set of measurements made with a sound-level meter equipped with octave filters. A convenient single number NCB-20 applies to this particular noise. If the NCB-20 contour had been specified as the highest permissible sound-pressure level in an air-conditioning contract, the above installation would just barely be acceptable.

Which contour should be selected as the allowable limit for background noise in a recording studio? This depends on the general studio quality level to be maintained, on the use of the studio, and other factors. There is little point in demanding NCB-15 from the air-conditioning system when intrusion of traffic and other noise is higher than this. In general, NCB-20 should be the highest contour that should be considered for a recording studio or listening room, and NCB-15 is suggested as a practical and attainable design goal for the average studio. NCB-10 would be excellent and it would probably take special effort and expense to reduce all noise to this level.

The advent of digital recording has changed our view of which NCB contour to select as a goal. Figure 5-12 showed that signal-to-noise ratios of more than 100 dB must be accommodated, for the top-flight studios at least. This means a lower noise floor. A lower noise floor means tighter construction practices and HVAC (heating, ventilating, and air-conditioning) contract noise specifications that substantially increase the cost.

If you want to know just how an NCB-15 or NCB-20 background noise really sounds, the following procedure is suggested. Beg, borrow, buy, or rent a sound-level meter with built-in octave filters. Measure

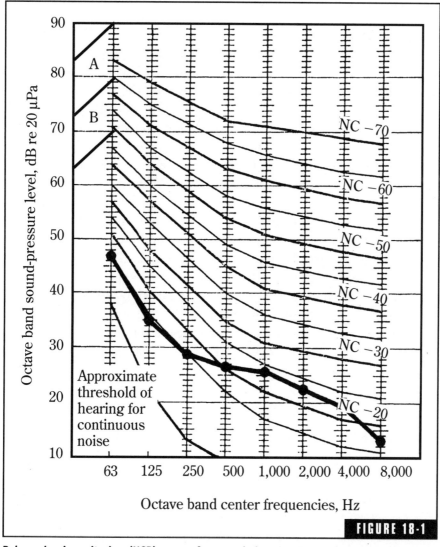

FIGURE 18-1

Balanced noise-criterion (NCB) curves for occupied rooms. Octave-band sound-pressure levels of the magnitudes indicated in the regions A and B may induce audible rattles or feelable vibrations in lightweight partitions and ceiling constructions. *After Beranek.*[1]

the sound levels in several studios that you suspect have high noise or studios you consider quite acceptable. By the time you have measured four or five such rooms with the HVAC turned on and off, you will have little NCB numbers in your head for ready recall in future discussions and you, too, will have become an expert.

Fan Noise

The fan is a chief contributor to HVAC noise in the studio, but it is by no means the only contributor. The sound power output of the fan is largely fixed by the air volume and pressure required in the installation, but there are certainly variations between the types of fans. Figure 18-2 gives the specific sound power output of just two types of fans: the airfoil centrifugal and the pressure blower. *Specific* sound power level means that the measurements have been reduced to the standard conditions of 1 cu ft per minute and a pressure of 1 in of water. On this basis, noise of various types of fans can be compared equitably.

The centrifugal fan is one of the quietest fans available. The amazing thing is that large fans are quieter than small fans. This is also true of the pressure blower fan as shown in Fig. 18-2.

Aside from the term *specific*, sound power in general is foreign to the audio field. All the acoustical power radiated by a piece of machinery must flow out through a hemisphere. The manufacturer is the source of noise data on any particular fan. Sound power is usually evaluated by sound-pressure level readings over the hemisphere. Sound power is proportional to sound pressure squared. Noise ratings of fans in terms of sound power can be converted back to sound-pressure levels applicable to a given room by means of the following formula:

Sound-pressure level = sound-power level $- 5 \log V - 3 \log f - 10 \log r + 25$ dB

where

V = room volume, cu ft
f = octave-band center frequency, Hz
r = distance source to reference point, ft

There is usually a tone generated by the fan with a frequency given by (rev/sec) (number of blades). This tone adds to the level of the octave band in which it falls. Three dB of the centrifugal fans and eight dB for the pressure fans should be added to the one octave band level to account for the contribution of the fan tone.

ASHRAE

It might sound like an Egyptian goddess, but our new word for today is *ASHRAE* (as in ashtray). It stands for the American Society of Heating, Refrigerating, and Air-Conditioning Engineers. Although its primary purpose is to enlighten and standardize its own engineers, this highly respected organization is a prolific source of help for the studio designer who is a novice in HVAC. In this brief chapter it is impractical to go deeply into specific design technologies, but the ASHRAE Handbooks can be faithful guides to those faced with such problems.[2] Chapter 7 of this handbook entitled *Sound and Vibration Fundamentals* (1985) introduces fundamental principles, including the source/path/receiver concept; basic definitions and terminology; and acoustical design goals.

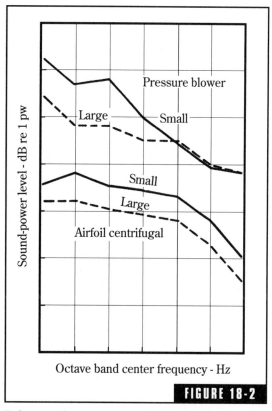

FIGURE 18-2

Noise sound-power output of different types of fans commonly used for air-conditioning. (Data are from Table 22-2.[2])

Chapter 32 (1984) covers *Sound and Vibration Control.* The objective of this chapter is to achieve a noise level appropriate for the functions of the space—not the lowest possible level. Overdesign is as unforgivable as underdesign. The material in chapter 32 includes: noise control for outdoor equipment installations; system noise control for indoor air-conditioning systems; general design considerations for good noise control; mechanical-equipment-room noise isolation; vibration isolation and control; and troubleshooting for noise and vibration problems.

These two chapters are excellent sources for the designer or manager of any kind of audio room. The *Handbook* is massive and expensive, but it is available in local libraries leaning toward service to industries. For those with some engineering training and a modicum of determination, these chapters in the Handbook can provide a background for intelligent dealing with the highest caliber of air-conditioning contractors. For the lower caliber type, this background is indispensable for avoiding big and expensive mistakes.

Machinery Noise

The first step in the reduction of HVAC noise in the studio is wisely locating the HVAC machinery. If this is left to chance, Murphy's Law will result in the equipment room being adjacent to or on the roof directly above the studio. The wall or roof panel, vibrating like a giant diaphragm, is remarkably efficient in radiating airborne noise into the studio. So, step number one is to locate the equipment as far removed from the sound-sensitive areas as possible.

The next step is to consider some form of isolation against structure-borne vibration. If the equipment is to rest on a concrete slab shared with the studio and plans are being drawn up, the machinery-room slab should be isolated from the main-floor slab. Compressed and treated glass fiber strips are available that are suitable for separating slabs during pouring. Other precautions would include proper vibration isolation mounts, designed accurately or they will be useless or downright damaging to the situation. Flexible joints in pipes and ducts where they leave the machinery room might be advisable.

Air Velocity

In air distribution systems, the velocity of air flow is a very important factor in keeping HVAC noise at a satisfactorily low level. Noise generated by air flow varies approximately as the 6th power of the velocity. As air velocity is doubled, the sound level will increase about 16 dB at the room outlet. Some authorities say that air-flow noise varies as the 8th power of the velocity and give 20 dB as the figure associated with doubling or halving the air velocity.

A basic design parameter is the quantity of air the system is to deliver. There is a direct relationship between the quantity of air, air velocity, and size of duct. The velocity of the air depends on the cross-sectional area of the duct. For example, if a system delivers 500 cu ft/minute and a duct has 1 sq ft of cross-sectional area, the velocity is 500 ft/min. If the area is 2 sq ft, the velocity is reduced to 250 ft/min; if 0.5 sq ft, velocity is increased to 1,000 ft/min. An air velocity maximum of 500 ft/min is suggested for broadcast studios, and this value is also about right for top-flight recording studios and other critical spaces. Specifying a low velocity eliminates many headaches later.

High-pressure, high-velocity, small-duct systems are generally less expensive than low-velocity systems. True budget systems commonly ensure noise problems in studios because of high air velocity and its resulting high noise. Compromise can be made by flaring out the ducts just upstream from the grille. The increasing cross-sectional area of the flare results in air velocity at the grille being considerably lower than in the duct feeding it.

Effect of Terminal Fittings

Even if fan and machinery noise are sufficiently attenuated, by the time the air reaches the sound-sensitive room, air turbulence associated with nearby 90° bends, dampers, grilles, and diffusors can be serious noise producers as suggested by Fig. 18-3.

"Natural" Attenuation

The designer must be careful to avoid expensive overdesign of an air distribution system by neglecting certain attenuation effects built into the system. When a plane wave sound passes from a small space, such as a duct, into a larger space, such as a room, some of the sound is reflected back toward the source. The effect is greatest for low-frequency sound. Recent research has also indicated that the effect is significant only when a straight section of ductwork, three to five diameters long, precedes the

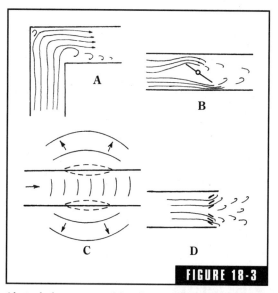

FIGURE 18-3

Air turbulence caused by discontinuities in the flow path can be a serious producer of noise. (A) 90-degree miter bend. (B) Damper used to control quantity of air. (C) Sound radiated from duct walls set into vibration by turbulence or noise inside the duct. (D) Grilles and diffusors.

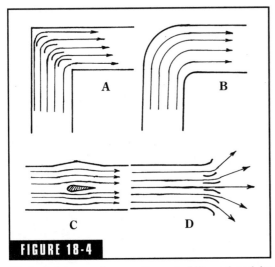

FIGURE 18-4

Air turbulence noise can be materially reduced by (A) use of deflectors, (B) radius bends, (C) airfoils, and (D) carefully shaped grilles and diffusors.

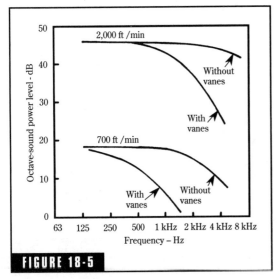

FIGURE 18-5

Noise produced by 12" x 12" square cross ssection 90-degree-miter elbow with and without deflection vanes and at 2,000 and 700 fpm air velocities. These curves have been calculated following the procedures in Ref.2.

duct termination. Any terminal device, such as a diffusor or grille, tends to nullify this attenuation effect. A 10-in duct dumping air into a room without a grille can give rise to a 15-dB reflection loss in the 63-Hz octave. This is about the same attenuation that a 50- to 75-ft run of lined ductwork would give. Figure 18-4 shows several methods of attenuation.

A similar loss occurs at every branch or takeoff. There is also an attenuation of noise in bare, rectangular sheet ducts due to wall flexure amounting to 0.1 to 0.2 dB per foot at low frequencies. Round elbows introduce an attenuation, especially at the higher frequencies. Right-angle elbows also introduce attenuation (see Fig. 18-5). All of these losses are built into the air-handling system and serve to attenuate fan other noise coming down the duct. It's there, it's free, so take it into consideration to avoid overdesign.

Duct Lining

The application of sound-absorbing materials to the inside surfaces of ducts is a standard method of reducing noise levels. Such lining comes in the form of rigid boards and blankets and in thicknesses of ½-inch to 2-inches. Such acoustical lining also serves as thermal insulation when it is required. The approximate attenuation offered by 1-in duct lining in typical rectangular ducts depends on the duct size as shown in Fig. 18-6. The approximate attenuation of round ducts is given in Fig. 18-7. Duct attenuation is much lower in the round ducts than in lined rectangular ducts comparable cross-sectional areas.

Plenum Silencers

A sound-absorbing plenum is an economical device for achieving significant attenuation. Figure 18-8 shows a modest-sized plenum chamber, which if lined with 2-in thickness of 3 lb/cu ft density glass fiber, will yield a maximum of about 21-dB attenuation. The attenuation characteristics of this plenum are shown in Fig. 18-9 for two thicknesses of lining. With a lining of 4-in of fiberboard of the same density, quite uniform absorption is obtained across the audible band. With 2 inches of glass fiberboard, attenuation falls off below 500 Hz. It is apparent that the attenuation performance of a plenum of given size is determined primarily by the lining.

Figure 18-10 gives actual measurements on a practical, lined-plenum muffler approximately the same horizontal dimensions as that of Fig. 18-8, but only half the height and with baffles inside. Attenuation of 20 dB or more above the 250-Hz octave was realized in this case, and it solved an otherwise intolerable problem.

Plenum performance can be increased by increasing the ratio of the cross-sectional area of the plenum to the cross-sectional area of the entrance and exit ducts, and by increasing the amount or thickness of absorbent lining. A plenum located at the fan discharge can be an effective and economical way to decrease noise entering the duct system.

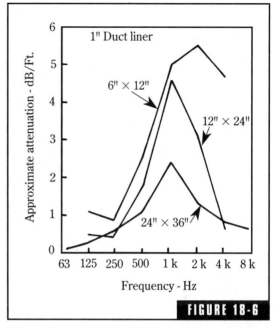

FIGURE 18-6

Attenuation offered by 1″ duct lining on all four sides of rectangular duct. Dimensions shown are the free area inside the duct with no air flows. This is a plot of the data from Ref.2.

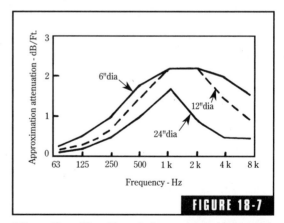

FIGURE 18-7

Attenuation offered by spiral wound round ducts with perforated spiral wound steel liner. Dimensions shown are the free area inside the duct, with no air flow. This is a plot of the data from Ref.2.

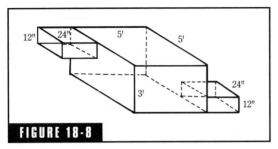

FIGURE 18-8

Dimensions of a plenum that will yield about 21 db attentution throughout much of the audible spectrum.

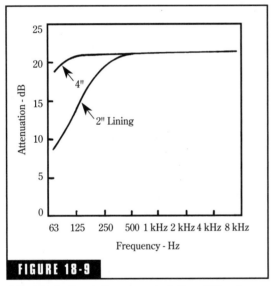

FIGURE 18-9

Calculated attenuation characteristics of the plenum of Fig. 18-8 lined with 2" and 4" glass of 3 lb/cu ft density. Calculations follow ASHRAE procedures.

Packaged Attenuators

Numerous packaged proprietary noise attenuators are available.[3] Cross sections of several types are shown in Fig. 18-11 with their performance plotted below. For comparison, the attenuation of the simple lined duct is given in curve A. Some of the other attenuators have no line-of-sight through them, i.e., the sound must be reflected from the absorbing material to traverse the unit and hence will have somewhat greater attenuation. The absorbing material is usually protected by perforated metal sheets in these packaged silencers. The attenuation of such units is very high at midband speech frequencies but not as good at low frequencies.

Reactive Silencers

Several passive, absorptive silencers have been considered that rely for their effectiveness on the changing of sound energy into heat in the interstices of fine glass fibers.[4] Another effective principle used in silencers is that of the expansion chamber as shown in Fig. 18-12. This type performs by reflecting sound energy back toward the source, thereby canceling some of the sound energy. Because there is both an entrance and exit discontinuity, sound is reflected from two points. Of course, the destructive interference (attenuation nulls) alternates with constructive interference (attenuation peaks) down through the frequency band, the attenuation peaks becoming lower as frequency is increased. These peaks are not harmonically related, therefore they would not produce high attenuation for a noise fundamental and all its harmonics, but rather attenuate slices of the spectrum. By tuning, however, the major peak can eliminate the fundamental while most of the harmonics, of much lower amplitude, would receive some attenuation. By putting two reactive silencers of this type in series and tuning one to fill in the nulls of the other, continuous attenuation can be realized throughout a wide frequency range. No acoustical material is required in this type of silencer, which operates like an automobile muffler.

Resonator Silencer

The resonator silencer illustrated in Fig. 18-13 is a tuned stub that provides high attenuation at a narrow band of frequencies. Even a small unit of this type can produce 40- to 60-dB attenuation. This type of silencer offers little constriction to air flow, which can be a problem with other types of silencers.

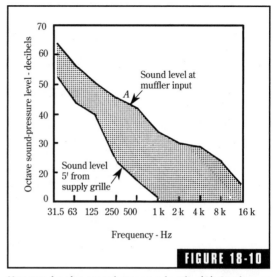

FIGURE 18-10

Measured noise sound pressure levels. (A) At plenum input. (B) In room 5 feet from the supply grilles. This plenum is approximately the size of that of Fig. 18-8 but only half the height and it contains baffles.

Duct Location

Why build an STC-60 dB wall between studio and control room, for instance, and then serve both rooms with the same supply and exhaust ducts closely spaced as in Fig. 18-14? This is a tactical error that results in a short path speaking tube from one room to the other, nullifying quite effectively the 60-dB wall. With an untrained air-conditioning contractor doing the work without a supervisor sensitive to the acoustical problem, such errors can easily happen. To obtain as much as 60-dB attenuation in the duct system to match the construction of the wall requires the application of many of the principles discussed earlier. Figure 18-15 suggests two approaches to the problem; to separate grilles as far as possible if they are fed by the same duct, or better yet, to serve the two rooms with separate supply and exhaust ducts.

Some Practical Suggestions

- The most effective way to control air-flow noise is to size the ducts so as to avoid high velocities. The economy of the smaller ducts, however, may be more than enough to pay for silencers to

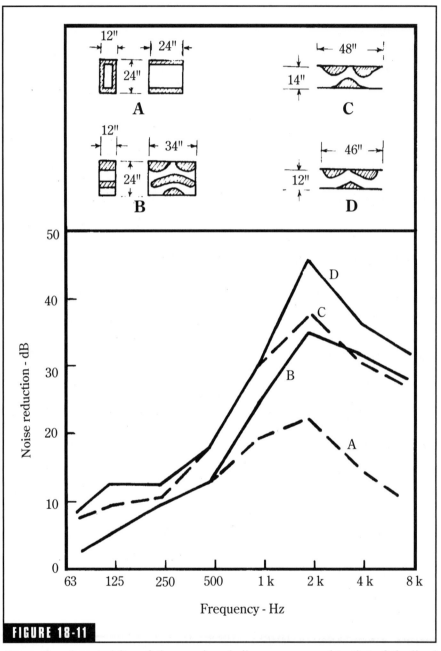

FIGURE 18-11

Attenuation characterisitcs of three packaged silencers compared to that of the lined duct, curve A. *Adapted from Doelling.*[3]

bring the higher noise to tolerable levels.

■ Right-angle bends, dampers, etc., create noise due to air turbulence. Locating such fittings 5 to 10 diameters upstream from the outlet allows the turbulence to smooth out.

■ Noise and turbulence inside a duct cause the duct walls to vibrate and radiate noise into surrounding areas. Rectangular ducts are worse offenders than round ones. Such noise increases with air velocity and duct size, but can be controlled with external treatment of thermal material.

■ Acoustical ceilings are not good sound barriers, hence in a sound-sensitive area, the space above a lay-in ceiling should not be used for high-velocity terminal units.

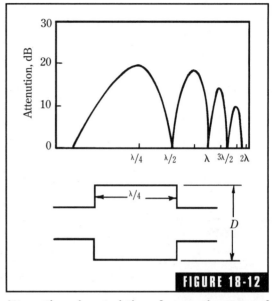

FIGURE 18-12

Attenuation characteristics of a reactive type of silencer, the expansion chamber. Sound is attenuated by virtue of the energy reflected back toward the source, canceling some of the oncoming sound. *Adapted from Sanders.*[4]

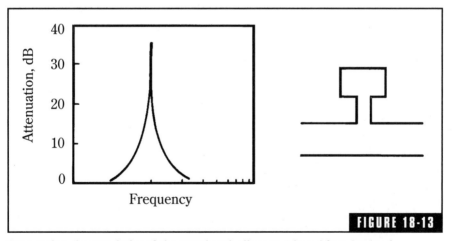

FIGURE 18-13

Attenuation characteristics of the tuned-stub silencer. *Adapted from Sanders.*[4]

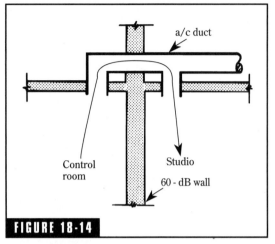

FIGURE 18-14

An expensive, high-transmission-loss wall can be bypassed by sound traveling over a short duct path from grille to grille.

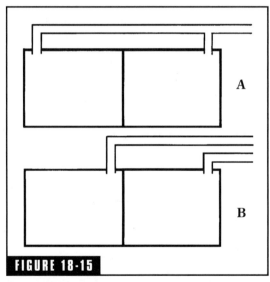

FIGURE 18-15

Two possible solutions to the problems of Fig. 18-14. (A) Separate grilles as far as possible. (B) Feed each room with a separate duct. Both approaches increase duct path length and thus duct attenuation.

■ The ear can detect sounds far below the prevailing NCB contour noise (see Fig. 18-1). The goal should be to reduce noise in the studio to a level at which it cannot be heard on a playback of a recording at normal level and without noise reduction.

■ Plenums are effective and straightforward devices adaptable to studio quieting programs, and they offer attenuation throughout the audible spectrum. They are especially effective at the fan output.

■ Some of the noise energy is concentrated in the highs, some in the lows. There must be an overall balance in the application of silencers so that the resulting studio noise follows roughly the proper NCB contour. Otherwise overdesign can result.

Endnotes

[1]Beranek, Leo L., *Balanced Noise Criterion (NCB) Curves*, J. Acous. Soc. Am., 86,2 (Aug 1989) 650-664.

[2]Anon., *Sound and Vibration Control*, ASHRAE Handbook, 1984 Systems, Chapter 32. Refer also to ASHRAE annual *Product Directories* and *Systems*. Published by Am Soc. Heating, Ventilating, and Air-Cond. Eng., 1791 Tullie Circle, N.E., Atlanta, GA 30329.

[3]Doelling, Norman, *How Effective Are Packaged Attenuators?*, ASHRAE Journal, 2, 2 (Feb 1960) 46-50.

[4]Sanders, Guy J., *Silencers: Their Design and Application*, Sound and Vibration, 2, 2 (Feb 1968) 6-13.

CHAPTER 19

Acoustics of the Listening Room

In the present context, "Listening Room" is taken as the normal home high-fidelity music center (or, possibly, the listening area in a professional work room). The control room or monitoring room of a recording studio is a very special type of listening room, which is treated in Chap. 21. This chapter considers that portion of a home set up for enjoyment of recorded music. Families having the means to dedicate a certain space exclusively for music are fortunate. The rest of us must be satisfied with a multipurpose space, typically a living room serving also as a listening room.

Ideally, all members of the household will be of one mind in such a project. Realistically, there must be compromises between the technical and the aesthetic. Just where the compromise point comes to rest (if ever) may turn out to be a measure of the relative force of personalities and strength of wills. If the persons involved share a genuine appreciation of good music, the war is over except for a few skirmishes along the way.

The Acoustical Link

The acoustics of the space is a vital part of both the recording and reproducing process. In every acoustical event, there is a sound source and some sort of receiving device with an acoustical link between the two.

Disc or tape recordings have the imprint of the acoustics of the recording environment recorded on them. If the sound source is a symphony orchestra and the recording is made in the performing hall, the reverberation of the hall is very much a part of the orchestral sound. If the reverberation time of the hall is 2 seconds, a 2-second tail is evident on every impulsive sound and sudden cessation of music, and it affects the fullness of all the music. In playing this record in our home listening area, what room characteristics will best complement this type of music?

Another recording may be of the popular type. This music was probably recorded in a very dead studio by a multitrack system. Basic rhythm sections playing in this very dead studio and well separated acoustically are laid down on separate tracks. During subsequent sessions the vocals and other instruments are recorded on still other tracks. Finally, all are combined at appropriate levels in a mix down, with a bit of "sweetening" added. The position in the stereo field of the sounds on each track is adjusted by turning a panpot knob. In the mix down, many effects, including artificial reverberation, are added. What listening room characteristics are best for playback of this recording?

If the taste of the high-fidelity enthusiast is highly specialized, the listening-room characteristics can be adjusted for relative optimum results for one type of music. If the taste is more universal, the acoustical treatment of the listening room might need trimming for different types of music.

The dynamic range of reproduced music in a listening room depends at the loud extreme on amplifier power, the power-handling ability of the loudspeakers, and the tolerance of family members and neighbors. The social limitation usually comes into play at a level far lower than the average electronic and transducer limitation. The low-level end of the dynamic range scale is limited by noise, environmental or electronic. Household noise usually determines the lower limit. The usable range between these two extremes is far, far less than, say, the range of an orchestra in a concert hall. Expanders, compressors, companders, etc., are capable of restoring some of the original range (see Fig. 5-12).

Peculiarities of Small-Room Acoustics

The ten-octave spread of the audible spectrum is so great that the acoustical analysis of small rooms is quite different from that for large rooms.

The reason is apparent when room size is considered in terms of the wavelength of sound. The 20-Hz to 20-kHz audio band covers sound wavelengths from 56.5 ft to 0.0565 ft (1¹⁄₁₆ inch). Below about 300 Hz (wavelength 3.8 ft), the average studio or listening room must be considered as a resonant cavity. It is not the studio that resonates, it is the air confined within the studio. As frequency increases above 300 Hz the wavelengths become smaller and smaller with the result that sound may be considered as rays and specular reflections prevail.

In this book, small rooms such as listening rooms, audio workrooms, control rooms, and studios are emphasized. The design of large rooms such as concert halls, theaters, and auditoriums is left to the professional acoustical consultant.

Reflections of sound from the enclosing surfaces dominate both the low- and high-frequency regions. At the lower frequencies reflections result in standing waves, and the room becomes a chamber resonating at many different frequencies. Reflections of sound from the room surfaces also dominate at the midband and higher audible frequencies, without the cavity resonances, but with specular reflections as the major feature.

For the professional acoustician and the critical audiophile, the listening room is as much of a challenge as the design of a professional recording studio, but it has not received as much attention. All the major acoustical problems are involved in the design of a listening room and any other small audio room. The acoustics of the listening room in this chapter are therefore considered as an introduction to the acoustics of other types of small audio spaces in following chapters.

Room Size

Problems are inevitable if sound is recorded or reproduced in spaces that are too small. Gilford[1] states that studio volumes less than approximately 1,500 cu ft are so prone to sound coloration that they are impractical. Rooms smaller than this produce sparse modal frequencies with exaggerated spacings, which are the source of audible distortions.

Room Proportions

Chapter 13, especially Table 13-2, lists the room proportions yielding the most favorable distribution of room modes. With new construction it is strongly advised to use these proportions as a guide, confirming any dimensions seriously considered by calculation and close study of the spacing of axial-mode frequencies.

In the home listening room the room shape and size in most cases are already fixed. The existing room dimensions should then be used for axial mode calculations after the pattern of Table 19-1. A study of these modal

Table 19-1 Axial modes.

	Room dimensions = 21.5' × 16.5' × 10.0'				
	Length L = 21.5 ft f_1 = 565/L Hz	Width W = 16.5 ft f_1 = 565/W Hz	Height H = 10.0 ft f_1 = 565/H Hz	Arranged in ascending order	Axial mode spacing Hz
f_1	26.3	34.2	56.5	26.3	7.9
f_2	52.6	68.4	113.0	34.2	18.4
f_3	78.9	102.6	169.5	52.6	3.9
f_4	105.2	136.8	226.0	56.5	11.9
f_5	131.5	171.0	282.5	68.4	10.5
f_6	157.8	205.2	339.0	78.9	23.7
f_7	184.1	239.4		102.6	2.6
f_8	210.4	273.6		105.2	7.8
f_9	236.7	307.8		113.0	18.5
f_{10}	263.0			131.5	5.3
f_{11}	289.3			136.8	21.0
f_{12}	315.6			157.8	11.7
				169.5	1.5
				171.0	13.1
				184.1	21.1
				205.2	5.2
				210.4	15.6
				226.0	10.7
				236.7	2.7
				239.4	23.6
				263.0	10.6
				273.6	8.9
				282.5	6.8
				289.3	18.5
				307.8	

Mean axial mode spacing = 11.7 Hz
Standard deviation = 6.9 Hz

frequencies will then reveal the presence of coincidences (two or more modes at the same frequency) or isolated modes spaced 25 Hz or more from neighbors. Such faults pinpoint frequencies at which colorations may occur.

Reverberation Time

Reverberation time has been demoted from being a primary determinant of acoustical quality of small rooms to a nonentity. The amount of overall absorbance in a listening room is still important in establishing the general listening conditions. If the room is excessively dead or too live and reverberant, listener fatigue might develop and music quality may deteriorate.

The old familiar Sabine equation for reverberation time (Eq. 7-1, of Chap. 7) makes possible an estimate of the amount of absorbing material required for a reasonable reverberant condition. Even though we place little importance on the reverberation time figure itself, it is expedient to assume a reasonable reverberation time, say about 0.3 sec, for the purposes of these calculations. From this, the total number of sabins of absorption can be estimated, which would result in reasonable listening conditions. In most home listening/living rooms, the structure and the furnishings often supply most of the basic absorbance required. Careful listening tests must determine the degree of room ambience most suitable for the favorite type of music.

The Listening Room: Low Frequencies

The bare room of Fig. 19-1 is the starting point. The room is 21.5 ft long, 16.5 ft wide, with a 10-ft ceiling. These dimensions fix the axial mode resonances and their multiples (harmonics). Following discussions in Chap. 15, axial mode effects will be emphasized and tangential and oblique modes will be neglected. Axial mode frequencies out to 300 Hz were calculated for the length, width, and height dimensions and tabulated in Table 19-1. These axial modal frequencies were then arranged in ascending order of frequency, irrespective of the dimensional source (length, width, or height). Spacings between adjacent modes were than entered in the right-hand column. No coincidences are noted; only a single pair are as close as 1.5 Hz. The dimensional ratios 1 : 1.65 : 2.15 are well within the "Bolt area" of Fig. 13-6.

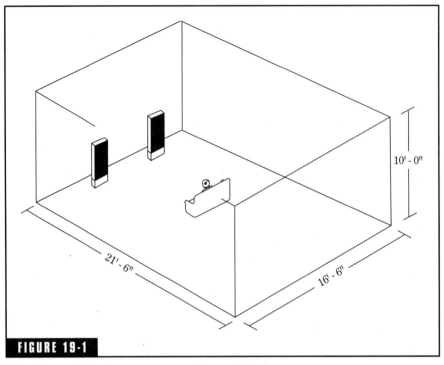

10' - 0"

21' - 6"

16' - 6"

FIGURE 19-1

Dimensions of listening room assumed for analysis.

The calculated axial modal frequencies listed in Table 19-1 must now be applied to the listening room space. This is done graphically in Fig. 19-2, following the "stylized impressions" of Toole.[5] The positions of the nulls locate null-lines, which are drawn through the listening room. Lines representing the length-mode nulls are drawn through both the elevation view and the plan view because these nulls actually form a null "sheet" that extends from floor to ceiling. In other words, the position of the listener's seat can be moved to avoid these particular nulls at 26, 53, and 79 Hz, but remember, there are 8 more below 300 Hz.

The three lowest axial-mode nulls associated with the height of the room (56, 113, and 170 Hz), are sketched on the elevation view of Fig. 19-2. These nulls are horizontal "sheets" at various heights. The head of the listener in the elevation view lies between two nulls and at the peak of the 79-Hz resonance.

The three axial modes of lowest frequency are sketched on the plan view. The nulls in this case are vertical "sheets" extending from floor

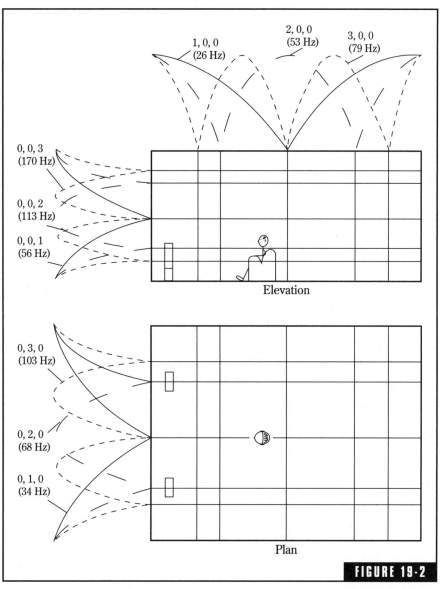

Plan and elevation views of listening room of Fig. 19-1 showing the low-frequency modal pressure distributions of the first three axial modes of Table 19-1.

to ceiling. The listener, situated on dead center of the room, intercepts the nulls of every odd axial mode.

The resonance nulls have been sketched because their location is definite, but between any two nulls of a given axial mode, a peak exists. Although nulls are capable of removing a sizeable chunk of spectrum, the low-frequency acoustics of the room is dominated by the wide, relatively flat peaks.

The extreme complexity of the modal structure of listening room acoustics is becoming apparent. Only the first three axial modes of length, width, and height have been shown. All of the axial modal frequencies listed in Table 19-1 have an important part to play in the low-frequency acoustics of the space. These axial modes exist only when they are excited by the low-frequency sound of the signal being reproduced in the room. The spectrum of the music is continually shifting, therefore, the excitation of the modes is also continually shifting. The length axial mode at 105.2 Hz (Table 19-1) comes alive only as spectral energy in the music hits 105.2 Hz. If only our eyes were sensitive to sound pressure (intensity) and frequency (color), what an awesome sight the constantly shifting room resonances would be as Beethoven's Eighth Symphony is reproduced! And all this represents only the sound energy below about 300 Hz.

Control of Modal Resonances

The low-frequency sound field at the listener's ears is made up of the complex vectorial sum of all axial, tangential, and oblique modes at that particular spot in the room. The loudspeakers energize the modal resonances prevailing at their locations. The modes that have nulls at a loudspeaker location cannot be energized, but those having partial or full maxima at this location will be energized proportionally. The interaction of low-frequency resonances in the listening room at the loudspeaker and listening positions is too complex and transient to grasp fully, but they can be understood if broken down into the contributions of individual modes.

Loudspeakers should be located as far away from reflecting surfaces as practicable. Loudspeaker positions should be considered tentative, moving them slightly if necessary to improve sound quality. The same is true for listening position.

With a basic understanding of the complexity of the low-frequency modal sound fields of a listening room, one must be amazed that sound quality is as good and natural as it is.

Bass Traps for the Listening Room

It is not practical to acoustically treat each mode separately. A general low-frequency treatment is usually sufficient to control "room boom" and other resonance anomalies. In fact, a room of wood-frame construction could very well have sufficient structural low-frequency absorption built in to provide all the general modal control necessary.

In addition to the adjustment of the general ambience of the room, low-frequency absorption in the two corners of the room near the loudspeakers can have an important effect on the stereo image. Figure 19-3 suggests four ways that such absorption can be obtained. The first, Fig. 19-3A, is a Helmholtz resonator trap, home-built in the corner. This could employ either a perforated face or spaced slats. Some design frequency must be assumed, perhaps 100 Hz, and an average depth of the triangular shape must be estimated. With these figures, Eq. 9-5 (for slats) or Table 9-4 (for perforations) can be used for completing the design. In fact, the design of Fig. 9-24 for a diaphragmatic absorber is still another one that can be used.

The easiest solution of the problem would be the selection of one of the three proprietary absorbers of Figs. 19-3B, C, or D. Tube Traps™, offered by Acoustic Sciences Corporation,[2] provide the necessary

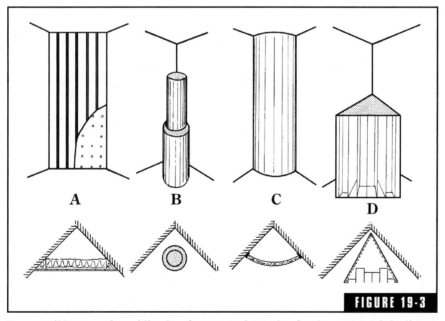

FIGURE 19-3

Four possible ways of providing low-frequency absorption for the corners of the listening room nearest the loudspeakers. (A) Home built Helmholtz resonator, (B) The use of Tube Trap™ or (C) Snap Traps™ available from Acoustic Sciences Corporation.[2] (D) The use of Korner Killers™ available from RPG Diffusor Systems, Inc.[3]

absorption by stacking a 9-in diameter unit on top of an 11-in unit, resulting in a total height of 6 ft. These traps are fibrous cylinders with wire mesh skeleton, together acting as resonant cavities. Half of the periphery is a reflective, limp-mass sheet. Low-frequency energy (below 400 Hz) readily penetrates this sheet while high-frequency energy is reflected from it. Used in a corner such as this, the cylinder would be rotated so that this reflective side faces the room. The cylinder thus contributes to diffusion of the room as well as to deep-bass absorption.

The corner treatment of Fig. 19-3C is based on another product of Acoustic Sciences Corporation[3] called the Snap Trap™. Mounting tracks of 1" × ½" J-metal are installed in the corner to hold the edges of the panel. The Snap Trap™ sheet is then bent and snapped in place. The air space behind the acoustic panel ensures good low-frequency absorption. A curved membrane reflector strip within the panel provides wide-angle reflection above 500 Hz.

Still another possible corner treatment offered by RPG Diffusor Systems, Inc.[3] is shown in Fig. 19-3D. The Triffusor™ is especially adapted to variable acoustic control with an absorbent side, a diffusing side, and a reflective side. An adaptation of the Triffusor™, called the Korner Killer™, has two absorptive sides and one quadratic residue diffusor side. With the absorptive sides into the corner for modal control, the diffusive side faces the room. This RPG™ diffusor face not only diffuses the sound energy falling on it, it also reduces the amplitude of the energy returned to the room. The nominal dimensions of the Korner Killer™: height, 4 ft, faces 24 in.

With a pair of any of the devices of Fig. 19-3 in the corners of the listening room near the loudspeakers, the chances are good that sufficient modal control is introduced to clean up any stereo image problems resulting from room resonances. There are two more corners of the room that could be treated similarly in the unusual case that more modal control is required.

Modal Colorations

The more obvious source of low-frequency colorations of sound results from the momentary deviations from flatness of the room response, which results from concentrations of modes or great spacings between modes (see Fig. 15-21). Transient bursts of music energy result in unequal, forced excitation of the modes. As the transient excitation is removed, each mode decays at its natural (and often different) frequency. Beats can occur between adjacent decaying modes. Energy at new and different frequencies is injected, which is a coloration of the signal.

The Listening Room: The Mid-High Frequencies

The propagation of sound of shorter wavelengths, above about 300 Hz, can be considered in the form of rays that undergo specular reflection. Figure 19-4 shows the same listening room and listener for the consideration of the mid-high-frequency reflections of the sound from the loudspeakers. Sound from the right loudspeaker is studied in detail as characteristic of the symmetrical room.

The first sound to reach the listener's ears is the direct sound, travelling the shortest distance. Reflection *F* from the floor arrives next. Reflections from the ceiling (*C*), the near side wall (*W*) and the far side wall (*W*) arrive later. One other early reflection is the one labelled *D* resulting from the diffusion of sound from the edges of the loudspeaker cabinet (Fig. 11-10). The reflection of this sound from the front wall is shown in the plan view of Fig. 19-4.

These constitute the early reflections, as contrasted to reflections from the rear surfaces of the room and general reverberation, arriving much later. The direct ray carries important information concerning the signal being radiated. If it is accompanied by the early reflections, the sharp stereo perception of the direct ray tends to be blurred.

The important research of Olive and Toole[4,5] is partially summarized in Fig. 19-5 (a repeat of Fig. 16-4 for convenience). The variables here are reflection level and reflection delay. It is instructive to compare the levels and delays of the early reflections of Fig. 19-4 with those of the Olive/Toole graphs of Fig. 19-5. Table 19-2 lists estimations of the level and delay of each of the early reflections identified in Fig. 19-4. These reflections, plotted on Fig. 19-5, all fall within the audible region between the reflection threshold and the echo-production threshold. The pure direct signal is immediately followed by a competitive swarm of early reflections of various levels and delays, producing comb-filter distortion.

The need is obvious: Reduce the level of the competing reflections so that the direct signal stands out in all its beauty—with one exception. Figure 19-5 is a study of a direct signal and a single lateral reflection.

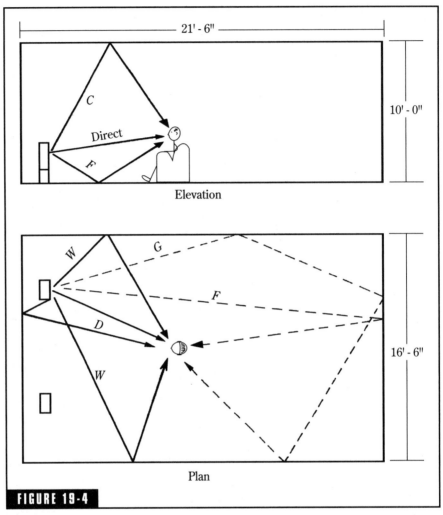

FIGURE 19-4

Plan and elevation views for listening room showing the early reflections from the floor, the ceiling, the side walls, and diffusion form edges of the loudspeaker cabinet. The later reflections *F* and *G* are the beginning of the reverberant component.

Allowing a single lateral reflection of adjustable level would place control of the spaciousness and image effects in the hands of the experimenter/listener! The potential of this concept is staggering. Therefore, the next step in improving the listening conditions in this listening room will be the effective elimination of all of the early reflections, except those lateral reflections off the left and right side wall that will be adjusted for optimum sound quality.

Identification and Treatment of Reflection Points

One method of reducing the levels of the early reflections is to treat the entire front portion of the room with sound absorbing material. This would also kill the lateral reflection and would probably make the room too dead for listening comfort. The principle recommended here is to add a minimum of absorbing material to treat only the specific surfaces responsible for the reflections.

Locating these reflection points is easy with a helper with a mirror. With the listener/experimenter seated at the "sweet spot" the assistant moves a mirror on the floor until the observer can see

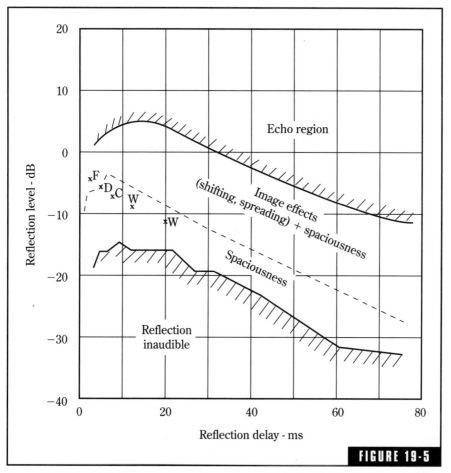

FIGURE 19-5

A repeat of Fig. 16-4 for more convenient reference. The small x's represent the early room reflections calculated in Table 19-2. *After Olive and Toole.*[5]

Table 19-2. Reflections: Their amplitudes and delays.

Sound paths	Path length (ft)	Reflection direct (ft)	Reflection level* (dB)	Reflection delay** (ms)
Direct	8.0	0.0	—	—
F (Floor)	10.5	2.5	−2.4	2.2
D (Diff)	10.5	2.5	−2.4	2.2
C (Cell)	16.0	8.0	−6.0	7.1
W (near wall)	14.0	6.0	−4.9	5.3
W (far wall)	21.0	13.0	−8.4	11.5
F (rear)	30.6	22.6	−11.7	20.0
G (rear)	44.3	36.3	−14.9	32.1

*Reflection level $= 20 \log \dfrac{\text{direct path}}{\text{reflected path}}$

**Reflection delay $= \dfrac{(\text{reflected path}) - (\text{direct path})}{1{,}130}$

(Assuming perfect reflection and inverse square propagation.)

the tweeter of the right loudspeaker reflected in it. This is the point on the floor where the floor reflection hits. This point is carefully marked and the procedure repeated for the tweeter of the left loudspeaker, and the second floor reflection spot is marked. A small rug covering these two marks should reduce the floor reflections to inaudibility.

The same procedure is carried out for locating the reflection points for the left and right side-wall and the ceiling reflections. Each of these points should also be covered with enough absorbing material to ensure ample coverage of the reflection points.

The point of reflection for the sound energy diffracted from the edges of the loudspeaker cabinet is more difficult to locate. Installing an absorber on the wall between the loudspeakers should subdue diffraction reflections.

When all the covered reflection points are in place (Fig. 19-6) try a preliminary listening test. The stereo image and the music will probably be much clearer and more precise now that the early reflections have been reduced.

Lateral Reflections: Control of Spaciousness

The lateral reflections from the side walls have been essentially eliminated by the absorbing material placed on the wall. The next listening test should be on the same music with the side-wall absorbers temporarily removed, but with the floor, ceiling, and diffraction absorbers still in place. The Olive/Toole graphs of Fig. 19-5 can now be tested. Does the fullstrength lateral reflection give the desired amount of spaciousness and image shifting or spreading? The lateral reflections can be reduced somewhat by hanging a cloth instead of the heavy absorber. The adjustment of the magnitude of

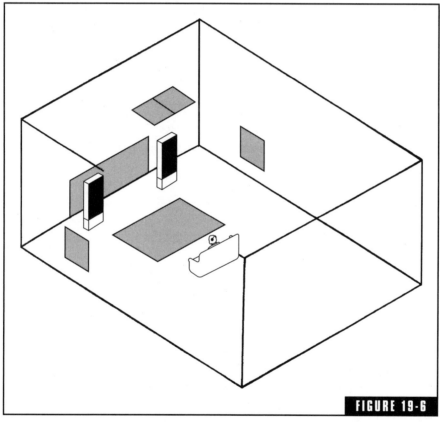

FIGURE 19-6

The room of Fig. 19-1 with minimum sound absorbing treatment to reduce the level of the early reflections from room surfaces. The reflectivity of the side-wall absorbers may be adjusted to control spaciousness and image effects in the listening room. Additional absorbing material may be needed to adjust the average reverberant character of the room for best listening.

the lateral reflections can be explored by using sound absorbers of varying absorbance (velour, heavy cloth, light cloth, oudoor carpet, indoor carpet) on the side-wall reflection points.

This is a new and highly promising field: The ability to adjust the lateral reflections to achieve the desired spaciousness and stereo image effect to suit the individual listener or to optimize conditions for different types of music.

Endnotes

[1]Gilford, C.L.S., *The Acoustic Design of Talks Studios and Listening Rooms*, Proc. Inst. Elect. Engrs., 106, Part B, 27 (May 1959), 245-258. Reprinted in J. Audio Eng. Soc., 27, 1/2 (1979) 17-31.

[2]Acoustic Sciences Corporation, P.O. Box 1189, Eugene, OR 97440, telephone 1-800-ASC-TUBE, FAX 503-343-9245.

[3]RPG Diffusor Systems, Inc., 651-C Commerce Drive, Upper Marlboro, MD 20772, telephone 301-249-0044, FAX 301-249-3912.

[4]Toole, Floyd E., *Loudspeaker and Rooms for Stereophonic Sound Reproduction*, Proc. Audio Eng. Soc. 8th International Conference, Washington, D.C. (1990) 71-91.

[5]Olive, Sean E., and Floyd E. Toole, *The Detection of Reflections in Typical Rooms*, J. Audio Eng. Soc., 37, 7/8 (July/Aug 1989) 539-553.

Acoustics of the Small Recording Studio

This is the day of the small recording studio. Musicians are interested in making demonstration records to develop their style and to sell their sounds. There are hundreds of small recording studios operated by not-for-profit organizations that turn out a prodigious quantity of material for educational, promotional, and religious purposes. Studios are required for the production of campus and community radio, television, and cable programs. All of these have limited budgets and limited technical resources. The operator of these small studios is often caught between a desire for top quality and the lack of means, and often the know-how to achieve it. This chap. is aimed primarily to those in these needy groups, although the principles expounded are more widely applicable.

What is a *good* recording studio? There is only one ultimate criterion—the acceptability of the sound recorded in it by its intended audience. In a commercial sense, a successful recording studio is one fully booked and making money. Music recorded in a studio is pressed on discs or recorded on tape and sold to the public. If the public likes the music, the studio passes the supreme test. There are many factors influencing the acceptability of a studio beside sound studio quality, such as the type of program and the popularity of the performers, but studio quality is vital, at least for success on a substantial, long-range basis.

Public taste must be pleased for any studio to be a success. Producing a successful product, however, involves many individuals along the way whose decisions may make or break a studio. These decisions may be influenced by both subjective and technical factors. The appearance of a studio, convenience, and comfort might outweigh acoustical quality, sometimes because the more tangible esthetic qualities are better understood than the intangible acoustical qualities. This chap. has little to say on the artistic, architectural, and other such aspects of a studio, but their importance cannot be denied. They just require a different kind of specialist.

Acoustical Characteristics of a Studio

Sound picked up by a microphone in a studio consists of both direct and indirect sound. The direct sound is the same as would exist in the great outdoors or in an anechoic chamber. The indirect sound, which immediately follows the direct, is the sound that results from all the various nonfree-field effects characteristic of an enclosed space. The latter is unique to a particular room and may be called *studio response*. Everything that is not direct sound is indirect, reflected sound.

Before dissecting indirect sound, let us look at the sound in its all-inclusive form in a studio, or any other room for that matter. Figure 20-1 shows how sound level varies with distance from a source, which could be the mouth of someone talking, a musical instrument, or a loudspeaker. Assume a pressure level of 80 dB measured 1 foot from the source. If all surfaces of the room were 100 percent reflective, we would have a reverberation chamber to end all reverberation chambers, and the sound pressure level would be 80 dB everywhere in the room because no sound energy is being absorbed. There is essentially no direct sound; it is all indirect. Graph B represents the fall off in sound

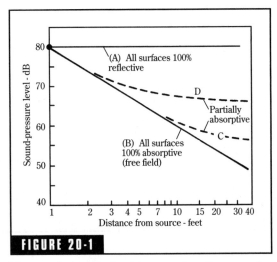

FIGURE 20-1

The sound-pressure level in an enclosed space varies with distance from the source of sound according to the absorbency of the space.

pressure level with distance from the sound source with all surfaces 100 percent absorptive. In this case all the sound is direct; there is no indirect component. The best anechoic rooms approach this condition. It is the true free field illustrated in Chap. 4, and for this condition the sound pressure level decreases 6 dB for each doubling of the distance.

Between the indirect "all reverberation" case of graph A of Fig. 20-1 and the direct "no reverberation" case of graph B lie a multitude of other possible "some reverberation" cases, depending on room treatment. In the area between these two extremes lies the real world of studios in which we live and move and have our being. The room represented by graph C is much more dead than that of graph D. In practical studios, the direct sound is observable a short distance out from the source, but after that the indirect sound dominates. A sudden sound picked up by a microphone in a studio would, for the first few milliseconds, be dominated by the direct component, after which the indirect sound arrives at the microphone as a torrent of reflections from room surfaces. These are spread out in time because of the different path lengths traveled.

A second component of indirect sound results from room resonances, which in turn are the result of reflected sound. The direct sound flowing out from the source excites these resonances, bringing into play all the effects listed in Chap. 15. When the source excitation ceases, each mode dies away at its own natural frequency and at its own rate. Sounds of very short duration might not last long enough to fully excite room resonances.

Distinguishing between reflections and resonances is an acknowledgment that neither a reflection concept nor a resonance concept will carry us through the entire audible spectrum. Resonances dominate the low-frequency region in which the wavelengths of the sound are comparable to room dimensions. The ray concept works for higher frequencies and their shorter wavelengths (Chap. 16). Around the 300- to 500-Hz region is a difficult transition zone. But with this reminder of the basic limitations of our method we can return to analyzing the components of sound in a small studio.

The third component of indirect sound is involved with the materials of construction—doors, windows, walls, and floors. These too are set into vibration by sound from the source, and they too decay at their own particular rate when excitation is removed. If Helmholtz resonators are involved in room treatment, sound not absorbed is reradiated.

The sound of the studio, embracing these three components of indirect sound plus the direct sound, has its counterpart in musical instruments. In fact, it is helpful to consider our studio as an instrument that the knowledgeable musician, technician, or engineer can play. It has its own characteristic sound, and a certain skill is required to extract from it its full potential.

Reverberation

Reverberation is the composite, average effect of all three types of indirect sound. Measuring reverberation time does not reveal the individual components of which reverberation is composed. Herein lies the weakness of reverberation time as an indicator of studio acoustical quality. The important action of one or more of the indirect components may be obscured by the averaging process. This is why it is said that reverberation time is *an* indicator of studio acoustical conditions, but not the only one.

There are those who feel it is improper and inaccurate to apply the concept of reverberation time to relatively small rooms. It is true that a genuine reverberant field may not exist in small spaces. Sabine's reverberation equation is based on the statistical properties of a random sound field. If such an isotropic, homogeneous distribution of energy does not prevail in a small room, is it proper to apply Sabine's equation to compute the reverberation time of the room? The answer is a purist "no," but a practical "yes." Reverberation time is a measure of decay rate. A reverberation time of 0.5 seconds means that a decay of 60 dB takes place in 0.5 seconds. Another way to express this is 60 dB/0.5 second = 120 dB/second decay rate. Whether the sound field is diffuse or not, sound decays at some particular rate, even at the low frequencies at which the sound field is least diffuse. The sound energy stored at the modal frequencies decays at some measurable rate, even though only a few modes are contained in the band being measured. It would seem to be a practical step to utilize Sabine's equation in small room design to estimate absorption needs at different frequencies. At the same time, it is well to remember the limitations of the process.

Studio Design

In a general book of this type, space is too limited to go into anything but basic principles. Fortunately, there is a rich literature on the subject, much of it written in easy-to-understand language. In designing a studio, attention should be given to room volume, room proportions, and sound decay rate, diffusion, and isolation from interfering noise.

Studio Volume

A small room almost guarantees sound colorations resulting from excessive spacing of room resonance frequencies. This can be minimized by picking one of the favorable room ratios suggested by Sepmeyer (see Fig. 13-6) 1.00 : 1.28 : 1.54, applying it to a small, a medium, and a large studio and seeing what happens. Table 20-1 shows the selected dimensions, based on ceiling heights of 8, 12, and 16 feet resulting in room volumes of 1,000, 3,400, and 8,000 cubic feet. Axial mode frequencies were then calculated after the manner of Table 15-5 and plotted in Fig. 20-2, all to the same frequency scale. As previously noted, the room proportions selected do not yield perfect distribution of modal frequencies, but this is of no consequence in our investigation of the effects of room volume. A visual inspection of Fig. 20-2 shows the increase in the number of axial modes as volume is increased, which of course results in closer spacing. In Table 20-2 the number of axial modes below 300 Hz is shown to vary from 18 for the small studio to 33 for the large. The low-frequency response of the large studio, 22.9Hz, is shown to be far superior to that of the two smaller studios at 30.6 and 45.9 Hz. This is an especially important factor in the recording of music.

We must remember that modes other than axial are present. The major diagonal

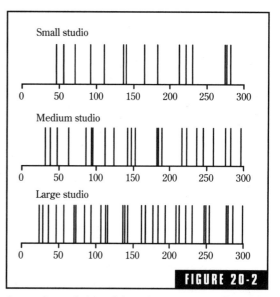

FIGURE 20-2

Comparison of the axial-mode resonances of a small (1,000 cu ft), a medium (3,400 cu ft), and a large (8,000 cu ft) studio all having the proportions 1.00: 1.28: 1.54.

Table 20-1. Studio dimensions.

	Ratio	Small studio	Medium studio	Large studio
Height	1.00	8.00 ft	12.00 ft	16.00 ft
Width	1.28	10.24 ft	15.36 ft	20.48 ft
Length	1.54	12.32 ft	18.48 ft	24.64 ft
Volume		1,000 cu ft	3,400 cu ft	8,000 cu ft

Table 20-2. Studio resonances in Hz.

	Small studio	Medium studio	Large studio
Number of axial modes below 300 Hz	18	26	33
Lowest axial mode	45.9	30.6	22.9
Average mode spacing	14.1	10.4	8.4
Frequency corresp. to room diagonal	31.6	21.0	15.8
Assumed reverb, time of studio, second	0.3	0.5	0.7
Mode bandwidth (2.2/RT60)	7.3	4.4	3.1

dimension of a room better represents the lowest frequency supported by room resonances because of the oblique modes. Thus, the frequency corresponding to the room diagonal listed in Table 20-2 is a better measure of the low-frequency capability of a room than the lowest axial frequency. This approach gives the lowest frequency for the large room as 15.8 Hz, compared to 22.9 Hz for the lowest axial mode.

The average spacing of modes, based on the frequency range from the lowest axial mode to 300 Hz, is also listed in Table 20-2. The average spacing varies from 8.4 Hz for the large studio to 14.1 Hz for the small studio.

The reverberation times listed in Table 20-2 are assumed, nominal values judged fitting for the respective studio sizes. Given these reverberation times, the mode bandwidth is estimated from the expression 2.2/RT60. Mode bandwidth varies from 3 Hz for the large studio to 7 Hz for the small studio. The advantage of closer spacing of axial modes in the large studio tends to be offset by its narrower mode bandwidth. So, we see conflicting factors at work as we realize the advantage of the mode skirts overlapping each other. In general, however, the greater number of axial modes for the large studio, coupled with the extension of room response in the low frequencies, produces a response superior to that of the small studio.

The examples of the three hypothetical studios considered above emphasize further the appropriateness of musical instrument analogy of a studio. We can imagine the studio as a stringed instrument, one string for each modal frequency. These strings respond sympathetically to sound in the room. If there are enough strings tuned to closely spaced frequencies, and each string responds to a wide enough band of frequencies to bridge the gaps between strings, the studio-instrument responds uniformly to all frequency components of the sound in the studio. In other words, the response of the studio is the vector sum total of the responses of the individual modes. If the lines of Fig. 20-2 are imagined to be strings, it is evident that there will be dips in response between widely spaced frequencies. The large studio, with many strings, yields the smoother response.

Conclusion: A studio having a very small volume has fundamental response problems in regard to room resonances; greater studio volume yields smoother response. The recommendation based on BBC experience still holds true, that coloration problems encountered in studios having volumes less than 1,500 cubic feet are severe enough to make small rooms impractical. For reasons of simplicity, the axial modes considered in the previous discussion are not the only modes, but they are the dominant ones.

Room Proportions

If there are fewer axial modes than are desired in the room under consideration, sound quality is best served by distributing them as uniformly as possibly. The cubical room distributes modal frequencies in the worst possible way—by piling up all three fundamentals, and each trio of multiples with maximum gap between. Having any two dimensions in multiple relationship results in this type of problem. For example, a height of

8 ft and a width of 16 ft means that the second harmonic of 16 ft coincides with the fundamental of 8 ft. This emphasizes the importance of proportioning the room for best distribution of axial modes.

The perfect room proportions have yet to be found. It is easy to place undue emphasis on a mechanical factor such as this. I urge you to be well informed on the subject of room resonances and to be aware of certain consequences, but let us be realistic about it—all of the recording that has ever taken place has been done in spaces less than perfect. In our homes and offices, conversations are constantly taking place with serious voice colorations, and we listen to and enjoy recorded music in acoustically abominable spaces. The point is that in striving to upgrade sound quality at every stage of the process, reducing sound colorations by attention to room modes is just good sense.

Reverberation Time

Technically, the term "reverberation time" should not be associated with relatively small spaces in which random sound fields do not exist. However, some first step must be taken to calculate the amount of absorbent needed to bring the general acoustical character of a room up to an acceptable level. While reverberation time is useful for this purpose, it would be unfortunate to convey the impression that the values of reverberation time so obtained have the same meaning as that in a large space.

If the reverberation time is too long (sound decays too slowly), speech syllables and music phrases are slurred and a definite deterioration of speech intelligibility and music quality results. If rooms are too dead (reverberation time too short), music and speech lose character and suffer in quality, with music suffering more. These effects are not so definite and precise as to encourage thinking that there is a specific optimum reverberation time, because many other factors are involved. Is it a male or female voice, slow or fast talker, English or German language (they differ in the average number of syllables per minute), a stand-up comic or a string ensemble, vocal or instrumental, hard rock or a waltz? In spite of so many variables, readers need guidance, and there is a body of experience from which we can extract helpful information. Figure 20-3 is an approximation rather than a true optimum—but following it will result in reasonable, usable conditions

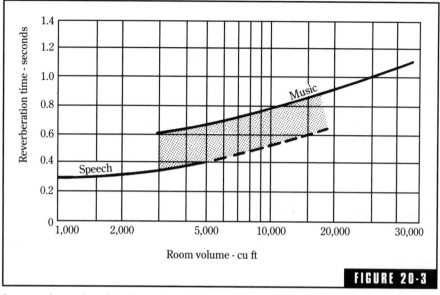

FIGURE 20-3

Suggested reverberation times for recording studios. The shaded area is a compromise region for studios in which both music and speech are recorded.

for many types of recording. The shaded area of Fig. 20-3 represents a compromise in rooms used for both speech and music.

Diffusion

Before the advent of the Schroeder (diffraction grating) diffusor, there was little advice to give regarding diffusion in the small studio. Splaying walls and the use of geometrical protuberances have only a modest diffusing effect. Distributing the absorbing material is a useful means of not only achieving some diffusion, but increasing the absorbing efficiency as well.

Modular diffusing elements are on the market that really diffuse as shown in Chap. 14. There are even 2 ft-x-4 ft-modular units that offer high-quality diffusion and excellent broadband absorption (0.82 coefficient at 100 Hz, for example), all within a 2-in thickness (the Abffusor™). The application of this new principle of diffusion, with or without the absorption feature, contributes a feeling of spaciousness through the diffusion of room reflections and the control of resonances.

Noise

Noise is truly something in the ear of the "behearer." One person's beautiful music is another person's noise, especially at 2 AM. It is a two-way street, and fortunately, a good wall that protects a studio area from exterior noise also protects neighbors from what goes on inside. The psychological aspect of noise is very important—acceptable if considered a part of a situation—disturbing if considered extraneous. Chap. 18 has already treated the special case of air-conditioning noise.

Studio Design Procedure

We have considered reverberation and how to compute it (Chap. 7), the reality of room resonances (Chap. 15), the need for diffusion (Chaps. 13 and 14), various types of dissipative and tuned absorbers (Chap. 9), and as mentioned, one of the most serious studio noise producers, the air-conditioning equipment (Chap. 18). All of these are integral parts of studio design. The would-be designer should also sample the literature to see how others have solved similar problems.[1-3]

Some Studio Features

A glance into other people's studios often makes one aware of "things I want to do" or "things I definitely don't like." Figures 20-4 and 20-5 show the treatment of a budget 2,500 cu ft studio. Built on the second floor of a concrete building with an extensive printing operation below, certain minimum precautions were advisable. The studio floor is ¾-inch plywood on 2-×-2-inch stringers resting on ½-inch soft fiberboard. Attenuation of noise through the double ⅝" drywall ceiling is augmented by a one inch layer of dry sand, a cheap way to get amorphous mass. The wall modules, containing a 4-inch thickness of Owens-Corning Type 703 Fiberglas (3 lb/cu ft density), help to absorb and diffuse the sound.

The studio of Fig. 20-6 with a volume of 3,400 cu ft has a couple of interesting features. The wall modules (Fig. 20-5) feature carefully stained and varnished frames and a neat grille cloth. Those of Fig. 20-7 are of two kinds, one sporting a very attractive fabric

View of a 2,500-cu ft voice studio looking into the control room. *World Vision, International.*

Rear view of the 2,500-cu ft voice studio of Fig. 20-4. Wall modules containing 4-inch thicknesses of dense glass fiber contribute to diffusion of sound in the room. *World Vision, International.*

FIGURE 20-6

A 3,400-cu ft studio used for both recording and editing voice tapes. Decorator-type fabric provides an attractive visual design for the absorber/diffusor wall modules. *Mission Communications Incorporated.*

FIGURE 20-7

Close view of the wall modules of Fig. 20-6. *Mission Communications Incorporated.*

design, the other more subdued. The studio of Fig. 20-6 has a rather high ceiling, hence a virtual, visual ceiling was established at a height of 8 feet. This consists of four 5-×-7-foot suspended frames as shown in Fig. 20-8, which hold fluorescent lighting fixtures and patches of glass fiber. The plastic louvers are acoustically transparent.

The voice studio of Fig. 20-9, with a volume of 1,600 cu ft, employs wall absorbing panels manufactured by the L.E. Carpenter Co. of Wharton, New Jersey. These panels feature a perforated vinyl wrapping and a ⅜-in rigid composition board backing. The concrete floor rests on soft fiberboard with distributed cork chips under it. The low-frequency deficiencies of carpet and wall panels require some Helmholtz correction, and

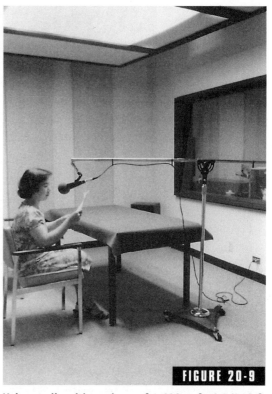

The high structural ceiling of the studio of Fig. 20-6 allows the use of four 5 X 7 ft suspended frames to bring the visual ceiling down to 8 ft and to support illumination fixtures and absorbing material. The plastic louvers are acoustically transparent. *Mission Communications Incorporated.*

Voice studio with a volume of 1,600 cu ft. A 7 X 10 ft suspended ceiling frame hides 13 Helmholtz resonators for low-frequency absorption. Wall modules are proprietary units covered with perforated vinyl. *Far-East Broadcasting Company.*

thirteen 20 × 40 × 8-inch boxes are mounted in the suspended ceiling frame out of sight.

Figure 20-10 is a 3,700-cu ft music studio that is also used for voice work. Low-frequency compensation is accomplished by the same Helmholtz boxes mentioned above, 14 of them in each of two suspended ceiling frames.

Elements Common to All Studios

Chap. 4 of Reference 1 treats sound lock treatment, doors and their sealing, wall constructions, floor/ceiling constructions, wiring pre-

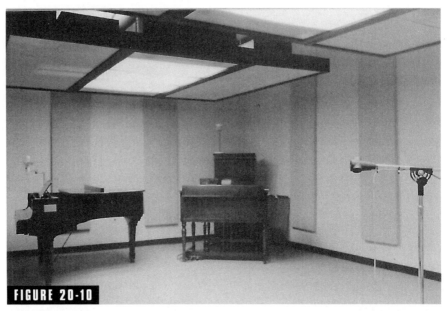

FIGURE 20-10

Music studio with a volume of 3,700 cu ft employing two 9 x 11 foot suspended ceiling frames that hold a total of 28 Helmholtz resonators. *Far East Broadcast Company*

cautions, illuminating fixtures, observation windows, and other things common to all studios and which can create serious problems if not handled properly.

Endnotes

[1]Ballou, Glen, ed., *Handbook for Sound Engineers—The New Audio Cyclopedia*, Indianapolis, IN, Howard W. Sams & Co., 2nd ed. (1991): Chap. 4, *Common Factors in All Audio Rooms*, by F. Alton Everest, p 67-101; Chap. 5, *Acoustical Design of Audio Rooms*, by F. Alton Everest, p. 103-141; Chap. 6, *Contempory Practices in Audio Room Design*, by F. Alton Everest, p. 143-170, Chap. 7, *Rooms for Speech and Music*, by Rollins Brook, p. 171-201.

[2]Allison, Roy F. and Robert Berkowitz, *The Sound Field in Home Listening Rooms*, J. Audio. Eng. Soc., 20, 6 (July/Aug 1972), 459-469.

[3]Kuhl, Walter, *Optimal Acoustical Design of Rooms for Performing, Listening, and Recording*, Proc. 2nd. International Congress on Acoustics, (1956) 53-58.

Acoustics of the Control Room

The acoustical design of control rooms has come of age during the past decade. Rather than discuss the numerous subjective opinions and approaches to control-room design that have characterized the literature of the past, we shall go directly to current practice, which promises even greater refinements in the future.

The Initial Time-Delay Gap

Every recording bears indelibly the marks of the room in which the sounds were recorded. Beranek[1] made an intensive study of concert halls around the world. He noted that those rated the highest by qualified musicians had certain technical similarities. Among them was what he called the *initial time-delay gap*. This is the time between the arrival of the direct sound at a given seat and the arrival of the critically important early reflections. He was impressed by the fact that halls rating high on the quality scale had a well-defined initial time-delay gap of about 20 milliseconds. Halls having this time delay gap confused by uncontrolled reflections were rated inferior by qualified listeners.

Davis's attention was directed to the initial time-delay gap of recording studios and control rooms through work with a newly introduced measuring technique known as *time-delay* spectrometry.[2] The factors generating the initial time delay of a typical recording studio are illustrated in Fig. 21-1A. The direct sound travels a short distance from source to microphone. Later, the sound reflected from the floor, ceiling, or nearby objects arrives at the microphone. This time gap between the arrival of the direct and reflected components is determined by the geometry of the particular setup in a particular studio. Although it varies with each setup, the delay gap for a studio typically falls within a fairly narrow range.

The operator in a conventional control room cannot hear the studio delay gap because it is masked by early control-room reflections. This

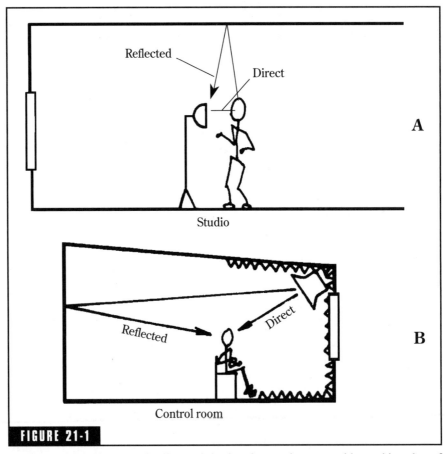

FIGURE 21-1

The wider time-delay gap of a live-end-dead-end control room avoids masking that of the studio. *After Davis.*[2]

means that the operator is deprived of an important component of the studio sound in his critical evaluation. Davis figured that the way to allow the operator to hear the studio delay gap was to eliminate or reduce the early reflections in the control room. Time-delay spectrometry dramatically revealed the comb-filter effects associated with early reflections from surfaces near the loudspeakers and from the console face. This clarification of the problem led directly to the solution of placing absorbing material on the surfaces surrounding the loudspeakers in the front part of the control room (Fig. 21-1B)

Chips Davis (not related to Don Davis), who was a party to this early experimentation and analysis in 1978 in a Syn-Aud-Con seminar and a man of action, decided to test the idea.[3,4] He mounted absorbing material on the surfaces of the front part of his control room in Las Vegas. The results were very encouraging; the sounds from the studio had improved clarity and the ambience of the control room took on a new spacious dimension.

The delay gap of the control room was now narrow enough to avoid masking that of the studio. Giving the control room a precise initial time-delay gap gave listeners the impression of a much larger room.

In simplified and idealized form, Fig. 21-2 shows the energy-time relationships essential for a properly designed and adjusted control room. At time = 0, the signal leaves the monitor loudspeaker. After an elapsed transit time, the direct sound reaches the ears of the operator. There follows some insignificant low-level "clutter" (to be neglected if 20 dB down), after which the first return from the rear of the room arrives. In the early days, emphasis was placed on the first, second, and third significant reflections. Suffice it to say that these important, prominent, delayed reflections constitute the end of the time-delay gap and the first signs of an exponential decay.

The Live End

The deadening of the front end of the control room near the observation window seemed like a fairly straightforward procedure—just cover the surfaces with absorbent. The resulting improvement of stereo image and sound quality gave no indication that other approaches to the treatment of the front end might work even better.

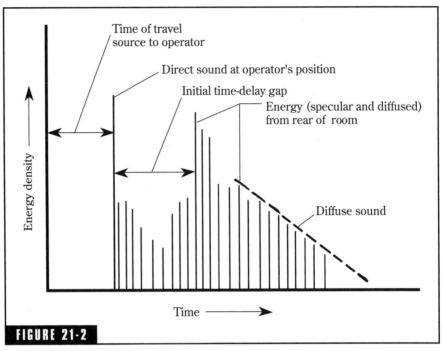

FIGURE 21-2

The definition of the initial time-delay gap for a control room.

Rather, once the front end is made absorbent, attention was naturally shifted to the rear, live end.

Delays offered by reflections from the rear wall dominated the thinking in the early days. Attention was correctly directed to the ability of the human auditory system to integrate these delayed reflections from the rear wall. The desire to make the rear wall diffusive was present in early thinking, but the only means at hand was to introduce the relatively ineffective geometrical irregularities. Had reflective phase-grating diffusors been available in 1978, they would have been eagerly utilized.

Specular Reflections vs. Diffusion

For the specular reflection of sound energy from the monitor loudspeakers from the rear wall (Fig. 21-3A), all of the acoustical energy from a given point on the reflector surface arrives in a single instant of time. If the same sound energy is incident on a reflection phase-grating

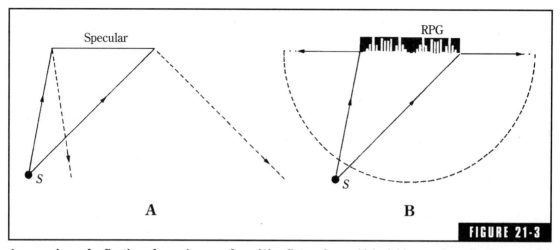

FIGURE 21-3

A comparison of reflection of sound energy from (A) a flat surface, which yields specular reflection, and (B) a reflection-phase grating, which yields energy dispersed in a hemidisc. *Peter D'Antonio, RPG Diffusor Systems, Inc., and Audio Engineering Society, Inc.*

diffusor on the rear wall (Fig. 21-3B), the back-scattered energy is spread out in time.[5] Each element of the diffusor returns energy, the respective reflected wavelets arriving at different times. This temporal distribution of reflected (diffused) energy results in a rich, dense, nonuniform mixture of comb filters that the human auditory system perceives as a pleasant ambience. This is in contrast to the sparse specular reflections that combine to form unpleasant wideband colorations.

With the reflection phase-grating diffusor, the reflected wavelets are not only spread out in time, they are also spread out in space. The one-dimensional diffusor spreads its reflected energy in the horizontal hemidisc of Fig. 14-7. By orienting other one-dimensional units, vertical hemidiscs of diffusion are easily obtained. This is in contrast to the specular panel, which distributes reflected energy in only a portion of half-space determined by the location of the source and the size of the panel.

Another feature of the reflection phase grating, illustrated in Fig. 21-4, makes it especially desirable for the live end of a control room. Let us consider the three black spots as side-wall reflections that impinge on the rear wall, returning energy to the operator, *O*. If the rear wall is specular, there is only one point on the surface returning

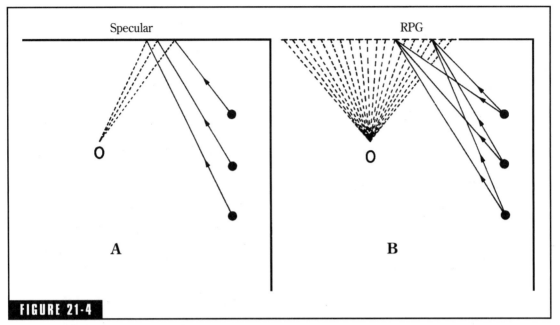

FIGURE 21-4

The three black spots represent side-wall reflection in a control room returning energy to the operator, *O*. If the rear wall is specular, only a single point on the rear wall returns energy to the operator. Each element of a reflection-phase-grating diffusor returns energy to the operator. *Peter D'Antonio, RPG Diffusor Systems, Inc., and Audio Engineering Society, Inc.*

energy from each source to the operator. In contrast, each element of the surface of the grating diffusor sends energy toward the operator. Energy from all sound sources (direct or reflected) falling on the diffusor are scattered to all observation positions. Instead of a single "sweet spot" at the console, a much wider range of good listening positions results.

Low-Frequency Resonances in the Control Room

The Techron™ instrument continues to turn out, under expert direction, new views of sound fields in enclosed spaces. I have devoted much space in this book to room modes, but visualizing them is made much easier by Fig. 21-5. This is a three-dimensional plot showing the relationship between time, energy, and frequency at the microphone position in a control room. The vertical scale is 6 dB between marks. This is

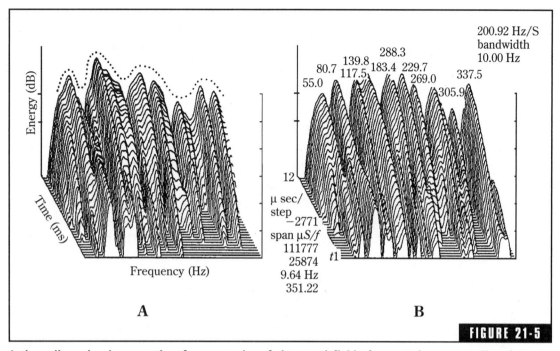

A three-dimensional energy-time-frequency plot of the sound field of a control room revealing the room modes and their decay. The (A) display is the "as found" condition. Display (B) reveals the cleaning up of modal interference by the installation of a low-frequency diffusor at the extreme rear of the control room. *Peter D'Antonio, RPG Diffusor Systems, Inc., and Audio Engineering Society, Inc.*

not sound-pressure level, but rather true energy. The frequency scale runs from 9.64 Hz to 351.22 Hz, a region expected to be dominated by modes. Time runs from the rear toward the reader, 2.771 msec per step, or about 0.1 second for the entire traverse.

The dramatic ridges in Fig. 21-5A are the modal (standing wave) responses of the control room, which together make up the low-frequency acoustical response of the room.[6] Without these modal resonance boosts, the control room would sound like the outdoors.

In addition to the modal response, this delay contains information concerning a second phenomenon of control rooms that must be taken into consideration. This is the interference between the direct low-frequency wave from the loudspeaker and its reflection from the rear wall. If the operator position is 10 feet from the rear wall, the reflection lags the direct sound by the time it takes the wave to travel 20 feet. The delay is t = (20 ft) / (1,130 ft/sec) = 0.0177 second. The frequency of the first comb-filter notch is *1/2t* or 1 / (2)(0.0177) = 28.25 Hz (see chapter 17). Subsequent notches, spaced 1/*t* Hz, occur at 85, 141, 198, 254 Hz, etc.

The depth of the notches depends on the relative amplitudes of the direct and reflected components. One way to control the depth of the notches is to absorb the direct wave so that the reflection is low. This removes precious sound energy from the room. A better way is to build large diffusing units capable of diffusing at these low frequencies. This is exactly what has been done in Fig. 21-5B—a 10-ft-wide, 3-ft-deep, floor-to-ceiling diffusor behind the midband diffusors.[6] The result is a very definite improvement in the response and a much smoother decay. These modes decay about 15 dB in the 0.1-second time sweep. This means a decay rate of roughly 150 dB/sec, which corresponds to a reverberation time of 0.4 second, but the variation of decay rate between the various modes is great.

Initial Time-Delay Gaps in Practice

The realization of the general applicability of the initial time-delay gap principle to different kinds of spaces was not long in coming. Beranek coined the term to apply to concert halls. It was then studied in recording studios, and now control rooms are being revolutionized in order for the operator to hear the delay gap of the studio. In the process, it was discovered that if the control room is altered to make the studio delay gap audible, some beautiful fallout in listening benefits are realized.

The Techron™ instrument's energy-time display makes possible a sophisticated, clinical evaluation of the distribution in time of acoustical energy in a room and the presence or absence of an initial time-delay gap. Figure 21-6 shows three energy-time displays of widely differing spaces. The display of Fig. 21-6A shows the response for the modern control room of Master Sound Astoria, Astoria, NY (see Fig. 21-9). The well-defined time gap (labeled ITD), the exponential decay (a straight line on a log-frequency scale), and the well-diffused reflections are the hallmarks of a well-designed room.

Figure 21-6B shows the energy-time display for the Concertgebouw, Haarlem, Netherlands. A well-defined time gap of about 20 msec qualifies it as one of Beranek's quality concert halls.

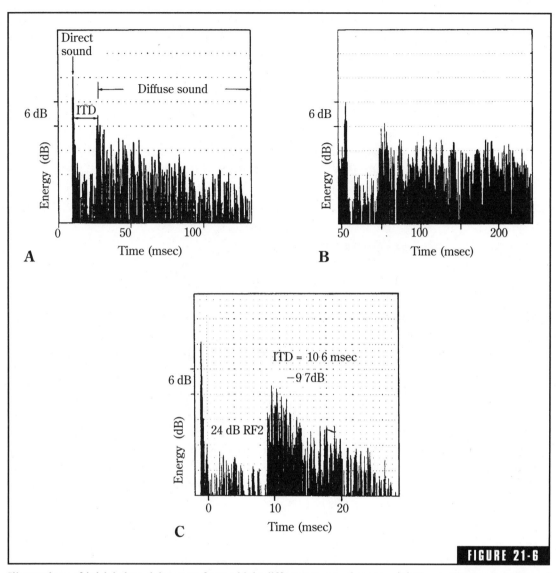

FIGURE 21-6

Illustrations of initial time-delay gaps from widely different types of spaces. (A) The quality-control room of Master Sound Astoria, NY (see Fig. 21-9). (B) A quality concert hall. Concertgebouw, Haarlem, The Netherlands. (C) A quality small listening room, Audio electronics Laboratory, Albertson, NY. *Peter D'Antonio, RPG Diffusor Systems, Inc., and Audio Engineering Society, Inc.*

There are many high-quality concert halls in the world, and in recent years, an ever-growing number of high-quality control rooms. However, there have been very few, if any, small listening or recording rooms that could be classified as quality spaces for the simple reason that they are dominated by normal modes and their associated problems. The physics of the situation seems to doom small rooms to inferior acoustical quality. But now physics, in another guise, comes to the rescue. The energy-time display of Fig. 21-6B gives hope that small listening rooms of genuine quality are now a possibility. This display is for a small listening room belonging to Audio Electronics Laboratory of Albertson, New York. A beautiful, initial time-delay gap of about 9 msec is made possible by a live-end-dead-end design in which the narrator and microphone are placed in the dead end of the room facing the live end. This end is dominated by reflection phase-grating diffusors designed by Peter D'Antonio of RPG Diffusor Systems, Inc.[7]

Managing Reflections

The management of reflections is a major concern in control rooms. Davis realized this and recommended making the entire front end dead by applying absorbent to the surfaces. More recently, Berger and D'Antonio have devised a method that does not depend on absorption but rather on the shaping of surfaces to nullify the bad effect of reflections.[8-11] This sounds much simpler than it really is. The placement of a loudspeaker close to solid boundaries can greatly affect its output. If it is placed close to an isolated solid surface, its power output into half space is doubled—an increase of 3 dB. If the speaker is placed close to the intersection of two such surfaces, there is an increase of 6 dB because the power is confined to quarter space. If placed close to the intersection of three such solid surfaces, the power radiated is increased 9 dB for one-eighth space.

Placing loudspeakers at some specific distance from the trihedral surfaces of a room has been common in home listening rooms and in control rooms. If the distances from the loudspeaker to the surfaces are appreciable in terms of wavelength of the sound, new problems are introduced. The overall power boost effect may be minimized, but frequency response might be affected due to the constructive and destructive combination of direct and reflected waves.

A point source in a trihedral corner has a flat response at an observation point if that observation point is in a reflection-free zone. There are no reflections to contribute to interference effects. D'Antonio extends this observation to the idea of placing control-room-monitor loudspeakers in trihedral corners formed from splayed surfaces. By splaying the room boundaries, a reflection-free zone can be created around the operator. By splaying the walls and ceiling, it is even possible to extend the reflection-free zone across the entire console, several feet above, and enough space behind to include the producer behind the mix position.

The Reflection-Free-Zone Control Room

In their 1980 paper,[12] Don Davis and Chips Davis specified that there should be "...an effectively anechoic path between the monitor loudspeakers and the mixer's ears." What they called "anechoic" is today called a *reflection-free zone*. The most obvious way to achieve an anechoic condition is through absorption, hence the "dead end" designation.

To design a control room with a reflection-free zone (RFZ™) the engineer must deal with the mathematics of image sources. The contribution of a reflection from a surface can be considered as coming from a virtual source on the other side of the reflecting plane on a line perpendicular to that plane through the observation point and at a distance from the plane equal to that to the observation point. With splayed surfaces in three dimensions, it is difficult to visualize all the virtual sources, but this is necessary to establish the boundaries of the reflection-free zone.

A floor plan of a reflection-free zone control room is shown in Fig. 21-7. The monitor loudspeakers are flush-mounted as close as possible to the trihedral corner formed with the ceiling intersection. Next, both the front side walls and the front ceiling surfaces are splayed accurately to keep reflections away from the volume enclosing the operator.[10] It is possible to create an adequate reflection-free zone at the operator's position by proper splaying of walls. In this way, an anechoic condition is achieved without recourse to absorbents.

If absorbent is needed to control specific reflections, it can be applied to the splayed surfaces.

The rear end is provided with a complete complement of reflection phase-grating diffusors. In Fig. 21-7 the self-similarity principle is employed in the form of fractals. A high-frequency quadratic residue

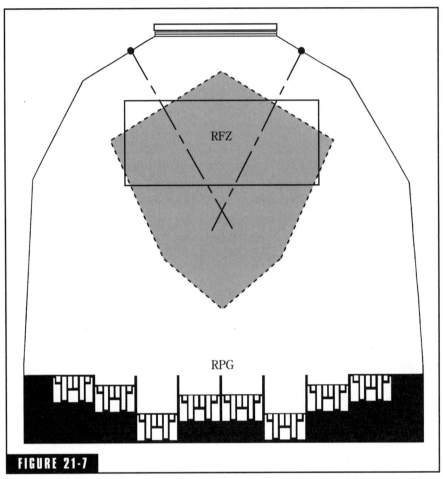

FIGURE 21-7

Plan view of a control room with a live, rather than a dead, front end. Front-end reflections in the operator's area are avoided by shaping the surfaces to create a reflection-free zone. The rear of the room is dominated by a mid- and high-band fractal diffusor and a low-frequency diffusor. *Peter D'Antonio, RPG Diffusor Systems, Inc., and the National Association of Broadcasters.*

diffusor is mounted at the bottom of each low-frequency diffusor well. Wideband sound energy falling on the rear wall is diffused and directed back to the operator with an appropriate time-delay gap. This sound is diffused both in space and time by the hemidisc pattern of the diffusors. An elevation view of the reflection-free zone control room is shown in Fig. 21-8.

Control-Room Frequency Range

Although this is not a treatise on control-room construction, some of the salient constructional features will be mentioned. The range of frequencies to be handled in the control room is very great, and every frequency-dependent component must perform its function over that range. The commonly accepted high-fidelity range of 20 Hz to 20 kHz is a span of 10 octaves, or 3 decades. This represents a range of wavelengths from about 57 feet to ⅝ inch. The control room must be built with this fact in mind.

The lowest modal frequency is associated with the longest dimension of a room, which may be taken as the diagonal. Below that frequency there is no modal resonance support for sound, much like the great outdoors. Needless to say, the room's response falls sharply below this frequency, which can be estimated by 1,130 / 2L, in which L = the diagonal distance in feet. For a rectangular room following Sepmeyer's proportions having dimensions of 15.26 × 18.48 × 12 feet, the

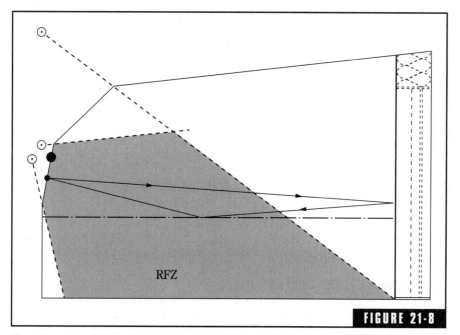

RFZ

FIGURE 21-8

Elevation view of the reflection-free zone of the control room of Fig. 21-7. *Peter D'Antonio, RPG Diffusor Systems, Inc., and the National Association of Broadcasters.*

diagonal is 28.86 feet. This places the low-frequency cutoff frequency for this particular room at 1,130/ [(2)(28.86)] = 21 Hz.

From 21 Hz up to about 100 Hz (for a room of this size), normal modes dominate and wave acoustics must be applied. From 100 Hz to about 400 Hz is a transition region in which diffraction and diffusion prevail. Above 400 Hz, true specular reflection and ray acoustics take over. These frequency zones determine construction of the control room—a massive shell to contain and distribute the low-frequency modal energy, and an inner shell for reflection control.

Outer Shell of the Control Room

The size, shape, and proportions of the massive outer shell of the control room determine the number of modal frequencies and their specific distribution as discussed in Chap. 15. There are two schools of thought: one prefers splaying of walls of the outer shell to "break up" modal patterns, and the other prefers the rectangular shape. Only a modest deviation from a rectangular shape toward a trapezoidal shape is feasible. Such a shape does not get rid of modal patterns, it just distorts them into an unpredictable form. Others feel that symmetry for both low-frequency and high-frequency sound better fits the demands of stereo.

To contain the low-frequency sound energy associated with control-room activities, thick walls, possibly 12-inch-thick concrete, are required.

Inner Shell of the Control Room

The purpose of the inner shell is, among other things, to provide the proper reflection pattern for the operator at the console. Consequently, its construction can be relatively light. For the inner shell, shape is everything.

Representative Control Rooms

An example of the application of reflection phase-grating diffusors in a studio is shown in Fig. 21-12. This is the Blue Jay Recording Studio of Carlisle, MA, also designed by Russell E. Berger. Figures 21-13 and

FIGURE 21-9

Control room of Master Sound Astoria, Astoria, NY, designed by Charles Bilello Associates utilizing RPG diffusors. *Photo by Kent Howard, Peter D'Antonio, RPG Diffusor Systems, Inc.*

FIGURE 21-10

Control room of Red Bus Studios of London, England, designed by Neil Grant of Harris Associates. RPG diffusors are used throughout. *Peter D'Antonio, RPG Diffusor Systems, Inc.*

FIGURE 21-11

Control room of Digital Services, Houston, TX, designed by Russell E. Berger, then of Pelton-Joiner-Rose Group of Dallas, TX. *Peter D'Antonio, RPG Diffusor Systems, Inc.*

21-14 are examples of low-frequency QRD™ diffusors utilizing the Diffractal™ principle, that is, diffusors within diffusors within diffusors to achieve wide-band effectiveness.

All of the diffusors shown in these photographs were supplied by RPG Diffusor Systems, Inc.[13] and are of the types discussed in Chap. 14.

Some European Designs

A glance at several Swiss control rooms and studios will broaden your view of the application of Schroeder diffusors. I am indebted to Helmuth Kolbe, a consultant in acoustics, for data on the rooms to be described, and which he designed.[7] The first example is Studio Sixty of Lausanne, Switzerland, owned by Wolfgang Ehrlich, which is busy around the clock on jobs from Paris and even from the United States.

A good example of the use of diffusors in a studio is shown in Fig. 21-15. The lower diffusor, a bit less than 7 feet high and 17.5 feet wide, is based on a primitive root-sequence. The wells are made from medium-density board with birch veneer of 1‌9⁄16-in width. The

FIGURE 21-12

Blue Jay Recording Studio of Carlisle, MA, showing extensive use of RPG diffusors on the ceiling and elsewhere. Designed by Russell L. Berger, then of the Pelton-Joiner-Rose Group of Dallas, TX. *Peter D'Antonio, RPG Diffusor Systems, Inc.*

maximum depth is 16.7 inches. The design frequency is taken as 400 Hz, which gives a calculated bandwidth of 300 to 4,305 Hz.

The upper diffusor is based on a quadratic residue sequence. A single period of 19 wells per period is used. Each well is 1 inch in width. Like the lower unit, no metal separators between the wells are used. The design frequency of the upper unit is 375 Hz, yielding a calculated bandwidth of 281.3 to 6,889 Hz.

Reactions of the musicians to the treatment of this studio are very positive: They like the crisp, natural-blended, lively sound it gives. Rear view of the control room for Studio Sixty is shown in Fig. 21-16.

FIGURE 21-13

Control room of Festival Records, Sydney, Australia. The low-frequency diffusor in the rear of the control room actually functions over a wide band because it is of the diffractal type which incorporates a diffusor within a diffusor within a diffusor. *Peter D'Antonio, RPG Diffusor Systems, Inc.*

The lower diffusor in the rear of this control room is based on a primitive root sequence. The design frequency of 400 Hz gives a calculated bandwidth of 300 to 6,624 Hz. It is made up of 1-inch medium-density board with a birch veneer facing. There are 67 wells per period and only a single period is used, which gives an overall width of 67.5 inches. Aluminum sheeting separates the wells.

The upper diffusor in Fig. 21-16 is based on a quadratic residue sequence. It has 37 wells of ¾ inch width, also made of medium-density board with birch veneer facing. The wells are separated by aluminum dividers, each a full 13 inches in depth. The design frequency in this case is 500 Hz, yielding a computed bandwidth of 375 to 8,611 Hz.

The plastic panels mounted on either side of the diffusors in Fig. 21-16 are special reflectors. The specular reflections from these panels are carefully placed in time either to shorten the initial time-delay gap, or as Kolbe wanted in this case, to sustain the important early reflections.

FIGURE 21-14

The rear of the control room of Winfield Sound, Toronto, Ontario, Canada. The low-frequency diffusor also features wide-band diffusion because of diffractals. *Peter D'Antonio, RPG Diffusor Systems, Inc.*

The small control room of A + D Studio, another Swiss studio, posed a number of problems and challenges to Kolbe. A direct view of the rear of this control room is shown in Fig. 21-17. A quadratic residue diffusor having a well width of 7¼ inches is located on either side of the door. The design frequency of this one is 500 Hz, giving a calculated bandwidth of 375 to 931 Hz. There are 19 wells per period with one period used. The total width of this diffusor is 11.5 feet.

The ceiling diffusors can be seen over and behind the operator's head position. The close-up of Fig. 21-18 reveals a center quadratic residue unit made up of boards of 0.45-in thickness with no metal

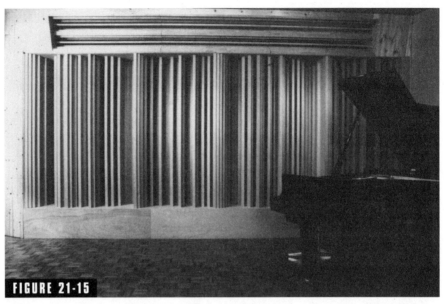

FIGURE 21-15

The recording studio of Studio Sixty, Lausanne, Switzerland. The large, lower diffusor, made of 1-⁹/₁₆ medium-density board with birch facing, is constructed on the basis of a primitive root sequence. The upper diffusor is based on a quadratic-residue sequence. *Acoustic Engineering H. Kolbe and Synergetic Audio Concepts.*

FIGURE 21-16

The rear of the control room at Studio Sixty, Lausanne, Switzerland designed by Helmut Kolbe. The lower diffusor is based on a primitive root sequence, the upper on a quadratic-residue sequence. The reflectors mounted on either side of the diffusors return specular reflections to the operator's position to support the diffused return. *Acoustic Engineering H. Kolbe.*

FIGURE 21-17

Control room of A+D Studios in Switzerland, designed by Helmut Kolbe. The rear, low-frequency diffusor is based on a quadratic-residue sequence, as are those on the walls and ceiling. *Acoustic Engineering H. Kolbe.*

dividers. Another quadratic residue diffusor made of $1\frac{1}{16}$-in boards is placed on either side of the center unit and on the walls (Fig. 21-19). These units have bandwidths that go up to 8,000 and 15,000 Hz.

In spite of the grave problems associated with such small control rooms and the difficulty of cramming so many diffusors into such a limited space, operators appreciate the fact that their room sounds as though it is much larger.

Figure 21-20 shows the control room of Studio Atmosphere between Zurich and Bern in Switzerland. The interesting feature here is the large concrete low-frequency diffusor out of sight behind the wide array of quadratic residue diffusors pictured. Together, a very wide frequency band is covered.

FIGURE 21-18

Closer view of the ceiling diffusors of the A+D control room of Fig. 17-16. *Acoustic Engineering H. Kolbe.*

Consultants

This chapter has only just scratched the surface of control-room design and has not even touched on the equally complicated design of recording studios. It is hoped that what has been covered will reveal some of the more recent developments that have contributed so much to sound quality in control rooms.

Anyone contemplating building, modernizing, or retrofitting a studio or control room should enlist the services of a qualified consultant in acoustics. How can a qualified consultant be found? One way is to make a point of visiting some of the better, more recent installations and evaluating them personally. Much can be learned from these owners about the consultants involved. One earmark of a qualified consultant is his dependence on modern instruments, such as the Techron™ instrument, in measuring and qualifying the space. It is a wise person who relies upon the knowledge and experience of a qualified consultant.

FIGURE 21-19

Another view of the control room A+D Studio in Switzerland. *Acoustic Engineering H. Kolbe.*

Endnotes

[1]Beranek, L.L., *Music, Acoustics, and Architecture*, New York, John Wiley and Sons, (1962).

[2]Davis, Don, *The Role of the Initial Time-Delay Gap in the Acoustic Design of Control Rooms for Recording and Reinforcing Systems*, 64th Convention, Audio Eng. Soc., (Nov 1979), preprint #1574.

[3]Davis, Chips and Don Davis, *Live-End-Dead-End Control Room Acoustics....(etc)*, Recording Eng/Prod, 10, 1 (Feb 1979) 41.

[4]Davis, Chips and G.E. Meeks, *History and Development of the LEDE™ Control-Room Concept*, 64th Convention, Audio Eng. Soc., Los Angeles, (Oct 1982), preprint #1954.

[5]D'Antonio, Peter and John H. Konnert, *The Role of Reflection-Phase-Grating Diffusors in Critical Listening and Performing Environments*, 78th Convention of the Audio Eng. Soc., Anaheim, CA, (May 1985), preprint #2255.

FIGURE 21-20

Closer view of the ceiling diffusors of the A+D control room of Fig. 17-16. *Acoustic Engineering H. Kolbe.*

[6]D'Antonio, Peter and John H. Konnert, *New Acoustical Materials and Designs Improve Room Acoustics*, 81st Convention, Audio Eng. Soc., Los Angeles, (Nov 1986), preprint #2365.

[7]Acoustic Engineering H. Kolbe, Zielacker Strasse 6, CH-8304 Wallisellen, Switzerland. Telephone: 01/830-10-39.

[8]Berger, Russell E., *Speaker/Boundary Interference Response (SBIR)*, Norman, IN, Synergetic Audio Concepts Tech Topics, 11, 5 (Winter 1984) 6 p.

[9]D'Antonio, Peter, *Control-Room Design Incorporating RFZ™, LFD™, and RPF™ Diffusors*, dB The Sound Eng. Mag., 20, 5 (Sept/Oct 1986) 47-55, 40 references.

[10]D'Antonio, Peter and John H. Konnert, *The RFZ™/RPG™ Approach to Control-Room Monitoring*, 76th Convention Audio Eng. Soc., New York, (Oct 1984), preprint #2157.

[11]Muncy, Neil A., *Applying the Reflection-Free Zone RFZ™ Concept in Control-Room Design*, dB The Sound Eng. Mag., 20, 4 (July/Aug 1986), 35-39.

[12]Davis, Don and Chips Davis, *The LEDE™ Concept for the Control of Acoustic and Psychoacoustic Parameters in Recording Rooms*, J. Audio Eng. Soc., 28, 9 (Sept 1980), 585-595.

[13]RPG Diffusor Systems, Inc., 651-C Commerce Drive, Upper Marlboro, MD 20772, Tel. 301-249-0044, FAX 301-249-3914.

Acoustics for Multitrack Recording

In the early days of recording, artists crowded around a horn leading to a diaphragm-driven stylus cutting a groove on a wax cylinder. In early radio dramatics, actors, actresses, and sound-effects persons moved in toward the microphone or faded back according to the dictates of the script. Greater freedom came as several microphones, each under separate control in the booth, were used. All this, of course, was monophonic.

Monophonic is still with us in commercial form, but the advent of stereophonic techniques has added a new dimension of realism and enjoyment to recordings, films, radio, etc. Stereo requires, basically, a dual pickup. This can be two separated microphones or two elements with special directional characteristics and electrical networks mounted close together. In broadcasting or recording a symphony orchestra, for example, it was soon found that some of the weaker instruments required their own microphones to compete with the louder instruments. The signal from these was proportioned between the left and right channels to place them properly in the stereo field. Here, again, we see a trend from two microphones to many.

Popular music has always been with us, but its form changes with time. Recording techniques came to a technological maturity just in time to be clasped to the breast of new-wave musicians and musical directors. Whether the Beatles were truly the vanguard of this new musical development or not will be left to historians, but their style spread like wildfire throughout the western world. "Good" sound quality in the traditional sense was not as much sought after as was a distinctive sound. Novelty effects, such as phasing and flanging, sold records by the millions. A new era of studio recording, variously called *multichannel*, *multitrack*, or *separation* recording, burst on the scene. It was beautifully adapted to the production of special effects, and the novel, distinctive sound, and it flourished.

Flexibility

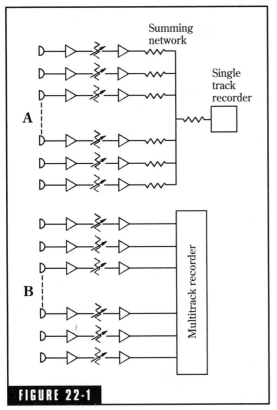

FIGURE 22-1

(A) The outputs of several microphones can be combined by a summing network and recorded on a single-track tape. (B) In basic multitrack recording, the output of each microphone is recorded on a separate track of the tape. The signals of the several tracks are combined later in a mix down session.

The key word is *flexibility*. Multitrack provides the means for recording one instrument or soloist at a time, if desired, as well as the introduction of special effects along the way. A production can be built up piece by piece and assembled later in the mix down. In Fig. 22-1A, signals from several microphones are combined in a summing network and fed to a single-track recorder. A variation is to use a two-track recorder, distributing signals from each microphone, partially or wholly, between the two tracks for artificial stereophonic recording and reproduction. In contrast, the signal of each microphone shown in Fig. 22-1B is recorded on a single track of a multitrack recorder. Many variations of this arrangement are possible. For example, a half-dozen microphones on the drums could be premixed and recorded on a single track, but mix down flexibility would be sacrificed in the process.

Once all the component parts of a musical production are recorded synchronously on separate tracks, they can then be mixed down to mono, stereo, or other multichannel form for release. Much attention can be lavished on each detail in the mix down, a stage that becomes a very important part in the production chain of events.

Advantages of Multitrack

Flexibility is the outstanding overall advantage of multitrack techniques, but to understand the true breadth of the word some supporting detail is offered. Multitrack recording makes possible the conquering of space and time. Suppose the drums and electric piano are recorded on separate tracks on Monday. On Tuesday, the guitar player is available, so a third track is recorded as he or she listens on headphones to a temporary mix of the first two. The tape can be shipped across the country to another studio to pick up a big-name female vocalist between engagements. In this way, a musical production can be built up a piece at a time. It might not be the best way, but it is possible.

Another big advantage of separation recording is the almost complete control it provides over the relative levels of each instrument and artist. Each track can be given just the equalization desired, often after considerable experimentation. Special effects can be injected at the mix down stage. Reverberation can be added in any desired amount.

It is expensive when artists, musicians, and technical crews are standing by for a retake of one performer or to argue the next step. The mix down is a calmer session than recording a group on a premix basis.

There is also a noise advantage in multitrack. In a "mix and record as the group plays" type of session, levels of the various instruments are adjusted, and the result, as recorded, is frozen as the final mix with no recourse. Some potentiometers are set high and some low depending on the requirements of the source, and the signal-to-noise ratio of each channel is thereby fixed. In separation-type recording, the standard practice is to record all channels at maximum level, which guarantees the best signal-to-noise ratio on the original tape. In the multichannel mix down, these levels are adjusted downward to achieve the desired balance, but there is still a significant noise advantage over the premixed case.

In addition to recording each track at maximum undistorted level, the bandwidth of some channels can be reduced without affecting the sound of the instruments. For example, violin sound has practically no energy below 100 Hz. By cutting the low-frequency end of that channel at 100 Hz, the noise is reduced with no noticeable degradation of quality. The sound of some instruments is essentially unaltered by cutting some low-frequency energy, some by cutting highs.

A pervasive argument for multichannel recording is that it is what the client wants, expects, and gets. Customer demand, in the final analysis, can be the greatest factor influencing the drift toward multichannel. Of course, the customer demands it because of the inherent flexibility, potential savings, and other virtues.

Disadvantages of Multitrack

In spite of the signal-to-noise ratio advantage, multitrack carries a disadvantage of noise buildup as the number of tracks combined is increased. When two tracks having equal noise levels are mixed together, the noise on the combined track is 3 dB higher than either original. If 32 tracks are engaged in a mix down, the combined noise is 15 dB higher than a single track. Table 22-1 lists noise buildup for commonly used track configurations. It is simply a matter of adding noise powers. Mixing eight tracks of equal noise powers means that the total noise is 10 log 8 = 9.03 dB higher than the noise of one track. If the noise of one track is −80 dB referred to the reference level, the noise of 16 tracks is −68 dB.

The *dynamic range* of a system is defined as the total usable range of audio level between the noise at the lower extreme and the full undistorted level at the upper extreme. The more tracks on a given width of tape, the narrower each track and the lower the reproduced signal level. Increased noise and decreased reproduced level spell narrower dynamic range.

Table 22-1 Multitrack noise buildup.

Number of tracks	Noise buildup above noise of one track, dB
2	3.01
4	6.02
8	9.03
16	12.04
24	13.80
32	15.05
48	16.81

The closer the spacing of tracks on a tape, the greater the crosstalk between adjacent tracks. Recording circumstances determine the magnitude of the resulting problem. For instance, if the two adjacent tracks are of two musical instruments recorded simultaneously in the studio while playing the same number, the congruity might make the cross talk acceptable. The degree of separation realized between microphones in the studio affects the judgment on the seriousness of tape crosstalk. If the material on adjacent tracks is unrelated (usually not the case in music recording) the crosstalk will be much more noticeable.

Artistic responsibility might become diffused in multitrack recording unless the musical director is intimately involved in both recording and mix down. The very nature of the mix down technique dictates seemingly endless hours of detailed comparison of tracks, recording pickups, and overdubs, which are the basic creative steps in a production. Often this meticulous duty falls on the recording engineer with only an occasional check by the music director. In contrast, the old style of premix recording session ends with an essentially completed product with the music director in full charge all the way.

Some separation recording sessions tend to separate musicians to the extent that spontaneous interaction is lost. Musicians respond to each other, and this desirable effect may or may not be maintained in the face of cueing by foldback headphones and as the musicians are physically isolated by baffles, screens, and isolation booths.

While we are considering the negative aspects of multitrack recording, the degradation of quality as the tape is run and rerun scores of times must be added to the rest. In what other endeavor does the original recording receive such treatment? With 2-inch magnetic tape, contact with the head becomes a problem as the tape is passed through the machine, especially the outer tracks. The wise recording engineer reserves for outside tracks those sounds least affected by loss in high-frequency response.

Achieving Track Separation

Achieving 15- to 20-dB intertrack separation requires intelligent effort and attention to detail. Without such separation, the freedom of establishing relative dominance in the mix down is sacrificed. The following methods are employed to yield the required separation:

- Adjusting the acoustics of the studio.
- Spacing the artists.
- Using microphone placement and directivity.
- Use of physical barriers.
- Use of gating techniques.
- Use of contact transducers or electrical instruments.

Studio Acoustics

Heretofore in considering studio acoustics, the criterion has been quality (naturalness and freedom from colorations) of the recorded signal. In multitrack recording, the emphasis is shifted to track separation, and the very meaning of quality tends to be dissipated in the process. Reflective surfaces in the studio are not ruled out, but they are generally localized for specific instruments while the general studio acoustics are made quite dead and absorptive. The number of musicians to be accommodated is limited, among other things, by the size of the studio. If the walls are highly absorptive, musicians can be placed closer to them, and more artists can be accommodated in a given space. Reverberation time rather loses its meaning in a studio specializing in separation recording, but if measured, it would tend to be quite short.

Distance between Artists

In an absorptive studio, increasing distance between the various instruments is a step toward track separation. Sound level falls off at a rate of 6 dB for each doubling of the distance in free field. While the falloff rate is less indoors, this is still a fair rule to use in estimating the separation that can be realized through spacing of musicians.

Microphone Management

This principle of separation by distance also applies to microphones. The placement of the microphone for musician A and the microphone for musician B must be considered along with the actual relative positions of A and B. In some cases, there is a directional effect associated with certain musical instruments that can be used to advantage, certainly the directional properties of microphones can be used to improve separation. The distance between adjacent musicians and the distance between microphones are obvious factors as well as the distance between each musician and his or her own microphone. There is an interplay between all these distance effects and microphone directivity. The nulls of a cardioid or bidirectional microphone pattern may save the day in controlling a troublesome crosstalk problem.

Barriers for Separation

Physical separation of musicians, absorbent studios, and proper selection, placement, and orientation of microphones still are limited in the degree of acoustical separation they can produce. Baffles (or screens, as they are sometimes called) are used to increase isolation of the sound of one musician from that of another. Baffles come in a great variety of forms: opaque and with windows, reflective and absorbent, large and small. And now, with the availability of quadratic residue diffusors, sound diffusion can become a feature of baffles. Extreme forms of barriers are nooks and crannies for certain instruments or a separate booth for drums or vocals.

The effectiveness is very low for baffles of any practical size at low frequencies. Once more we come up against the basic fact of physics that an object must be large *in terms of the wavelength of the sound* to be an effective obstacle to the sound. At 1 kHz, the wavelength of sound is about 1 ft, hence a baffle 6-ft wide and 4-ft tall would be reasonably effective. At 100 Hz, however, the wavelength is about 11 feet and a sound of that frequency would tend to flow around the baffle, regardless of the thickness or material of the baffle.

Electronic Separation

Some use, although not extensive, has been made of electronic gating circuits to improve separation between sources. These circuits reject all signals below an adjustable threshold level.

Electronic Instruments and Separation

Contact pickups applied to almost any musical instrument with a special adhesive can transform an acoustical instrument to an amplified instrument. In addition, there are many electrical instruments that are completely dependent on the pickup transducer and amplification. The electrical output from such instruments can be fed to the console, providing dependence on an electrical signal rather than a microphone pickup of an acoustical signal. The separation between two such tracks can be very high. Amplified instruments with their own loudspeakers in the studio can be picked up by microphones placed close to the loudspeakers. Even though the quality of the sound is degraded, this approach has its enthusiastic followers.

The Future of Multichannel

Multichannel recording techniques have production advantages that promise to be a permanent part of the recording scene of the future to one degree or another. The audio control room of a network television production center probably has a mixing console with at least 48 channels and multitrack recording facilities to match. Recording a symphony orchestra today commonly requires as many channels and as many tracks. In this case, it is not true separation recording, but rather single-point recording for ambience with augmentation of certain instruments, sections, and soloists as required. At this time, there seems to be a trend away from the artificialities of strict separation recording in some areas.

Automation

The number of knobs, switches, buttons, and VU meters on the average recording console is enough to dazzle the uninitiated and impress the musically oriented client. In fact, it becomes something of a problem for an operator equipped with only one pair of eyes, one brain, two hands, and a normal reaction time to operate all these controls and still have time for the more creative aspects of his or her job. Mini-computer automation control relieves the operator of much of the tedium and releases the operator for more creative work. Automatic computer-controlled adjustment of pot settings, equalization, etc., in a mix down, especially, is a boon to the operator, making him or her a far more productive worker. Development in the direction of greater automation is inevitable for the future.

Audio/Video Tech Room and Voice-Over Recording

Post recording; digital sampling; midi; editing; sound effects; (foley) dialog replacement; voice over; sound processing; digital/analog recording; synthesizers; composing; video production; equipment testing/evaluation.

A separate room for each function? Hardly practical. One room for all these functions? Possible, with some compromises, but this is the working hypothesis for this chapter.

With so many functions, the only recourse is to consider the acoustical factors common to all of the functions:

GENERAL ROOM ACOUSTICS Any function requiring an open microphone demands reasonably good general acoustics.

LISTENING Every audio function must eventually be checked by the human ear. This means that loudspeakers will be required as well as adhering to the principles of Chap. 19.

NOISE Especially quiet conditions are required if digital recording standards are to be met. Noise from the outside as well as from within the room must be considered. The noise from many pieces of production equipment (fans, etc.) must be considered.

Selection of Space: External Factors

So many noise problems can be solved at the outset by choosing a space in a quiet location. Is it close to a railroad switch yard or busy intersection? Under the approach path (or worse yet, the take-off pattern) of the local airport? The maximum external noise spectrum must be reduced to the background NC noise goal within the space by the transmission loss of the walls, windows, etc.

Selection of Space: Internal Factors

If the space being considered is within a larger building, other possible noise and vibration sources within the same building must be identified and evaluated. Is there a printing press on the floor above? A machine shop? Is there a noisy elevator? A reinforced concrete building efficiently conducts noises throughout via structural paths.

Audio/video work space might require support activities such as duplication, accounting, sales, telephone, each with its own noise generation potential. Serious review of such noise sources concentrates attention on transmission loss characteristics of internal walls, floors, and ceilings.

Work Space Treatment

A bare work space obviously needs some sort of treatment. The structure of the walls determines the degree to which external noise is attenuated (Chap. 8). This is equally true of the floor and the ceiling structures in regard to attenuation of noise from below or from above. The bare work space is merely a container of air particles responsible for the propagation of sound (good for breathing, too). Practically all attenuation of sound (signals as well as noise) takes place at the boundaries of this air space. The absorption of sound by the air itself is negligible, except at the higher audio frequencies. Carpet on the floor boundary, lay-in panels on the ceiling boundary, and absorbent on the wall surfaces will reduce sound energy in the room upon each reflection. The example to follow illustrates how each of these needs can be carefully optimized.

Many of the points to be applied to the audio/video work place have already been covered in depth earlier in this book. To avoid excessive redundancy, frequent references to appropriate earlier coverage is made. A review of this referenced material is very much a part of this chapter.

Audio/Video Work Place Example

It has been stated earlier that an audio room smaller than 1,500 cu ft is almost certain to produce sound colorations. A medium-sized work place (selected from Table 20-1) will be considered having dimensions of 12' × 15'-4" × 18'-6" (Fig. 23-1), yielding a floor area of 284 sq ft and a volume of 3,400 cu ft. The proportions are as close to "optimum" as is feasible (Table 13-2 and point B, Fig. 13-6).

Appraisal of Room Resonances

Even though the room proportions are favorable and highly recommended, it is well to verify axial mode spacing. Only axial modes are considered because the tangential modes are 3 dB down and the oblique modes are 6 dB down with respect to the more powerful axial modes (Ref. 3 in Chap. 15). All of the length, width, and height axial mode resonance frequencies to 300 Hz are listed in Table 23-1. These constitute the low-frequency acoustics of this room. The spacings of these modes determine the smoothness of the low-frequency room response. Spacings of 25.8 and 30.6 are the only ones exceeding Gilford's suggested limit of 20 Hz (Ref. 4 in Chap. 15). Figure 15-21 suggests that these two are possible sources of speech and music coloration. This is by no means a major flaw of the room. Favorable room proportions only minimize the potential for modal problems, not eliminate them.

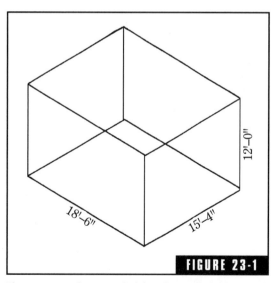

FIGURE 23-1

The space to be occupied by the audio/video work place example. The proportions are selected to distribute room mode resonances favorably.

Table 23-1 Axial modes of trial audio/video work place.

	Length L = 18.5' $f_1 = 565/L$ (Hz)	Width W = 15.3' $f_1 = 565/W$ (Hz)	Height H = 12.0' $f_1 = 565/H$ (Hz)	Arranged in ascending order	Axial mode spacing (Hz)
	Room dimensions = 18.5' × 15.3' × 12.0'				
f_1	30.6	36.8	47.1	30.6	6.2
f_2	61.2	73.6	94.2	36.8	10.3
f_3	91.8	110.4	141.3	47.1	14.1
f_4	122.4	147.2	188.4	61.2	12.4
f_5	153.0	184.0	235.5	73.6	18.2
f_6	183.6	220.8	282.6	91.8	2.4
f_7	214.2	257.6	329.3	94.2	16.2
f_8	244.8	294.4		110.4	12.0
f_9	275.4	331.2		122.4	18.9
f_{10}	306.0			141.3	5.9
				147.2	5.8
				153.0	30.6
				183.6	0.4
				184.0	4.4
				188.4	25.8
				214.2	6.6
				220.8	14.7
				235.5	9.3
				244.8	12.8
				257.6	17.9
				275.4	7.2
				282.6	11.8
				294.4	11.6
				306.0	

Mean axial mode spacing = 12.0
Standard deviation = 7.2

Control of Room Resonances

The control of room resonances is now approached with some understanding. Some bass trap absorption effective in the 50–300 Hz region will be necessary. If the audio/video work space example is to be of frame construction, an impressive amount of low-frequency sound absorption is built into the structure. The floor and wall diaphragms vibrate and absorb low-frequency sound in the process. This is borne out in calculations coming up. Additional low-frequency absorption might be needed.

Treatment of Work Place

For a rough estimate of the absorption units (sabins) needed in the room we fall back on Eq. 7-1:

$$\text{Reverberation Time} = \frac{(0.049)\,(V)}{\text{Total Absorption}}$$

in which V = volume of the space in cu ft

$$= 3,400 \text{ cu ft}$$

Assuming a reverberation time of 0.3 second, Eq. 7-1 becomes:

$$\text{Total Absorption} = \frac{(0.049)\,(3,400)}{0.3}$$

$$= 555 \text{ absorption units (sabins)}$$

This is an approximate absorption that will result in a reasonable acoustical situation that can be trimmed at a later time to meet specific needs.

Calculations

The proposed treatment of the audio/video work place example is shown in Fig. 23-2. This is a "fold-out" plan in which the four walls, hinged along the edges of the floor, are laid out flat. Table 23-2 shows each step of the calculations using absorption coefficients for the different materials found in the appendix. It is assumed that the room is of frame construction, with a wooden floor, ½-in drywall on all walls, and carpet covering the floor. The drop ceiling panels are Owens-Corning Frescor. These are the common elements of treatment of the room.

The less common elements include polycylindrical diffusing/absorbing units, a bass trap under the polys, and Abffusor™ units in the ceiling and on one wall.

Even though they are very old-fashioned, polys are inexpensive and impressive in appearance. Their simple construction is detailed in Figs. 9-25 through 9-28. As for effectiveness as a diffusor, Fig. 14-21 compares the diffusion of the poly directly with the diffusion of a flat panel

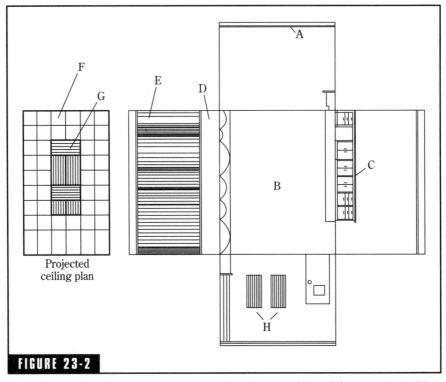

FIGURE 23-2

Possible treatment of the example of an audio/video work place. **(A)** Suspended ceiling, **(B)** carpet on the floor, **(C)** suggested work table and storage space, **(D)** low-frequency Helmholtz resonator bass trap, **(E)** polycylindrical sound absorbers and diffusors, **(F)** ceiling lay-in panels, **(G)** Abffusors™, absorbers and quadratic residue diffusion combined, **(H)** Abffusor™ panels. (RPG Diffusor Systems, Inc.)

(reference only, no one claims it to be a good diffusor), distributed absorption, and the quadratic residue diffusor. The poly having the smaller chord gives a normal incidence diffusion pattern somewhat similar to the quadratic residue diffusor.

Instead of polys, this wall could very well be covered with Abffusors™, which would give almost perfect diffusion plus excellent absorption. Cost should determine which to use. If labor costs to build the polys are too high, Abffusors™ would be ideal. Abffusors are laid into the suspended ceiling framework (50 sq ft) and two panels (16 sq ft) are mounted on the wall near the door to discourage lengthwise flutter.

The bass trap beneath the polys is a simple perforated panel Helmholtz-type of absorber tuned to give peak absorption at the low

Table 23-2 Reverberation calculations trail audio/video work place.

Material	S Sq. ft.	125 Hz a	Sa	250 Hz a	Sa	500 Hz a	Sa	1 kHz a	Sa	2 kHz a	Sa	4 kHz a	Sa
Drywall	812	0.10	81.2	0.08	65.0	0.05	40.6	0.03	24.4	0.03	24.4	0.03	24.4
Wood floor	284	0.15	42.6	0.11	31.2	0.10	28.4	0.07	19.9	0.06	17.0	0.07	19.9
Drop ceiling	234	0.69	161.5	0.86	201.2	0.68	159.1	0.87	203.6	0.90	210.6	0.81	189.5
Carpet	284	0.08	27.7	0.24	68.3	0.57	161.9	0.69	196.0	0.71	201.6	0.73	207.3
Polys	148	0.40	59.2	0.55	81.4	0.40	59.2	0.22	32.6	0.20	29.6	0.20	29.6
Bass trap	37	0.65	24.1	0.22	8.1	0.12	4.4	0.10	3.7	0.10	3.7	0.10	3.7
Abffusor™	50	0.48	27.8	0.98	49.0	1.2	60.0	1.1	55.0	1.08	54.0	1.15	57.5
Total absorption, sabins			424.1		504.1		513.6		535.2		540.9		531.9
Reverberation time, sec.			0.39		0.33		0.32		0.31		0.31		0.31

end of the audio band. This can be covered with an open weave material for appearance, with minimum compromising of its operation. Full design particulars for the perforated panel bass trap resonator are given in Chap. 9.

This is not necessarily the end of the acoustic treatment of the room. The bass trap might not be sufficient to control boom in the room. If more bass absorption is needed, the two corners opposite the door could be treated in one of the ways described in Fig. 19-4.

The Voice-Over Booth

The old telephone-booth-sized voice recording spaces, common during the early days of radio, were notorious for their abominable acoustics. The surfaces were treated with "acoustic tile" or some such material that absorbs high-frequency energy well, but low-frequency energy practically none at all. This means that important voice frequencies were over-absorbed and that the low-frequency room modes were left untouched. Because of the small dimensions of the "booth," these room modes were obnoxious because they were few and widely spaced.

Dead-End Live-End Voice Studio

To replace the "voice booth" a bit more space is needed, although not much. The object is to obtain a clean direct sound free from early reflections followed by a normal ambient decay. One approach is that of Fig. 23-3. The microphone is so placed with respect to the absorbent end that no reflections reach it except those diffused from the live end of the room. All walls of the dead end must be absorptive as well as the floor. A small early time gap will exist between the time of arrival of the direct sound and the arrival of the first highly-diffused sound from the live end. A very clean voice recording should be obtained with this arrangement. This type of treatment could also be used for isolation purposes.

Voice-Over Booths

Making a small booth highly absorbent, for both the high and low frequencies, produces an eerie effect on the narrator. Such an anechoic

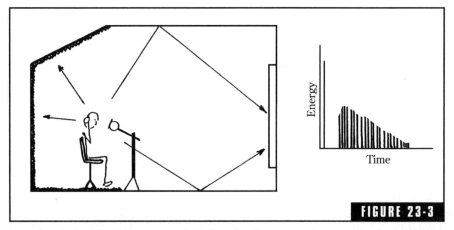

FIGURE 23-3

Voice booth for high-quality sound. Early reflections are eliminated by the absorbent rear end and the diffusor assures a dense ambient decay.

space gives essentially no acoustic feedback, which the narrator needs for orientation and voice adjustment. The use of headphones only contributes to the sense of isolation as the direct voice sound is cut off. The only sound feedback is through bone conduction. For these reasons, and others, an anechoic space is unsuitable for voice-over or announce booths.

The Quick Sound Field™

The concept of a "Quick Sound Field™" (i.e., live, but dry), which comes from Acoustic Sciences Corporation (Ref. 3 in Chap. 19) can be applied to the small (e.g. 4 × 6 ft or 6 × 8 ft) voice recording room. The realization of a practical quick sound field is based on a multiplicity of half-round Tube Traps™ as shown in Fig. 23-4. A quarter-round Trap is mounted in each wall-wall and ceiling-wall intersection to provide absorption to control the normal modes. Half-round traps alternate on the walls and ceiling with hard, reflective drywall

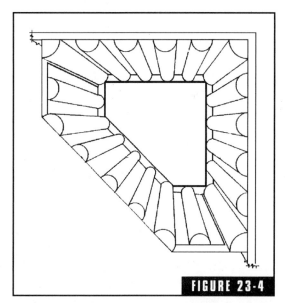

FIGURE 23-4

View of small voice-over studio based on the Quick Sound Field™ idea. The ceiling, walls, door, and the observation window are all covered with spaced half-round Tube Traps™. *Acoustic Corporation.*

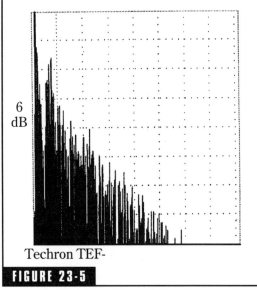

Techron TEF-

FIGURE 23-5

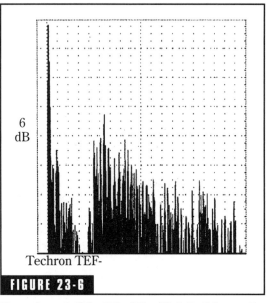

Techron TEF-

FIGURE 23-6

Energy-time-curve (ETC) of the Quick Sound Field™ studio. About 1,000 reflections per second are produced in such a space resulting in a dense, but rapidly decaying ambient field. The horizontal time base goes to 80 ms. *Acoustic Sciences Corporation.*

The early part of the ETC of Fig. 23-5. In this case the expanded time base goes to 20 ms. A well-defined early time gap of about 3 ms is revealed. *Acoustic Sciences Corporation.*

strips as wide as the traps (9-, 11-, or 16-inch). Even the door and window need to be covered in this way to avoid compromise of the sound field. Limited visibility is available through the window glass strips of reflecting surface. The floor remains untreated.

In the usual small recording room there are too few early reflections that result in noticeable comb-filter colorations. Reflections off the script stand or window, which normally cause comb-filter colorations, are overwhelmed by the swarm of diffuse reflections from the many reflecting surfaces of the room. The broad-band absorption of the half-round traps combines with the reflections from the hard areas between the traps to produce a dense, diffuse group of reflections immediately following the direct signal.

An energy-time-curve (ETC) of the room's response is shown in Fig. 23-5. The horizontal time axis goes to 80 ms in this case. The spike at the left extreme is the direct signal, followed by a smoothly decaying ambient sound field. This decay is quite fast, corresponding to a reverberation time of 0.08 sec (80 ms). The first 20 ms of the same ETC is shown in Fig. 23-6. The early time gap of 3 ms is distinct and clear, followed by a smooth, dense fill of reflections.

The time-energy-frequency (TEF) "waterfall" record of the complete performance of the room is shown in Fig. 23-7. The vertical amplitude scale is 12 dB per division. The horizontal frequency scale runs from 100 to 10,000 Hz. The diagonal scale is time, running from zero at the rear, to 60 ms at the extreme toward the observer. This remarkable measurement shows no prominent room modal resonances, only a dense series of smooth decays with time throughout the entire 100 Hz to 10 kHz frequency range.

The Quick Sound Field™ voice-over studio just described yields voice recordings that are said to be accurate and clean-sounding. Further, recordings made at one time match those made at other times, even if different microphone positions are used. Moving the microphone only changes the fine structure of the ambient (some thousand reflections per second) and has little effect on the quality of sound. A photograph of a Quick Sound Field™ studio is shown in Fig. 23-8.

Endnote

[1]Noxon, Arthur M., *Sound Fusion and the Acoustic Presence Effect*, Presented at the 89th Convention of the Audio Engineering Society, Los Angeles (Sept 1990), preprint #2998.

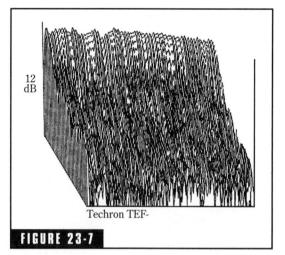

12 dB

Techron TEF-

FIGURE 23-7

"Waterfall" recording of the Quick Sound Field™ of the studio of Fig. 23-4. The vertical scale is energy (12 db per division), the horizontal scale is frequency (100 to 10,000 HZ), and the diagonal scale is time (zero at the back to 60 ms closest to reader). The broadband decay of sound is smooth and balanced and very dense (compare Fig. 21-5). *Acoustic Sciences Corporation.*

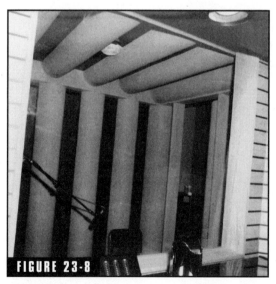

FIGURE 23-8

A studio constructed according to the Quick Sound Field™ principles. (Tube Traps™ have been removed from the window for the photograph.) *Acoustic Sciences Corporation.*

Adjustable Acoustics

If an audio space is used for only one purpose, it can be treated with some precision. Even though a multipurpose audio space carries with it some compromises, economics may dictate that a given space must serve more than one purpose. It is necessary to weigh the acoustical compromises against the ultimate sound quality of the product.

For this chapter it is desirable to cast aside the impression that acoustical treatment is an inflexible exercise and to consider means of introducing the element of adjustability.

Draperies

As radio broadcasting developed in the 1920s, draperies on the wall and carpets on the floor were almost universally used to "deaden" studios. During this time there was remarkable progress in the science of acoustics. It became more and more apparent that the old radio studio treatment was quite unbalanced, absorbing middle- and high-frequency energy but providing little absorption at the lower frequencies. As proprietary acoustical materials became available, hard floors became common and drapes all but disappeared from studio walls.

A decade or two later the acoustical engineers, interested in adjusting the acoustical environment of the studio to the job to be done, turned with renewed interest to draperies. A good example of this early return to draperies was illustrated in the rebuilding of the old Studio 3A of the National Broadcasting Company of New York City in 1946. This studio was redesigned for optimum conditions for making records for home use and transcriptions for broadcast purposes. The acoustical criteria for these two jobs differ largely as to the reverberation-frequency characteristic. By the use of drapes and hinged panels (considered later), the reverberation time was made adjustable over more than a two-to-one range. The heavy drapes were lined and interlined and were hung some distance from the wall to make them more absorbent at the lower frequencies (see Figs. 9-13 through 9-16). When the drapes were withdrawn, polycylindrical elements having a plaster surface were exposed. (Plywood was in critical supply in 1945–46.)

If due regard is given to the absorption characteristics of draperies, there is no reason, other than cost, why they should not be used. The effect of the fullness of the drape must be considered. The acoustical effect of an adjustable element using drapes can thus be varied from that of the drape itself when closed (Fig. 24-1) to that of the material behind when the drapes are withdrawn into the slot provided. The wall treatment behind the drape could be anything from hard plaster for minimum sound absorption to resonant structures having maximum absorption in the low-frequency region, more or less complementing the effect of the drape itself. Acoustically, there would be little point to retracting a drape to reveal material having similar acoustical properties.

Adjustable Panels: Absorption

Portable absorbent panels offer a certain amount of flexibility in adjusting listening room or studio acoustics. The simplicity of such an arrangement is illustrated in Fig. 24-2A and B. In this example a

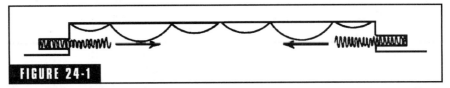

FIGURE 24-1

The ambience of a room may be varied by pulling absorptive drapes in front of reflective areas.

FIGURE 24-2A

(A). The simplest and cheapest way to adjust the reverberation characteristics of a room is to use removable panels. *These photographs were taken in the studios of the Far East Broadcasting Company, Hong Kong.*

FIGURE 24-2B

(B). Close-up of hanging detail.

perforated hardboard facing, a mineral fiber layer, and an air cavity constitute a low-frequency resonator. Hanging such units on the wall adds low-frequency absorption, and contributes somewhat to sound diffusion. There is some compromising of the effectiveness of the panels as low-frequency resonators in that the units hang loosely from the mounting strip. "Leakage" coupling between the cavity and the room would tend to slightly destroy the resonant effect. Panels may be removed to obtain a "live" effect for instrumental music recording, for example, or introduced for voice recording.

Free standing acoustical flats are useful studio accessories. A typical flat consists of a frame of 1×4 lumber with plywood back filled with a low density (e.g., 3 lb per cu ft) glass fiber board faced with a fabric such as muslin or glass fiber cloth to protect the soft surface. Arranging a few such flats strategically can give a certain amount of local control of acoustics.

Adjustable Panels: The Abffusor™

Combining broadband absorption in the far field with horizontal or vertical diffusion in the near field down to 100 Hz for all angles of incidence is the accomplishment of RPG Diffusor Systems[3] in their Abffusor™. The Abffusor™ panel works on the absorption phase grating principle using an array of wells of equal width separated by thin dividers. The depth of the wells is determined by a quadratic residue sequence of numbers to diffuse what sound is not absorbed.

The Abffusor™ panels are approximately 2×4 and 2×2 ft. They can be mounted in ceiling grid hardware or as independent elements. Figure 24-3 describes one method of mounting the panels on the wall with beveled cleats. The panels can be easily removed by lifting off the cleats. The sectional drawing in Fig. 24-3 reveals the construction of the unit.

The absorption characteristics of the Abffusor™ are shown in the graphs of Fig. 24-4 for two mountings. Mounted directly on a wall an absorption coefficient at 100 Hz of about 0.42 is obtained. With 400 mm of air space between the Abffusor™ and the surface, the coefficient is doubled. The latter is approximately the performance with the Abffusor™ mounted in a suspended ceiling grid. Near perfect absorbance is obtained above 250 Hz. The idea of obtaining such

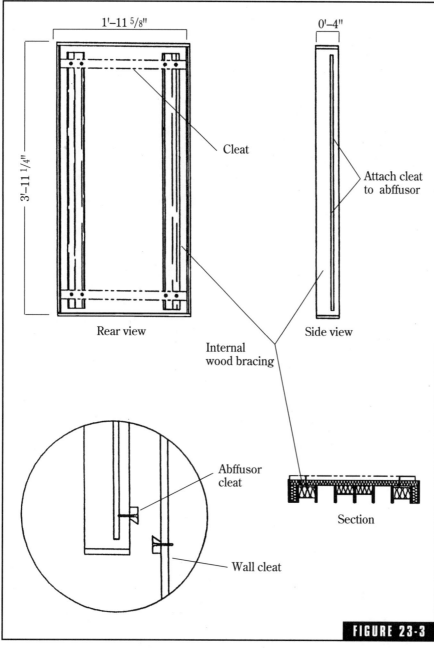

1'-11 5/8"

3'-11 1/4"

Cleat

Rear view

0'-4"

Attach cleat
to abffusor

Side view

Internal
wood bracing

Abffusor
cleat

Section

Wall cleat

FIGURE 23-3

Abffusor™ wall mounting detail.

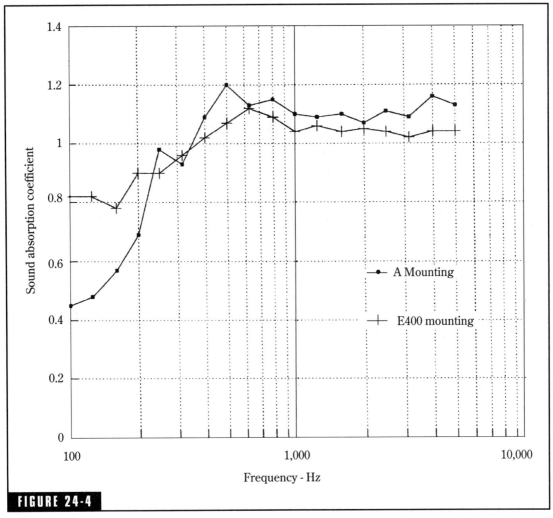

FIGURE 24-4

Abffusor™ absorption coefficients.

wideband sound absorption plus sound diffusion is very appealing to the designer.

Hinged Panels

One of the least expensive and most effective methods of adjusting studio acoustics is the hinged panel arrangement of Figs. 24-5A and B. When closed, all surfaces are hard (plaster, plasterboard, or plywood). When

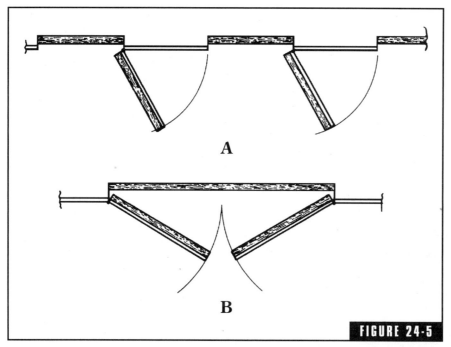

FIGURE 24-5

An inexpensive and effective method of incorporating variability in room acoustics is through the use of hinged panels, hard on one side and absorbent on the other.

opened, the exposed surfaces are soft. The soft surfaces can be covered with 3 lb/cu ft density glass fiber boards 2 to 4 inches thick. These boards could be covered with cloth for the sake of appearance. Spacing the glass fiber from the wall would improve absorption at low frequencies.

Louvered Panels

The louvered panels of an entire section can be adjusted by the action of a single lever in the frames commonly available for home construction, Fig. 24-6A. Behind the louvers is a low-density glass fiber board or batt. The width of the panels determines whether they form a series of slits, Fig. 24-6B, or seal tightly together, Fig. 24-6C. In fact, opening the louvers of Fig. 24-6C slightly would approach the slit arrangement of Fig. 24-6B acoustically, but it might be mechanically difficult to arrange for a precise slit width.

The louvered panel arrangement is basically very flexible. The glass fiber can be of varying thickness and density and fastened

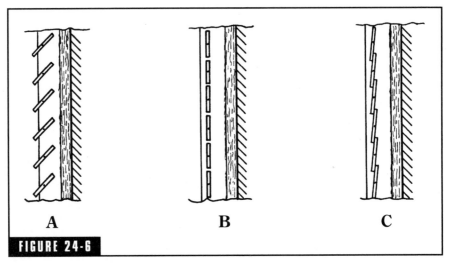

FIGURE 24-6

Louvered panels may be opened to reveal absorbent material within, or closed to present a reflective surface. Short louvers can change from a slat resonator (closed) to reveal absorbent material within (open).

directly to the wall or spaced out different amounts. The louvered panels can be of hard material (glass, hardboard) or of softer material such as wood and they can be solid, perforated, or arranged for slit-resonator operation. In other words, almost any absorption-frequency characteristic shown in the graphs of earlier chapters can be matched with the louvered structure with the added feature of adjustability.

Variable Resonant Devices

Resonant structures for use as sound absorbing elements have been used extensively in the Danish Broadcasting House in Copenhagen[1]. One studio used for light music and choirs employs pneumatically operated hinged perforated panels as shown in Fig. 24-7A. The effect is basically to shift the resonant peak of absorption as shown in Fig. 24-7B. The approximate dimensions applicable in Fig. 24-7A are: width of panel 2 ft, thickness 3/8 in., holes 3/8 in. diameter spaced 1-⅜ inch on centers. A most important element of the absorber is a porous cloth having the proper flow resistance covering either the inside or outside surface of the perforated panel.

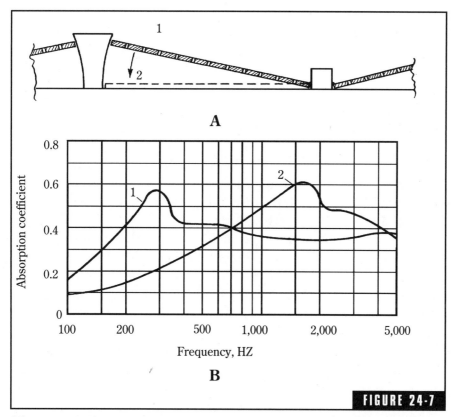

(A) Pneumatically operated hinged, perforated panels used to vary reverberation in the Danish Broadcasting House in Copenhagen. An important element not shown is a pourous cloth of the proper flow resistance covering one side of the perforated panel. **(B)** Changes in absorption realized by shifting the panel of (A) from one extreme to another.

When the panel is in the open position the mass of the air in the holes and the "springiness" (compliance) of the air in the cavity behind act as a resonant system. The cloth offers a resistance to the vibrating air molecules, thereby absorbing energy. When the panel is closed the cavity virtually disappears and the resonant peak is shifted from about 300 to about 1,700 Hz (Fig. 24-7B). In the open condition the absorption for frequencies higher than the peak remains remarkably constant out to 5,000 Hz.

A studio designed by the late William B. Snow for the sound mixing-looping stage at Columbia Pictures Corporation studios in Hollywood used another interesting resonant device.[2] Sound mixing requires good listening conditions; looping requires variable voice recording conditions to simulate the many acoustic situations of motion picture scenes portrayed. In short, reverberation time had to be adjustable over about a two-to-one range for the 80,000 cu. ft. stage.

Both side walls of the stage were almost covered with the variable arrangement of Fig. 24-8 which is a cross section of a typical element extending from floor to ceiling with all panels hinged on vertical axes. The upper and lower hinged panels of 12 ft length are hard on one side (2 layers of ⅜ in. plaster board) and soft on the other (4 in. fiberglass). When open they present their soft sides and reveal slit resonators (1 × 3 slats spaced ⅜ in. to ¾ in. with mineral fiber board behind), which utilize the space behind the canted panels. In some areas glass fiber was fastened directly to the wall. Diffusion is less of a problem when only highly absorbent surfaces are exposed but when the hard surfaces are exposed, the hinged panels meet, forming good geometric diffusing surfaces.

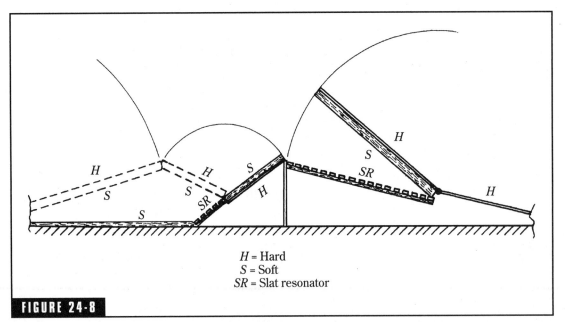

H = Hard
S = Soft
SR = Slat resonator

FIGURE 24-8

Variable acoustical elements in the mixing-looping stage at Columbia Pictures Corporation, Hollywood. Reflective areas are present when closed, absorbent areas and slat resonators are presented when the doors are opened. After Snow.[2]

The Snow design illustrates the extreme flexibility offered in combining many types of absorbers in an effective yet inexpensive overall arrangement.

Rotating Elements

Rotating elements of the type shown in Fig. 24-9 have been used in radio station KSL in Salt Lake City, Utah. In this particular configuration the flat side is relatively absorbent and the cylindrical diffusing element is relatively reflective. A disadvantage of this system is the cost of the space lost, which is required for rotation. The edges of the rotating element should fit tightly to minimize coupling between the studio and the space behind the elements.

At the University of Washington a music room was designed with a series of rotating cylinders partially protruding through the ceiling. The cylinder shafts were ganged and rotated with a rack-and-pinion drive in such a way that sectionalized areas of the cylinder exposed gave moderate low-frequency absorption increasing in the highs, good low-frequency absorption decreasing in the highs, and high reflection absorbing little energy in lows or highs. Such arrangements, while interesting, are too expensive and mechanically complex to be seriously considered for most studios.

A truly elegant solution to the rotating type of adjustable acoustics element is the Triffusor™, another product of RPG Diffusor Systems, Inc.[3] shown in Fig. 24-10. The Triffusor™ is a rotatable equilateral-triangular prism with absorptive, reflective, and diffusive sides. A nonrotating form of Triffusor™ is available with two absorptive sides and one diffusive side, especially adapted for use in corners. The nominal dimensions of the Triffusor™ are: height 4 ft, faces 2 ft across. In a normal mounting the edges would be butted and each unit supplied with bearings for rotation. In this way an array of these units

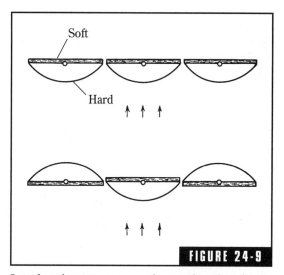

FIGURE 24-9

Rotating elements can vary the reverberation characteristics of a room. They have the disadvantage of requiring considerable space to accommodate the rotating elements.

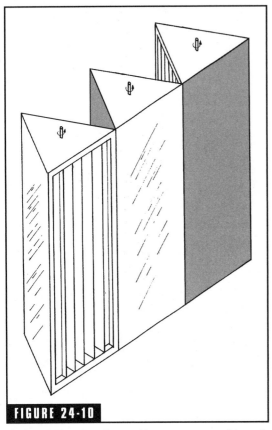

FIGURE 24-10

The Triffusor™ may be used in groups to provide variable acoustics in a space. Rotation of the individual units can bring diffusing, absorbing, or reflecting surfaces into play.

could provide all absorptive, all diffusive, all reflective, or any desired combination of the three surfaces.

Portable Units: The Tube Trap™

A proprietary, modular, low-frequency absorber with a number of interesting features has been introduced by Acoustic Sciences Corporation.[4] Known as the Tube Trap™, it is a cylindrical unit available in 9-, 11-, and 16-inch diameters and 2- and 3-foot lengths. An 11-inch unit with a 9-inch unit stacked on top of it is shown in the corner of a listening room in Fig. 24-11. A quarter-round adaptation of the same idea is shown in Fig. 24-12. The construction of the trap is shown in Fig. 24-13. It is basically a simple cylinder of 1-inch glass fiber given structural strength by an exoskeleton of wire mesh. A plastic sheet designated as a "limp mass" covers half of the cylindrical surface. For protection and appearance, a fabric cover is added.

Absorption coefficients are used in the familiar equation:

$$(area)\ (coefficient) = sabins\ absorption$$

With the Tube Traps™, it makes more sense to list directly the sabins of absorption contributed to a room by each tube. The absorption characteristics of the three-foot long Tube Traps™ and of the 9-, 11-, and 16-inch-diameter models are shown in Fig. 24-14. Appreciable absorption, especially with the 16-inch trap, is achieved below 125 Hz.

There is another benefit to be derived from stacking a couple of Tube Traps™ in each corner behind the loudspeakers. The limp mass, which covers only half of the area of the cylinder, provides reflection for midrange and higher frequencies. This limp mass, however, does not prevent low-frequency energy from passing through and being absorbed. By reflection of the mid-to-high frequencies, it is possible to control the "brightness" of the sound at the listening position. Figure 24-15 shows the two positions of the tubes. If the reflector faces the room (Fig. 24-15B), the tube fully absorbs the lower frequency energy while the listener receives the brighter sound. It is reported that most users prefer the brighter sound. The mid- and high-frequency sound is diffused by its cylindrical shape. If less bright sound is preferred, the reflective side is placed to face the wall. This might possibly introduce colorations resulting from the cavity formed by the intersecting wall surfaces and the cylindrical reflective panel. By placing absorptive panels on the wall surfaces, as indicated in Fig. 24-15A, this coloration can be controlled.

FIGURE 24-11

An 11-inch-diameter Tube™ with a 9-inch one on top of it. When placed in room corners, they can provide significant control of room modes. *Acoustic Sciences Corporation.*

Tube Traps™ can be placed in the rear two corners of the room if experimentation indicates it is desirable. Two Tube Traps™ may be stacked in a corner, the lower, larger one absorbing lower frequencies and the upper, smaller one absorbing moderate lows and midrange energy. Half-round units are also available that can be used to control sidewall reflections or provide general absorption elsewhere. Whether the Tube Traps™ plus carpet, furnishings, structural (wall, floor, ceiling) absorption, etc., combine to provide the proper overall decay rate (liveness, deadness) or not must be determined either by listening, calculation, or measurement. If flutter echoes are detected, steps must be taken to eliminate them. There is nothing as effective as an experienced ear in fine tuning the listening room.

Portable Units: The Korner Killer™

The special form of RPG Diffusor System's Triffusor™ with two absorptive sides and one diffusive side is called the Korner Killer™.

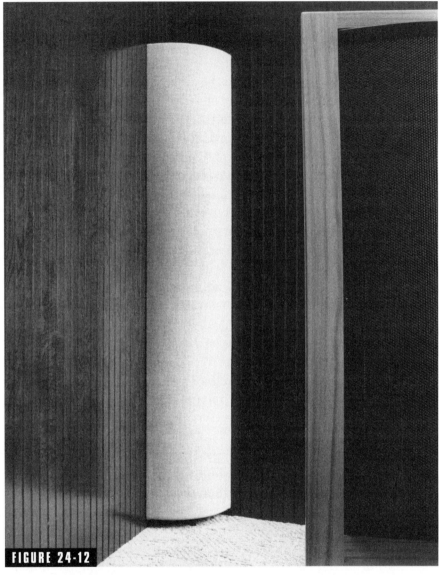

FIGURE 24-12

A quarter-round adaptation of the basic Tube-Trap™. *Acoustic Sciences Corporation.*

Placed with the absorptive faces into the corner it directs the diffusive face toward the room. This not only helps control normal modes, it adds important diffusion to the room. The diffused reflections are reduced 8 to 10 dB in the diffusion process, which would keep them from contributing to perceptual confusion of the stereo image. This is

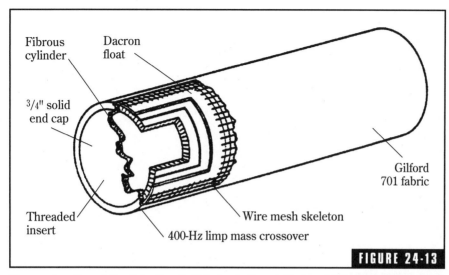

FIGURE 24-13

The construction of the Tube Trap™. It is basically a cylinder of 1-inch glass fiber with structural support. A plastic "limp mass" covers half the cylindrical surface, which reflects and diffuses sound energy above 400 Hz. *Acoustic Sciences Corporation.*

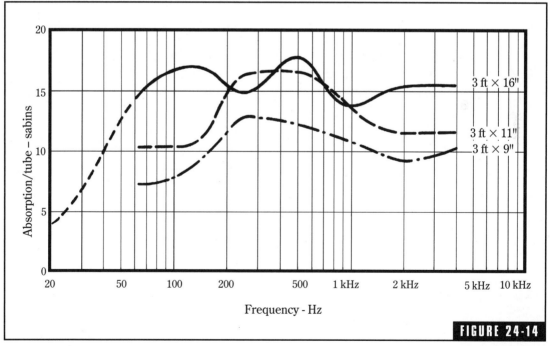

FIGURE 24-14

Absorption characteristics of three sizes of Tube Traps™. The 16-inch unit provides good absorption down to about 50 Hz.

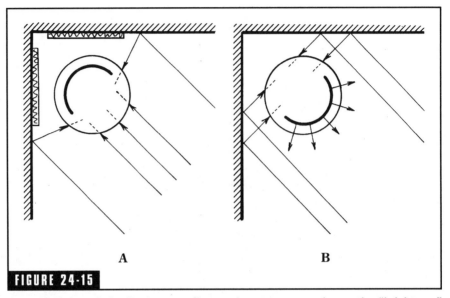

A   B

FIGURE 24-15

The positioning of the limp mass reflector gives some control over the "brightness" of the sound in the room. (A) With the limp mass facing the corner, the absorbent side of the cylinder absorbs sound over a wide range. (B) If the limp mass faces the room, highs above 400 Hz are reflected.

in contrast to the higher level limp mass reflections of the Tube Trap™ in Fig. 24-15B, which as early reflections, would tend to confuse the stereo image.

Endnotes

[1]Brüel, Per V., *Sound Insulation and Room Acoustics*, London, Chapman and Hall (1951), p. 242.

[2]Snow, William B., *Recent Application of Acoustical Engineering Principles in Studios and Review Rooms*, J. Soc. Motion Picture and Television Eng., 70 (Jan 1961) 33-38.

[3]RPG Diffusor Systems, Inc., 651-C Commerce Drive, Upper Marlboro, MD 20774, Telephone: 301-249-0044, FAX: 301-249-3912.

[4]Acoustic Sciences Corporation, 4275 West Fifth Ave., Eugene, OR 97402, Telephone: 1-800-ASC-TUBE or 503-343-0727.

CHAPTER 25

Acoustic Distortion*

There have been extraordinary advances in the quality of the hardware in our audio systems, but less progress has been made in improving the quality of the acoustical pathway which all sound must travel from the loudspeaker to our ears. Once the sound has reached our ears there are still many psychoacoustical factors that determine how we perceive the sound. Much psychoacoustical research is now being vigorously pursued to understand how the mind perceives sound. Between the hardware and the perception of sound there is still that analog sound-transmission path in which many distortions arise. The remainder of this book is devoted to acoustic distortions and the new tools and computer programs which have appeared to measure and model them.

Acoustic Distortion and the Perception of Sound

There are three psychoacoustic perceptions that are affected by acoustic distortions: frequency response or timbre, imaging, and spatial impression. Frequency response should be familiar by now. Good, clean high and low frequencies with a uniform response may be a goal, but *timbre,* or overall recognition and appreciation of the harmonic

*Contributed by Peter D'Antonio, RPG Diffusor Systems, Inc., Upper Marlboro, Maryland 20774.

content is that for which we are striving. As we listen to the musicians we form a mental image of those producing the music. The image can be very vivid when things are adjusted properly. The size and shape of the sonic source are pictured as well as its height, depth, and width. New discoveries relate image formation to specific early reflections of the sound from the sidewall in an almost magical way.

Sources of Acoustic Distortion

Acoustic distortion in the medium itself is difficult to visualize but easy to hear. The hearing ability of the listener and the distortion of the amplifiers and loudspeakers are outside the scope of this discussion of sources of acoustic distortion. There are many factors in adjusting the acoustics of a room; one of them is the matter of room proportions, which was covered in Chap. 13. The treatment of room surfaces and the diffusion and absorption of sound have also been covered in earlier chapters. The four remaining sources of acoustic distortion, covered in this chapter, are (1) room modes, (2) speaker-boundary interference response, (3) comb filtering, and (4) poor diffusion.

Coupling of Room Modes

There are many resonances in a normal room (see Chapt. 15). Specifically, the mathematics of the situation require three modes for a rectangular space, the axial, the tangential, and the oblique. The axial modes are the result of normal (right-angle) reflections between the end surfaces of the room, the sidewall surfaces, and for the third pair, added floor and ceiling reflections. Any sound in this room would excite these three resonances, or modes. To complicate matters, each fundamental resonance has a series of harmonics, which are also resonances in every sense of the word. These modes are the acoustics of this space and are the source of acoustic distortion as the individual modes interact with each other. Figure 27-1 is what could be called a frequency-response measurement of the room filled with axial, tangential, and oblique modes. Three axial modes are prominently identified, but the rest of the graph represents the vector sum of all the other modes. Figure 27-2 shows first-, second-, and third-order modes of a room. These modes vary from zero to maximum sound pressure and would have a major effect on the production or reproduction of sound in this room. Every room has great fluctuations of sound pressure from one point to another, and these are a source of acoustic distortion.

Sound waves are longitudinal waves; that is, they actually oscillate (expand and contract) in the direction of propagation. As the sound waves expand and contract, they cause high-pressure regions and low-pressure regions. The instantaneous pressure on opposite sides of a pressure minimum has opposite polarity. The pressure on one side is increasing while the pressure on the other side is decreasing. The position of a loudspeaker and the listener's ear with respect to these pressure variations will determine how they couple to the room.

Speaker-Boundary Interference Response

The next type of acoustic distortion is due to the coherent interference between the direct sound of a loudspeaker and the reflections from the room, in particular the corner immediately surrounding it (Fig. 27-3). This distortion occurs across the entire frequency spectrum but is more significant at low frequencies. It is called the speaker-boundary interference response, or SBIR. The room's boundaries surrounding the loudspeaker mirror the loudspeaker, forming virtual images. When these virtual loudspeakers (reflections) combine with the direct sound, they can either enhance or cancel it to varying degrees depending on the amplitude and phase relationship between the reflection and the direct sound at the listening position.

In Fig. 27-5, a loudspeaker is located 3 ft from each room surface with coordinates (3,3,3). The four virtual images on opposite sides of the main room boundaries that are responsible for first-order reflections are also shown. A virtual image is located an equivalent distance on the opposite side of a room boundary. The distance from a virtual source to the listener is equal to the reflected path from source to listener. In addition to the four virtual images shown, there are 7 more. Three virtual images and 1 real image exist in the speaker plane and 4 virtual images of these above the ceiling and floor planes. Imagine that the walls are removed and 11 additional physical speakers are located at the virtual image positions. The resultant sound at a listening position would be equivalent to the sound heard from one source and 11 other loudspeakers!

The effect of the coherent interference between the direct sound and these virtual images is illustrated in Fig. 27-3. For display, the SBIR is averaged over all listening positions with the loudspeaker located 4 ft from one, two, and three walls surrounding the loudspeaker. It can be seen that as each wall is added, the low-frequency response increases by 6 dB and the notch, at roughly 100 Hz, gets deeper. It is important to note at this point that once this notch is created, due to poor placement, it is virtually impossible to eliminate it without moving the listener and loudspeaker, since it is not good practice to electronically compensate for deep notches. Thus the boundary reflections either enhance or cancel the direct sound depending on the phase relationship between the direct sound and the reflection at the listening position. At the lowest frequencies the direct sound and reflection are in phase and they add. As the frequency increases the phase of the reflected sound lags the direct sound. At a certain frequency the reflection is out of phase with the direct sound and a cancellation occurs. The extent of the null will be determined by the relative amplitudes of the direct sound and reflection. At low frequencies there is typically very little absorption efficiency on the boundary surfaces and the notches can be between 6 and 25 dB! One important conclusion that can be drawn from this is never place a speaker's woofer equidistant from the floor and two surrounding walls!

The low-frequency rise in Fig. 27-3 illustrates why one can add more bass by moving a loudspeaker into the corner of a room. Actually, one has two choices. Move the loudspeaker either as close to the corner as possible or as far away from the corner as is physically practical. As you move the speaker closer to the corner, the first cancellation notch moves to higher frequencies, where it may be attenuated with porous absorption. This can be seen in Fig. 27-3 for the last condition in which the loudspeaker is positioned 1 ft from the floor, rear, and sidewalls. In addition, the loudspeaker's own directivity pattern diminishes the backward radiation, thus reducing the amplitude of the reflection relative to the direct sound. This principle is the basis for flush mounting loudspeakers in a corner soffit. The bad news is that with the loudspeaker in the wall-ceiling dihedral corner or wall-wall-ceiling trihedral corner, the loudspeaker very efficiently couples with the room modes. If the dimensional ratios are poor, leading to overlapping or very widely spaced modal frequencies, there will be significant modal emphasis. Many loudspeaker manufacturers now provide a roll-off equalization to compensate for the added emphasis of flush mounting the loudspeaker.

We can also move the loudspeaker farther away from the adjacent corner. In this case the first cancellation notch moves to a lower frequency, hopefully below the lower cutoff frequency of the loudspeaker or the hearing response of the listener. To obtain a 20-Hz first cancellation notch one needs to position the loudspeaker 14 ft from the rear wall.

Clearly, some happy medium must be found, and with the myriad of alternatives available the most effective approach is to start with a multidimensional computer loudspeaker-listener placement optimization program (described in Chap. 27) and adjust to taste.

Comb Filtering

Another form of acoustic distortion introduced by room reflections is comb filtering. It is due to coherent constructive and destructive interference between the direct sound and a reflected sound. In critical listening rooms we are primarily concerned with the interaction between the direct sound and the first-order (i.e., single-bounce) reflections. Reflections cause time delays, because the reflected path length between the listener and source is longer than the direct sound path.

In this way, when the direct sound is combined with the reflected sound, we experience notches and peaks referred to as comb filtering. The reflections enhance or cancel the direct sound to varying degrees depending on the phase (path length) difference between the reflection and the direct sound at the listening position. An example of comb filtering between the direct sound and a reflection delayed by 1 ms is shown in Fig. 25-1. Four conditions are illustrated. 0 dB refers to the theoretical situation in which the reflection is at the same level as the direct sound. The remaining three interference curves indicate situations in which the reflection is attenuated by 3, 6, and 12 dB. In Fig. 25-2 the locations of the first 5 interference nulls are indicated as a function of total delay. A delay of 1 ms or 1.13 ft produces a first null at 500 Hz with subsequent notches 1,000 Hz apart. The constructive interference peaks lie midway between successive nulls. When a reflection is at the same level as the direct sound the nulls theoretically extend to infinity.

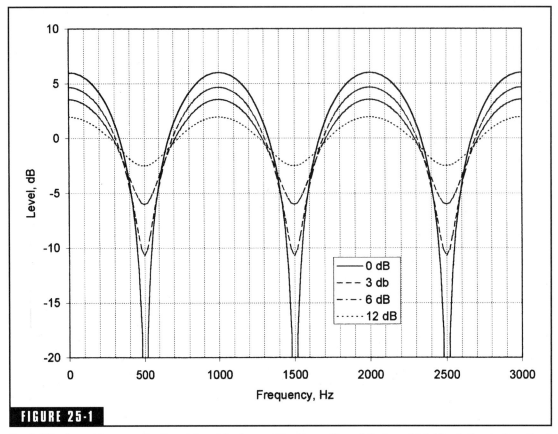

FIGURE 25-1

Comb filtering with one reflection delayed by 1 ms with attenuation of 3, 6, and 12 dB relative to the direct sound.

This name evolved because comb filtering, on a linear frequency graph, resembles a series of equally spaced notches like the teeth of a comb. The location of the first notch is given by the speed of sound divided by two times the total path length difference. The spacing between subsequent notches is twice this frequency. In Fig. 25-3 the comb filtering due to a series of regularly spaced reflections separated by 1 ms (flutter echo) is shown. Note how the peaks are much sharper than the single reflection.

The audible effect of comb filtering is easy to experience using a delay line. If you combine a signal with a delayed version, you will experience various effects referred to as chorusing or flanging, depending on the length of the delay and the variation of the delay

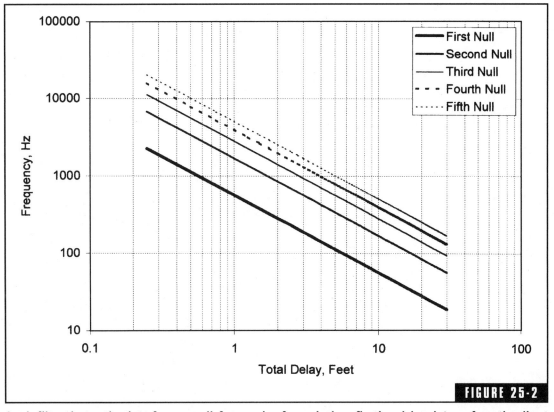

Comb filter destructive interference null frequencies for a single reflection delayed 1 ms from the direct sound. Constructive interference peaks occur midway between adjacent nulls.

with time. Shorter delays have wider bandwidth notches and thus remove more power than longer delays. This is why microsecond and millisecond delays are so audible.

The effect of a reflection producing comb filtering is illustrated in Figs. 25-4 and 25-5. Figure 25-4 illustrates the time and the frequency response of the right speaker only. The upper curve shows the arrival time and the lower curve shows the free-field response of the loudspeaker. Figure 25-5 shows the effect of adding a sidewall reflection to the sound of the right loudspeaker. The upper curve shows the arrival time of both the direct sound and the reflected sound. The lower curve shows the severe comb filtering that a single reflection introduces. If a loudspeaker has a free-field response like this lower curve, it would probably be rejected by most listeners. Yet many rooms are designed

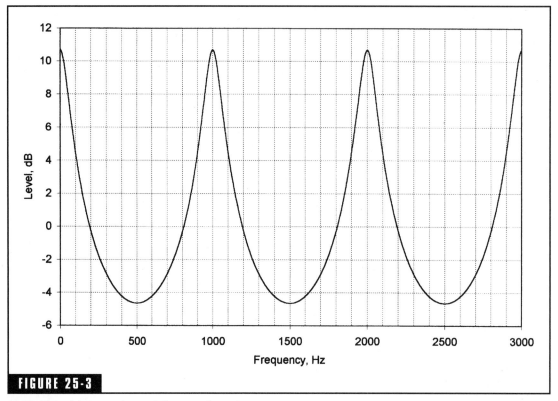

FIGURE 25-3

Comb filter due to equispaced flutter echoes 1 ms apart.

without reflection control. In reality our ear-brain combination is more adept at interpreting the direct sound and reflection than the FFT (fast Fourier transform) analyzer, so that the perceived effect may be somewhat less severe.

Comb filtering is controlled by attenuating the room reflections or by controlling the loudspeaker's directivity to minimize boundary reflections. If the loudspeaker has constant directivity as a function of frequency, then broad-bandwidth reflection control is necessary. Since the directivity of conventional loudspeakers increases with frequency, low-frequency reflection control is important. For this reason, one would not expect to control low-frequency comb filtering with a thin porous acoustical foam or panel.

Comb filtering can be controlled by using absorption, which removes sound energy from the room, or diffusion, which distributes

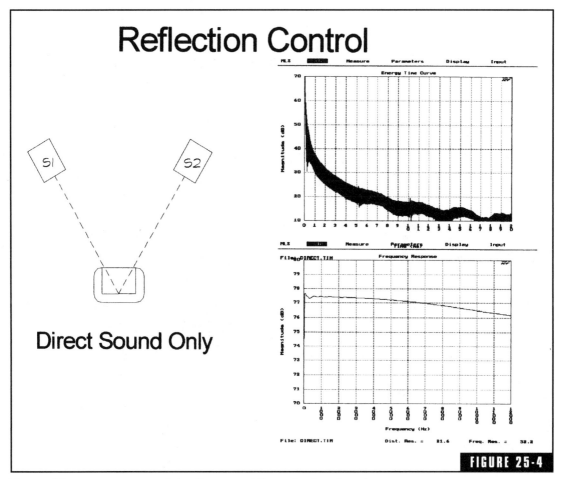

Figure 25-4

Time and frequency response of the direct sound from a loudspeaker.

the reflection over time without absorption. Both approaches are valid and produce different psychoacoustical reactions. Using absorption to reduce the effect of a specular reflection will produce pinpoint spatial phantom images. On the other hand, diffusion will produce sonic images with more spaciousness (width, depth, and height). The degree of this effect can be controlled.

Comb filtering usually results in image and timbre corruption. The effect of comb filtering at low frequency from the loudspeaker's constructive interference with the surfaces surrounding it has been discussed in the speaker-boundary interference section.

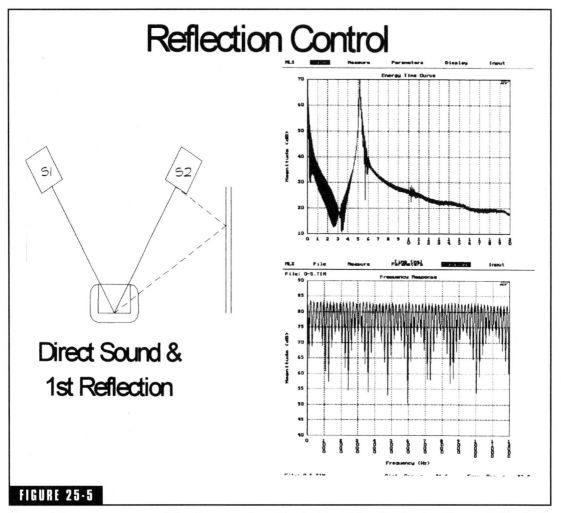

FIGURE 25-5

Time and frequency response of the direct sound combined with a sidewall reflection. The comb filtering is apparent.

Poor Diffusion

In a performance facility the acoustical environment contributes to the overall impression of the sound. The role of diffuse reflections in concert hall acoustics has been of interest for many years, this interest being partly fueled by the fact that some of the oldest halls with the best reputations are highly diffuse. In halls like the Grosser Musikvereinssaal the architectural style of the period led to the highly diffuse surface finishings. Due to rising building costs, increased seating requirements, and changes in interior design fashions, flat plaster, concrete, drywall, and cinder block surfaces have become all too commonplace, replacing ornate designs. Acoustic requirements have, however, led to highly modulated surface designs in some more modern halls with a primary aim of creating a highly diffuse sound field.

Recently, with the knowledge of subjective research since the 1960s, rather than use blanket diffusion, it has become more common to use diffusing elements more selectively. Diffusors can also be valuable in promoting lateral reflections to produce a sense of envelopment or spatial impression in rooms.

Very recent research by Haan and Fricke[1,2] showed that the surface diffusivity index (SDI), which is a qualitative characterization of surface diffusion, correlates very highly with an acoustic quality index (AQI) in the most highly acclaimed concert halls in the world. SDI was determined from a visual inspection of the surface roughness. While this research requires further confirmation, it bodes well for installations where a large number of diffusors have been used. In addition to geometrical shapes and relief ornamentation, number-theoretic diffusors are finding widespread application in performance spaces.

Small critical listening environments are spaces such as recording and broadcast studios, residential listening rooms, and music practice rooms. In concert halls, these spaces are usually designed to be neutral, adding little of the signature of the room to pre-encoded reproduced sound. A method to achieve this was described by Davis and Davis in their live end–dead end control room design concept.[3] An abbreviated list of two of the room criteria follows:

> There (should be) an effectively anechoic path between the monitor loudspeakers and the mixer's ears for at least 2–5 ms beyond the studio's initial time-delay gap. There (should be) a highly diffused sound field present during the initial onset of the so-called Haas effect.

In 1984, D'Antonio et al.[4] extended the ideas of Davis and Davis[3] by introducing practical ways to provide a temporal reflection-free zone (RFZ) surrounding the listening position using flush-mounted loudspeakers and a diffuse sound field using reflection phase grating (RPG) diffusors on the rear wall. The RFZ is created by splaying massive, speaker boundary surfaces, which may also contain porous absorption to further minimize the speaker-boundary interference. The RFZ also creates an initial time delay before the onset of the diffuse energy from the rear wall. The rear wall diffusors create a passive surround sound and a diffuse sound field for the room. They also widen the "sweet spot" and provide low-frequency absorption below the diffusion design frequency. Over the past decade this design has proved to be a useful approach to designing listening facilities and has been used in over a thousand facilities.

In small critical listening rooms, the geometry, close proximity, and surface topology of the boundary surfaces to the listening position may result in a very sparse temporal and spatial distribution of reflections. The nonuniform reflection pattern can result in specularity effects, such as comb filtering and image shifting, as well as nonuniform sound distribution throughout the room.

Conclusion

In conclusion we can see that acoustic distortions are related to the positions of listeners and loudspeakers in a room of reflecting surfaces. Some excellent computer programs are now available to study the acoustics of a room and to solve the problem of positioning loudspeakers and listeners. See Chaps. 26, 27, and 28.

Endnotes

[1] Haan, C. N. and F. R. Fricke, *Surface Diffusivity as a Measure of the Acoustic Quality of Concert Halls,* Proc. of Australia and New Zealand Architectural Science Association Conference, Sydney (1993).

[2] Haan, C. N. and F. R. Fricke, *The Use of Neural Network Analysis for the Prediction of Acoustic Quality of Concert Halls*, Proc. of WESTPRAC V '94, 543–550, Seoul (1994).

[3] Davis, D. and C. Davis, *The LEDE™ Concept for the Control of Acoustic and Psychoacoustic Parameters in Recording Control Rooms,* J. Audio Eng. Soc. 28, 585—595 (1980).

[4] D'Antonio, P. and J. Konnert, *The RFZ/RPG Approach to Control Room Monitoring,* Audio Eng. Soc. Convention Preprint 2157 (I-6), October 1984.

Room Acoustics Measurement Software*

I often say that when you measure what you are speaking about, can express it in numbers, you know something about it; but when you cannot express it in numbers, your knowledge is of a meager and unsatisfactory kind. It may be the beginning of knowledge but you have scarcely, in your thought, advanced to the stage of Science whatever the matter may be.—Lord Kelvin.

Measurements have always been critical to science as they allow us to put assumptions, theories, and equations to the test of reality. As a result, scientists have always been looking for ways to improve the availability, accuracy, and precision of measurements. The science of room acoustics is no exception.

High-fidelity sound reproduction involves the frequency range from 20 to 20,000 Hz. This range of frequencies includes acoustic waves with wavelengths ranging from a single centimeter to nearly 20 meters. As the previous chapters of this book have shown, the dimensions of rooms and loudspeakers also fall within this range, making them subject to various acoustic effects due to diffraction, resonances, boundary effects, and reflections. Given the extremely complex nature of acoustics, these effects are often very difficult to model with a useful level of precision. Therefore, measurements are the most reliable way to accurately quantify room acoustics.

*Contributed by Geoff Goacher, Acoustical Research Associates, Irvine, California, and Doug Plumb, AcoustiSoft, Peterborough, Ontario, Canada.

Fortunately, acoustical measurement tools are now more available and affordable than ever. Much of this is due to the recent popularity of the personal computer and the evolution of digital signal processing. In this chapter, we will take a look at how computers and digital signal processing have opened up all sorts of improved measurement capabilities for audio enthusiasts.

The Evolution of Measurement Technologies

Around the turn of the 20th century, a Harvard physics professor named Wallace Clement Sabine painstakingly discovered the fundamental relationship between the size of a room, its sound absorption characteristics, and its reverberation time. From that knowledge, he figured out the equation (now called the Sabine equation) for calculating the reverberation time of a room. Sabine is considered by many to be the first "scientific" acoustician.

Interestingly, Sabine's only tools for this research were his ears, a portable wind chest, some organ pipes, a stopwatch, and a bunch of seat cushions that he borrowed from a local theater. With these tools, he played sound into a room, measured the decay rate of its reverberation with a stopwatch and his ears, and modified the decay rate with the seat cushions he had borrowed. While these experiments may now seem primitive, Sabine's experiments were important because they took the poorly understood subject of room acoustics out of the dark ages and launched it into a respected field of study. People began to discover and quantify a great variety of acoustical parameters involved in "good sound"—a work in progress, which continues to this day.

Perhaps the most amazing advancement of acoustics in the modern era is the availability of good measurement capabilities for the common audio enthusiast, thanks to the rampancy of personal computers and the advancements of digital signal processing techniques. A "multimedia" computer can now serve as acoustic measurement hardware. This generally makes a PC-based measurement system much less expensive than a dedicated measurement system because the hardware is already at our fingertips—all we need is a test microphone and microphone preamp to record signals into the computer and software to perform the analysis functions. With a laptop computer, the user has an added advantage that the measurement system is also very portable, making it much easier to take measurements in different locations.

Another significant feature of software-based measurement systems is that the software can account for the imperfections in the hardware (such as sound card frequency response anomalies) and can compensate for them. The time-delay spectrometry (TDS) and maximum length sequence (MLS) test signals (which we will discuss later) used with many of the modern measurement programs are very insensitive to the nonlinearities present in low-cost computer sound cards.

This is not to say that software-based measurement programs are better than dedicated measurement systems or that they will definitely replace them altogether. This is simply meant to encourage people to explore the conveniences and benefits of the computer revolution.

Acoustical measurement software programs on the market can measure a variety of acoustical parameters, many of which have been discussed in the previous chapters. With the software that is currently on the market, you can commonly measure:

- Room reverberation time (without stopwatches or organ pipes!)
- Pseudo-anechoic loudspeaker frequency response in normal rooms
- Reverberant loudspeaker frequency response
- Energy-time curves (specific reflection levels)
- Room resonances
- Impulse response
- Loudspeaker delay times
- Many other critical audio and acoustic parameters involved in high fidelity

As the previous chapters of this book have illustrated, these measurements are invaluable for analyzing, troubleshooting, and perfecting room acoustics and sound reproduction. In order to give you a better understanding of the sophistication of some of these programs, let's take a look at the basic principles of how some of the more advanced ones work.

Building a Better Analyzer

Most audio enthusiasts are familiar with "sound-level meters." They are probably one of the most basic sound measurement devices and are useful for a variety of tasks in which sound pressure levels at a particular location must be measured. However, a sound-level meter does not give an indication of the "sound quality" which is typically sought in most audio applications. Measuring sound "quality," at a minimum, also requires the measurement of frequency response. There are a number of instruments and methods for measuring frequency response. Another important requirement for sound "quality" measurements is time response. This is frequently used for measuring reverberation time, speech intelligibility, and other time-domain phenomena. Again, there are a variety of instruments used for measuring this factor of sound.

One major problem that has traditionally plagued the measurement of these parameters of room acoustics is the presence of background noise. Acoustical background noise typically resembles pink noise, which has a power spectrum of -3 dB/octave across the frequency spectrum. Random background noise is typically more prevalent at lower frequencies because they travel through walls and structures easily.

Over the years, it was found that this noise could only be overcome in room acoustics measurements by feeding the room with stimulus energy so great that the background noise becomes insignificant. In the past, this was accomplished by using gunshots as a test signal for impulse response and reverberation time measurements. Nevertheless, this method had some technical flaws and provided only a rudimentary look at how sound actually behaved in rooms. Ultimately, it was concluded that to effectively study room acoustics and the correlation between certain acoustical parameters and "good sound," an instrument with both time and frequency measurement capabilities and significant noise immunity was required.

Time-Delay Spectrometry (TDS) Measurement Techniques

In response to this need for better measurement techniques, the late Richard C. Heyser refined a measurement method known as time-delay spectrometry (TDS) in the late 1960s. By the early 1980s, a TDS measurement system was implemented commercially by Techron in their TEF Analyzer. This small but very powerful digital instrument was implemented on the platform of an industrial portable PC. TEF used time-delay spectrometry to derive both time and frequency response information from measurements.

The basic operating principle of TDS resides in a variable-frequency sweep excitation signal and a receiver with sweep tuning synchronized with that signal. A critical third element is an offset facility that can introduce a time delay between the swept excitation signal and the receiver. The latter is not needed in electrical circuits in which signals travel near the speed of light, but it is vital in acoustical systems in which sound travels at a leisurely 1,130 ft/s.

The coordination of the outgoing sine-wave sweep and the sharply tuned tracking receiver is shown in Fig. 26-1. In this illustration, the horizontal axis is time and the vertical axis is frequency. The signal begins at A (t_1, f_1) and sweeps linearly to B (t_3, f_2). After a time delay (t_d), the receiver sweep begins at C (t_2, f_1) and tracks linearly at a rate identical to that of the signal to D (t_4, f_2). The receiver, being delayed by the time t_d, is now perfectly tuned to receive the swept-signal sine wave after it has traveled in air for the time t_d. At any instant during the sweep, the receiver is offset f_0 Hz from the signal.

The advantages of this arrangement of swept exciting signal and offset-swept tuned receiver are manifold. For example, it takes a certain amount of time for the signal to reach a wall, and it takes a definite amount of time for the reflection to travel back to the measuring microphone. By offsetting the receiver an amount equal to the sum of the two, the receiver "looks" only at that specific reflected component. The offset may be viewed as a frequency offset. When the offset in the previous example is right to accept the reflection from the wall, all the unwanted reflections arrive at the microphone when the analyzing receiver is tuned to some other frequency; hence they are rejected. In this way the receiver is detuned for noise, reverberation, and all unwanted reflections in the measurement. The frequency response of only that particular desired reflection or sound is displayed for study.

Also note that the full signal energy is applied over the testing time throughout the swept spectrum. This is in stark contrast to applying a

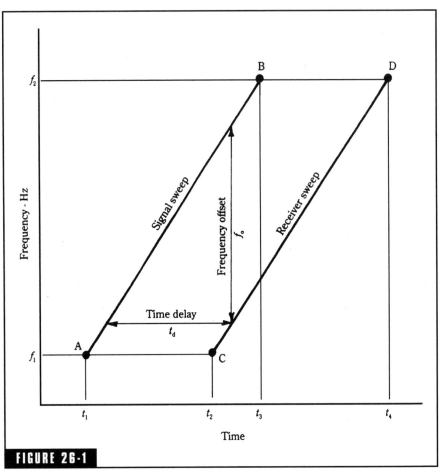

FIGURE 26-1

The basic principles of Heyser's time-delay spectrometry measurements. The outgoing signal is swept linearly from A to B. After a delay appropriate to select out the reflection desired, the receiver is swept from C to D. Only energy from the desired reflection is accepted; the receiver is detuned to all others.

brute-force signal to a system, which often results in driving elements into nonlinear regions.

The TEF Analyzer was a tremendous breakthrough in measurement tools for evaluating the effects of room acoustics on perceived sound quality. One of its primary benefits is that the original TDS signal from a test can be stored and then repeatedly modulated with delayed versions of the sine and cosine chirp to get multiple "snapshots" of the time vs. frequency content. These frequency-response "slices" allow for generation of the beautiful three-dimensional waterfall graphs of energy vs. time vs. frequency that have been shown in this book.

Another one of the most well-loved features of its measurements was that it allowed for pseudo-anechoic measurements in ordinary rooms by only measuring the initial sound plus the first few (or no) reflections. This technique could be used to take loudspeaker measurements and ignore the room effects by removing reflections from the measurement. The effect of reflections on the elusive "good sound" could therefore be studied.

This ability to remove undesired information from the time-response measurements is often referred to as "gating" in sound measurement. Gating permits audio analyzers to remove reflections from a measurement that occur beyond a certain point in time after the room has initially been stimulated, and hence the nickname "pseudo-anechoic" measurements.

One disadvantage of gating is the loss of resolution that occurs from truncating (cutting off) the time response. For example, room reflections often occur within 5 ms of the direct sound in a typical room with 10-ft ceilings. In a pseudo-anechoic measurement made in such a room all data that is taken 5 ms after the initial stimulus would have to be removed, limiting the resolution of the measurement. For instance, a 5-ms gate limits the lowest frequency that can be measured to 1 / 5 ms = 200 Hz. Frequency-response curves are smoothed as a result and frequency effects that appear with resolution finer than 200 Hz appear with reduced detail.

Overall, TDS swept tone measurements have two major advantages. First, the postprocessing causes unwanted harmonics to be rejected from the measurement, making the measurement accuracy less dependent on system linearity. Devices that add a large distortion component can be reliably measured for frequency response. Second, the swept tone can be used over long durations of time, injecting a lot of energy into the room. This has the effect of increasing measurement signal-to-noise ratio.

The major disadvantage of TDS is that a new measurement must be taken each time a change in the time resolution is desired. If it were desired to take the above measurement again with a time window of say 10 ms, another measurement would have to be taken using a different sweep rate.

Despite the tremendous advantages that were brought about by TDS measurement capabilities, it wasn't long before acousticians were yearning for a measurement system having the above TDS advantages with the additional capability that one single measurement could later be reprocessed with a different time window. Therefore, only one physical measurement would be required for a complete picture of the response. And along came "MLS" measurement techniques.

Maximum-Length Sequence (MLS) Techniques

The maximum-length sequence (MLS) measurement technique was refined in the 1980s and has been found by many to be a superior measurement method for room acoustics. MLS measurements use a pseudo-random binary sequence to excite a system and/or room with a test signal resembling wideband white noise. The noise-rejection capabilities of MLS are such that the noise excitation can be played at a low level while maintaining a good signal-to-noise ratio and measurement accuracy. An MLS analyzer can also play a test signal for a long period of time, injecting enough energy into a room to reduce the effects of acoustical background noise to an acceptable level. Like TDS, MLS measurements also have very good distortion immunity.

The binary sequence used to produce the noise signal is precisely known and generated from a logical recursive relationship. Recordings of the test signal played through a system are used to generate an impulse response of the system using a fast method known as the "fast Hadamard transformation."

The ideal impulse can never be realized in practice, but the situation can be approximated perfectly up to a set frequency limit. This limit is due to the "aperture effect." An example would be a system excited by a pulse with time duration of 1/1,000 seconds. The aperture effect will reduce the frequency content above 1,000 Hz when the frequency response is calculated from the recorded data.

Fortunately, sampling rates of 44,100 Hz and higher in digital measurements prevent the aperture effect from affecting measurements in the audible frequency range. Shannon's sampling theorem proves that all of the information that is contained below 22,500 Hz can be recovered completely. This ideal is realized in practice with modern PC-based measurement tools and very well approximated with real-time digital components such as CD players and computer sound cards.

There is absolutely no reason or need to increase this sampling rate when taking measurements for the audible frequency range.

The fact that the impulse response of the system can be determined when using MLS-type stimulus represents a distinct and very significant advantage of MLS systems over TDS systems. The impulse response can be used to calculate all remaining linear parameters of a system. These include the various forms of frequency-response measurement as well as time measurement. Knowledge of the impulse response of a system permits the construction of a square-wave response, triangular-wave response, or any other form of time response required. All intelligibility calculations can easily be done with only knowledge of the impulse response. When the complete time response of a system is known, the frequency response can be calculated directly using a Fourier transform. Frequency response, by definition, is the Fourier transform of a system time response obtained by impulse excitation.

Many companies have taken advantage of the superb measurement capabilities of MLS techniques and have created software that can be used to turn a multimedia computer into an exceptional electroacoustic measurement system. There are now several reputable acoustic measurement programs on the market and more will surely become available in the future. For the sake of demonstration, one of these programs and its measurement capabilities is highlighted below.

AcoustiSoft's ETF Program

In 1996, AcoustiSoft introduced their "ETF Loudspeaker and Room Acoustics Analysis Program." The program was one of the first software-based measurement programs intended to turn an ordinary multimedia PC into a proquality measurement tool.

At the time of this writing, the program is in its fifth version and is an MLS-based measurement program. The program contains calibration features for lab-quality precision and accuracy.

The way the program works is that the operator connects a test microphone and microphone preamplifier to the computer's sound card line input (any sound card will work so long as it is a "full duplex" card, meaning it can play and record simultaneously). A test signal is then played from the sound card's line output through the system (loudspeaker and/or room combination) and is simultaneously recorded through the sound card's line input and stored as a wave file on the computer for analysis. The software automatically processes the recording and then generates an impulse response of the system being measured, which is stored in the computer for postprocessing. Therefore, all other quantities can be analyzed in postprocessing after the measurement takes place. Figure 26-2 illustrates the process in which the test signal is recorded and converted into an impulse response.

As shown in Fig. 26-2, the ETF system is a two-channel analyzer. The left channel of the input signal is a reference channel and the right channel of the input signal is the test channel. This is why the program

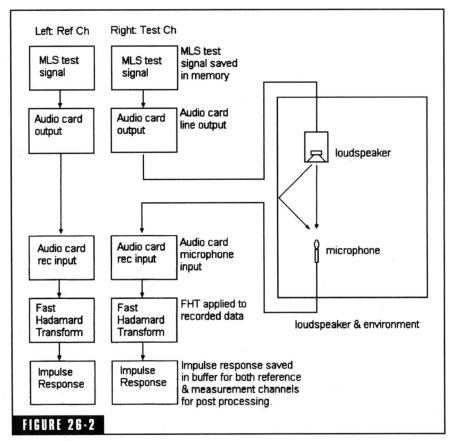

FIGURE 26-2

Block diagram of ETF 5's impulse response measurement methodology. *(Courtesy of Doug Plumb, AcoustiSoft.)*

can give precise and accurate results with any full-duplex sound card—the two-channel system subtracts any response anomalies caused by the sound card so that what is left is actually caused by the system under test.

The measured impulse response of a system can be very revealing, in and of itself. Figure 26-3 is an impulse response of a speaker that was measured at 1 meter distance, on axis with the speaker.

As shown in Fig. 26-3, placing the microphone close to and directly in front of the loudspeaker diminishes the effect of room reflections on the measurement. As will be shown in frequency-response measurements later, this type of impulse response will also be equated with the cleanest frequency response.

Figure 26-4 contains more room information. It is an impulse response of a loudspeaker that was also measured at a close distance to the loudspeaker, but with a single reflecting surface located nearby.

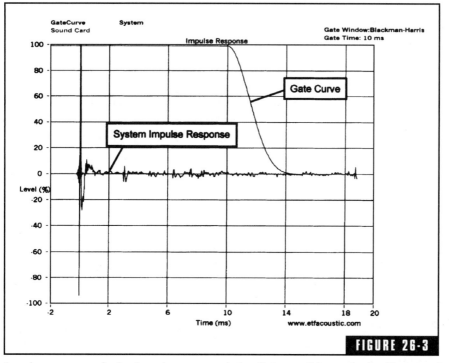

FIGURE 26-3

Impulse response of a loudspeaker measured at a distance of 1 m, on axis. *(Courtesy of Doug Plumb, AcoustiSoft.)*

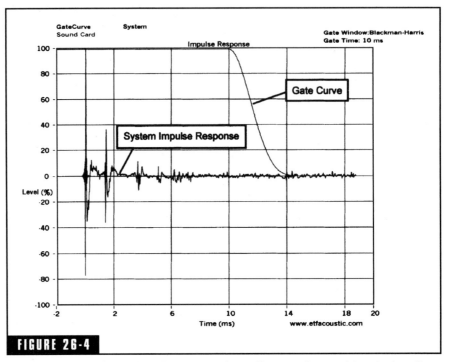

FIGURE 26-4

Impulse response of a loudspeaker measured at a distance of 1 m, with a single reflecting surface nearby. *(Courtesy of Doug Plumb, AcoustiSoft.)*

As you can see in Fig. 26-4, the single wall reflection shows up clearly. A little further, you will see that these types of reflections also generate a very visible comb-filtering effect in the frequency response. Another interesting feature about reflections in the impulse-response graphs is that reflections containing high-frequency content are clearly visible on the impulse-response measurement. Reflections containing only midfrequency and low-frequency content can be difficult or impossible to see in an impulse response.

Figure 26-5 shows the impulse response of a loudspeaker in a room with the test microphone located approximate 3 meters from the loudspeaker. It clearly shows the effect of multiple reflections in an impulse-response measurement. This type of impulse response generally contains a much more sporadic frequency response due to the interference effects of the room reflections on the direct sound from the loudspeakers.

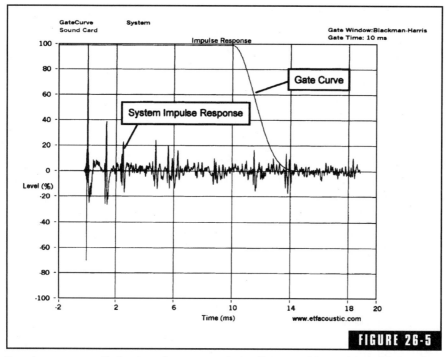

Impulse response of a loudspeaker measured at a distance of 3 m, with multiple reflections. *(Courtesy of Doug Plumb, AcoustiSoft.)*

Now let's take a look at how the program can translate these impulse responses into a variety of other types of useful measurements.

Frequency-Response Measurements

As previously mentioned, the impulse response measured by the ETF program is saved in buffers and can be used to compute a variety of results, otherwise called "postprocessing." Frequency response is one of these postprocessing options. Since the gate times used for the frequency-response measurements can be changed as desired in postprocessing, measurements do not have to be repeated nearly as often as with other types of measurement systems. Figure 26-6 shows the steps ETF uses to convert the measured impulse response of a system into its equivalent frequency response.

The next three figures show the frequency-response equivalents for the three impulse-response measurements above. Figure 26-7 is a sample of a pseudo-anechoic frequency-response measurement taken in an ordinary room. The test setup is the same one used to generate the impulse-response measurement shown in Fig. 26-3. The ability of the gating to cut off room reflections allows for the measurement of the direct response of the loudspeaker, minus the effect of the room. This is similar to what is possible to measure in an anechoic test chamber with no significant reflections (hence the name "pseudo-anechoic" measurement).

In this pseudo-anechoic frequency-response measurement, the impulse response was gated to eliminate information that was recorded after 2 ms of the direct sound so that an accurate picture of the loudspeaker's response could be seen without the effect of room reflections. This frequency response is typical of near-field listening in acoustically treated rooms. Pseudo-anechoic measurements are help-

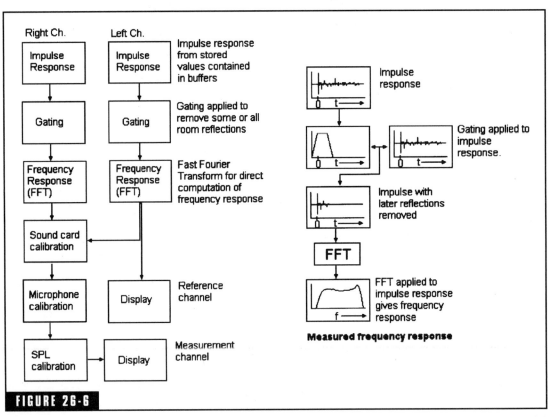

FIGURE 26-6

Block diagram of ETF 5's frequency response measurement methodology. *(Courtesy of Doug Plumb, AcoustiSoft.)*

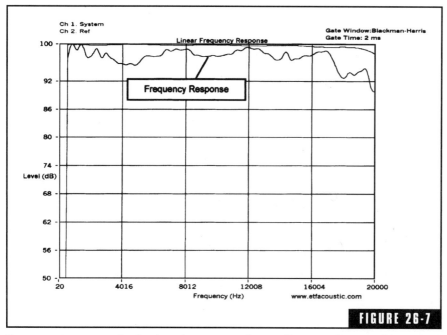

Frequency response of a loudspeaker measured at a distance of 1 m, with minimal reflections. *(Courtesy of Doug Plumb, AcoustiSoft.)*

ful for speaker design and are also helpful for finding the actual response of a loudspeaker's "direct" sound so that it can be differentiated from the response caused by the room. This information can be used by audio enthusiasts to equalize out undesirable response problems that are caused by the direct sound of loudspeakers.

Both linear and logarithmic frequency-response graphs are included in the ETF program. However, a linear frequency-response graph was used here to make it easier to diagnose the presence of comb filtering. None is seen in the above graph.

Figure 26-8 is a measurement of a loudspeaker at fairly close range, like that of Fig. 26-7. This measurement, however, has a single reflection that is fairly close in both time and level to that of the direct sound. The effects of this reflection can be seen in Fig. 26-8.

The single reflection obviously comb-filters the response. This results in the equally spaced peaks and dips shown in Fig. 26-8. This frequency response is typical of near-field listening in acoustically untreated rooms with nearby reflecting surfaces.

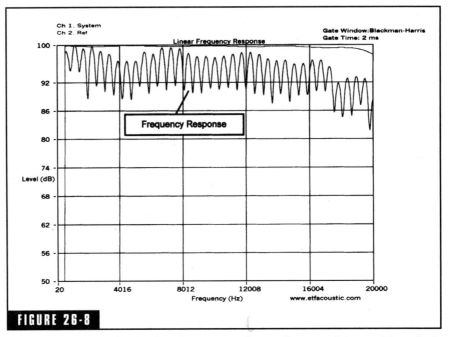

FIGURE 26-8

Frequency response of a loudspeaker measured at a distance of 1 m, with a single reflecting surface nearby. *(Courtesy of Doug Plumb, AcoustiSoft.)*

Figure 26-9 shows the frequency response for a measurement position that is roughly 3 meters from the loudspeaker. This measurement has many reflections in it, and a longer gate time has been selected, which is similar to that of the integration time of the human ear-brain system.

This measurement shows the effect of multiple room reflections in the frequency-response calculation. This would be a typical frequency response for far-field listening in an acoustically untreated room. The effects of comb filtering are obviously present, but much less idealized than they are in the preceding response graph. Multiple reflections causing these sharp dips and peaks can have a negative effect on stereo imagery. The addition of absorbers to surfaces causing reflections will tend to smooth this curve.

Compared to the other two measurements, the "in-room" response graph shown in Fig. 26-9 is of little value in an environment with multiple reflections. A more meaningful response graph for this type of condition can be achieved in ETF by further postprocessing the results to yield a fractional octave response (such as 1/3 octave) that more accurately reflects how we would subjectively perceive this frequency

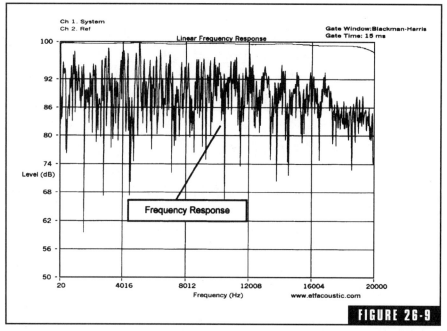

Frequency response of a loudspeaker measured at a distance of 3 m, with multiple reflections. *(Courtesy of Doug Plumb, AcoustiSoft.)*

response in such a listening environment. This will be further discussed under Fractional-Octave Measurements below.

Pinpointing the level and delay times of room reflections, with respect to the direct sound, is best evaluated using a filtered energy-time curve. ETF also automatically generates these energy-time curves for analysis of individual reflection delay times and levels. This is discussed in greater detail below.

Resonance Measurements

ETF 5 can also help pinpoint resonances in the frequency-response measurements. This is done by examining the later portions of the impulse response for ringing. This ringing results from sharp resonance in a low-frequency room response or in loudspeaker cabinets and drivers. The ringing appears as a sharp peak in the frequency response and can be more easily seen with delayed response graphs, otherwise called "time slices." Figure 26-10 shows the steps ETF uses to convert the measured impulse response of a system into a time slice frequency-response graph showing resonances.

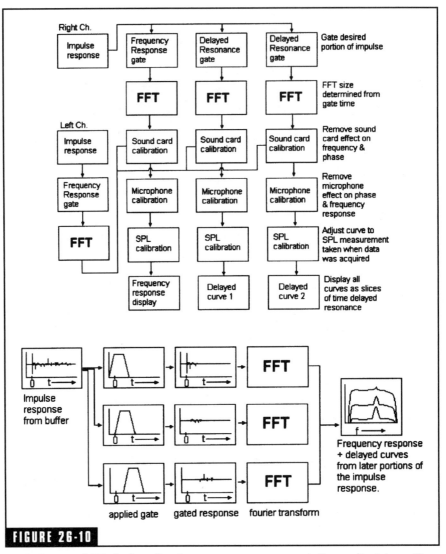

FIGURE 26-10

Block diagram of ETF 5's time slice frequency response computation methodology. *(Courtesy of Doug Plumb, AcoustiSoft.)*

Figure 26-11 shows a low-frequency room response measurement with time slices. This type of graph is particularly useful for evaluating the effect of room resonances on the system's low-frequency response.

The frequency response $t = 0$ (top) curve in Fig. 26-11 shows the overall response but does not allow the viewer to easily differentiate between boundary effects and resonances. Boundary effects are the effect of reinforcement (reflections) from surfaces near the microphone or loudspeaker. Resonances are frequencies for which the room energetically supports vibrations when excited by a sound source. Room resonances are easy to detect in a graph such as this because the peak structures will have the same shape in each time slice if they are truly resonant—boundary effects will not.

The later slices of the response of Fig. 26-11 show resonances at roughly 35, 65, and 95 Hz. Noise appears to have caused the sharp peaks at 120 Hz because the delayed slices all show the peak at the same level (roughly 68 dB in the graph). It is therefore not decaying, as a resonance would. Measuring resonances in this manner is helpful for optimizing the placement of loudspeaker and listeners in listening rooms. It is also helpful for finding the specific frequency and bandwidth of troublesome resonances so that you can tune a bass trap or parametric equalizer to bring them under control.

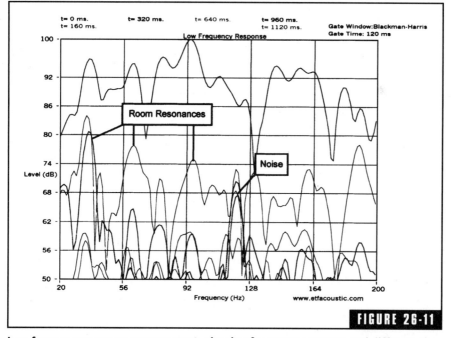

FIGURE 26-11

Low-frequency response measurement, showing frequency response and different time intervals ("time slices"). Resonances can be identified as peaks that appear every time slice, with varying level. Peaks that do not vary in level can be assumed to be noise. *(Courtesy of Doug Plumb, AcoustiSoft.)*

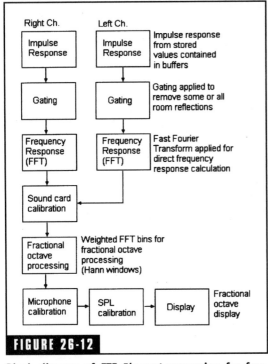

Right Ch.   Left Ch.

| Impulse Response | Impulse Response | Impulse response from stored values contained in buffers |

| Gating | Gating | Gating applied to remove some or all room reflections |

| Frequency Response (FFT) | Frequency Response (FFT) | Fast Fourier Transform applied for direct frequency response calculation |

Sound card calibration

| Fractional octave processing | Weighted FFT bins for fractional octave processing (Hann windows) |

| Microphone calibration | SPL calibration | Display | Fractional octave display |

FIGURE 26-12

Block diagram of ETF 5's post-processing for fractional octave displays. *(Courtesy of Doug Plumb, AcoustiSoft.)*

Fractional-Octave Measurements

ETF also averages the frequency-response curves into fractional octaves such as 1/3 octave, 1/6 octave, and so on. These fractional-octave measurements can also be gated so the curve contains only certain portions of the measurement information. This type of gating and frequency averaging allows for measurements that are highly correlated with our perceptions of sound in various circumstances. Figure 26-12 illustrates the process by which ETF 5 creates these fractional-octave displays.

The 1/3-octave frequency response is one of the best indicators of subjective frequency balance as it approximates the bandwidth accuracy of our hearing. By converting a high-resolution frequency-response measurement into 1/3 octave, it is easy to see that the sharp dips and peaks previously found in the frequency response are greatly diminished. Only the larger variations in frequency response remain. In the mid and high frequencies, reflections are therefore more the cause of stereo imaging problems rather than major subjective frequency balance changes.

Figure 26-13 shows a 1/3-octave "in room" frequency response. The gate time is set to 20 ms. This measurement would be a good indicator of how a listener would perceive the tonality of the system as the 20-ms gate eliminates what our ears would mostly perceive as "room sound" and the 1/3-octave averaging is close to the critical bandwidth of the ear.

In some instances, higher levels of frequency resolutions are preferred in fractional-octave measurements such as 1/10 or 1/12 octave. For these cases, higher-resolution fractional-octave displays are possible with ETF 5. The information gained from this measurement can be used for troubleshooting loudspeaker designs. With an equalizer

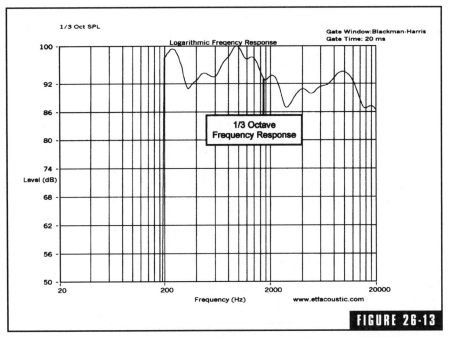

One-third octave frequency response measurement, with a 20-ms gate. *(Courtesy of Doug Plumb, AcoustiSoft.)*

patched into a system, this measurement can also be used to aid in the "flattening" of a system's response.

Energy-Time Curve Measurements

ETF also generates energy-time curves which are useful for determining the time and level of reflections versus that of the direct sound. Energy-time curves show the level as a logarithmic quantity, making lower-level behavior more visible. In ETF, the energy-time curves can be viewed as "full-range" levels, or they can be broken into various octave bands for checking the levels of specific ranges of frequencies. Figure 26-14 shows the process by which ETF 5 generates energy-time curves from the measured impulse response. It is interesting to compare and contrast energy-time curves to their equivalent impulse response. Figure 26-15 shows the energy-time curve for the same measurement as the impulse response shown in Fig. 26-3. This is a measurement of a loudspeaker taken at 1 meter distance, on axis with the speaker.

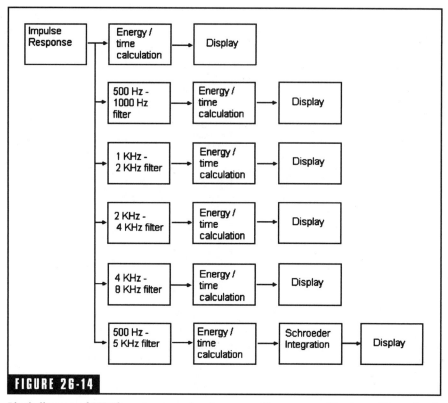

Block diagram of ETF 5's post-processing for energy time curves. *(Courtesy of Doug Plumb, AcoustiSoft.)*

Figure 26-15 shows very little room reflected energy. It is clearly present in the energy-time curves and can be seen to be about 20 dB below the initial sound. These energy-time curves can be compared to the results of Olive and Toole's research on the psychoacoustic effects of reflections, which have been summarized in Chaps. 3 and 19 of this book. Based on their findings, a listener would not hear much (if any) of the room reflections when positioned this close to the loudspeaker. Near-field monitors take advantage of the principle that room reflections become less significant if the listener is positioned close to the loudspeaker.

Figure 26-16 shows an energy-time curve for a loudspeaker that was also measured at a close distance to the loudspeaker, but with a reflecting surface located nearby. This is the same measurement as the one depicted in Fig. 26-4.

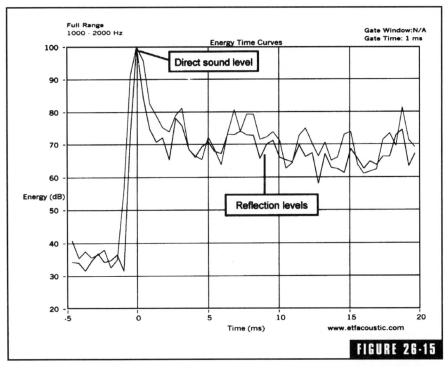

Energy-time curve of a loudspeaker measured at 1 m distance with minimal reflections. *(Courtesy of Doug Plumb, AcoustiSoft.)*

The energy-time curve of Fig. 26-16 contains high-level reflections that occur only a few milliseconds after the arrival of the direct sound. Also, there are some lower-level reflections visible in this energy-time curve that were not visible in the impulse-response measurement shown in Fig. 26-4.

Figure 26-17 is the equivalent energy-time curve for the impulse response shown in Fig. 26-5. This is a measurement of a loudspeaker with the test microphone located approximately 3 meters from the loudspeaker.

This figure shows a typical room response measurement for a room with untreated acoustics. Ideal room treatment for the most stable and precise sound reproduction would reduce all early reflections occurring within roughly 15 to 20 milliseconds of the direct sound to a level that is 15 to 20 dB below the level of the direct sound. This has the effect of smoothing out the phase and frequency response of the perceived sound, and it also enables the listener to more clearly hear the

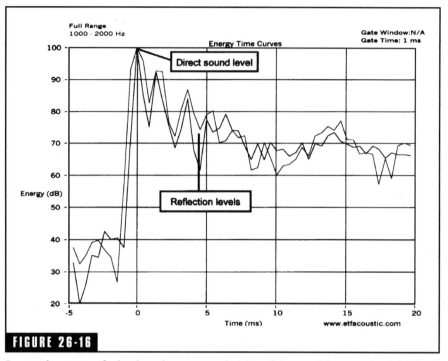

FIGURE 26-16

Energy-time curve of a loudspeaker measured at 1 m distance with one or two high-level reflections. *(Courtesy of Doug Plumb, AcoustiSoft.)*

direct sound and the acoustics of the space without being masked by listening-room reflections. The removal of early reflections also provides an improvement to stereo imagery by improving the coherence of the signal at the listener's ears.

Later reflected sound is usually desired to provide ambience. The longer time delay and mixing of the later reflections causes them to be decorrelated from the original sound and heard as ambience. Loudspeakers in a surround sound system can be also be used to provide ambience in an acoustically "dead" room.

Reverberation Time

ETF 5 also calculates the reverberation time, or "decay rate," of sound in rooms (RT60) from the measured impulse response. Much of the information about a room's ambience, mentioned above, can be inferred from the room's reverberation time. As shown in Chap. 7, different types of rooms and room sizes require different reverberation

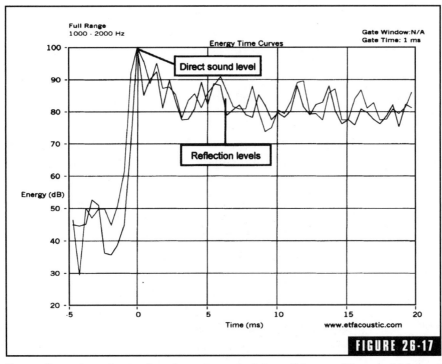

Energy time curve of a loudspeaker measured at 3 m distance with many high-level reflections. *(Courtesy of Doug Plumb, AcoustiSoft.)*

times for optimum capabilities. Figure 26-18 illustrates how reverberation time is calculated from the impulse response.

Figure 26-19 is a sample RT/60 measurement taken with ETF 5. In Fig. 26-19, the RT/60 values average 0.5 second across most of the audible band, rising to around 1 second at the lowest frequencies. The range of measurable RT60 values with ETF 5 is from 0 to 3 seconds. The information gained from a measurement such as this can be compared to the ideal reverberation time charts in Chap. 7, and any necessary acoustical modifications for improved RT/60 suitability will be apparent.

There are many other acoustical parameters of sound systems that this particular program is capable of measuring. However, the above measurement samples and graphs provide a glimpse at the wide range of useful measurement capabilities that this type of program can provide for anyone with a need to critically analyze audio systems and their acoustical environments. The Endnotes for this chapter give

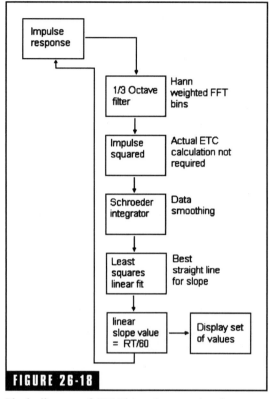

```
┌──────────────┐
│ Impulse      │
│ response     │
└──────────────┘

┌──────────────┐  Hann
│ 1/3 Octave   │  weighted FFT
│ filter       │  bins
└──────────────┘

┌──────────────┐  Actual ETC
│ Impulse      │  calculation not
│ squared      │  required
└──────────────┘

┌──────────────┐  Data
│ Schroeder    │  smoothing
│ integrator   │
└──────────────┘

┌──────────────┐  Best
│ Least        │  straight line
│ squares      │  for slope
│ linear fit   │
└──────────────┘

┌──────────────┐     ┌──────────────┐
│ linear       │────▶│ Display set  │
│ slope value  │     │ of values    │
│ = RT/60      │     └──────────────┘
└──────────────┘
```

FIGURE 26-18

Block diagram of ETF 5's post-processing for reverberation time computations. *(Courtesy of Doug Plumb, AcoustiSoft.)*

details on some good resources for finding more information on all the latest measurement systems.

Conclusion

The development of advanced acoustic measurement techniques like MLS, and the capability of implementing them through software on a common PC, has made it possible for professional consultants and audio enthusiasts alike to have better measurement capabilities than ever before. The types of measurements these programs provide are an invaluable aid to any sound contractor, acoustical consultant, audiophile, recording engineer, or speaker builder, and they are now well within the means of almost everyone.

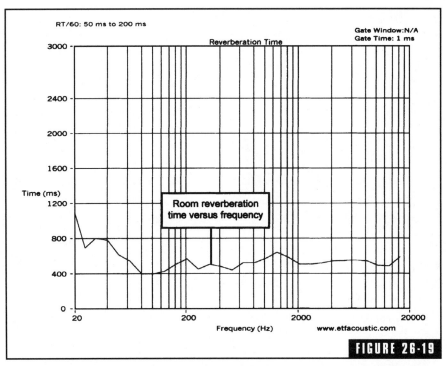

Room reverberation chart showing measured RT/60 values at various frequencies. *(Courtesy of Doug Plumb, AcoustiSoft.)*

Endnotes

[1]Rife, Douglas D., *Transfer Function Measurement with Maximum Length Sequences*, J. Audio Eng. Soc., Vol. 37, No 6, June 1989.

[2]Dunn, Chris, and Malcolm O. Hawksford, *Distortion Immunity of MLS Derived Impulse Measurements*, J. Audio Eng. Soc., Vol, 41, No. 5, May 1993.

[3]Vanderkooy, John, *Another Approach to Time Delay Spectrometry*, J. Audio Eng. Soc., Vol. 34, No. 7/8, July/August 1986.

[4]Heyser, R. C., *Acoustical Measurements by Time Delay Spectrometry*, J. Audio Eng. Soc., Vol, 15, p. 370, 1967.

[5]Toole, F. E., *Loudspeaker Measurements & Their Relationship to Listener Preferences*, Part 2, J. Audio Eng. Soc., Vol. 34, pp. 227–235, April 1986.

[6]Haykin, Simon, *An Introduction to Analogue and Digital Communications*, Copyright 1989 John Wiley & Sons, Inc., TK5101.H373 1989.

[7]Toole, Floyd E., *Loudspeakers and Rooms for Stereophonic Sound Reproduction*, National Research Council of Canada, AES 8th International Conference Preprint, 1989.

[8]Davis, Don, and Carolyn Davis, *Sound System Engineering, Second Edition*, Indianapolis: Sams, 1987, pp. 168–169.

[9]White, Glenn, *The Audio Dictionary, Second Edition*, Seattle, University of Washington Press, 1995.

[10]ETF is available directly from AcoustiSoft, 463 King Street Upper, Peterborough, Ontario, Canada, K9J-2T1, (800) 301-1423, doug@etfacoustic.com.

[11]Many additional types of acoustical measurement software programs are available through Old Colony Sound Laboratory, P.O. Box 876, Peterborough, New Hampshire 03458-0876, (888) 924-9465, custserv@audioxpress.com.

Room Optimizer*

Introduction

The sound that we hear in a critical listening room is determined by the complex interaction among the quality of the electronics, the quality and the placement of the loudspeakers, the hearing ability and placement of the listener, the room dimensions (or geometry if non-cuboid), and the acoustical condition of the room's boundary surfaces and contents. All too often these factors are ignored and emphasis is placed solely on the quality of the loudspeakers. However, the tonal balance and timbre of a given loudspeaker can vary significantly, depending on the placement of the listener and loudspeaker and the room acoustic conditions. The acoustic distortion introduced by the room can be so influential that it dominates the overall sonic impression.[1,2] The two causes of this acoustic distortion are the acoustical coupling between the loudspeakers and listener with the room's modal pressure variations or room modes and the coherent interaction between the direct sound and the early reflections from the room's boundaries.

Critical listeners have invested considerable time in trial-and-error attempts to minimize these effects. However, no automated method to search for the optimum locations has been proposed. With the advent

*Contributed by Peter D'Antonio, RPG Diffusor Systems, Inc., Upper Marlboro, Maryland 20774.

of 5.1 home theater and multichannel music, physical trial-and-error approaches become even less feasible. The task of optimally locating five loudspeakers and multiple subwoofers presents a significant challenge. In addition to optimizing the low-frequency response via optimum listener and loudspeaker placement, one must also address imaging[3,4] and the influence of acoustical surface treatment on the size and location of sonic images, as well as the sense of envelopment or spaciousness experienced in the listening room.

Therefore, to address these acoustical issues we describe an automatic computerized simulation program that suggests optimum locations for loudspeakers, listener, and acoustical surface treatment. In addition, the program can also help with new room design, by optimizing the room dimensions. For this discussion, we will focus on the optimum placement of loudspeakers and listener in a room.

Modal Response

All mechanical systems have natural resonances. In rooms, sound waves coherently interfere as they reflect back and forth between hard walls. This interference results in resonances at frequencies determined by the geometry of the room. In lossless cuboid rooms, where the normal component of the particle velocity is zero at the surface, the modal frequencies are associated with the eigenvalues of the wave equation. These modal frequencies are distributed among axial modes involving two opposing surfaces, tangential modes involving four surfaces, and oblique modes involving all surfaces. For an axial mode between two opposite boundaries, this frequency is equal to the speed of sound c divided by twice the room dimension in that direction. For example, for $c = 344$ m/s, a 4.57-m wall-to-wall dimension results in a first-order fundamental room mode of 37.6 Hz. As an example, Fig. 27-1 shows the measured modal frequency response of a room with a 4.57-m dimension. The loudspeaker was located in a corner and the microphone was placed against a wall perpendicular to the 4.57-m dimension, in order to record all axial modes. The first-order (100), second-order (200), and third-order (300) modes are identified in Fig. 27-1 at 37.6, 75.3, and 113 Hz, respectively. The pressure of the eigenfunctions or normal modes in a rectangular parallelepiped is given in references 5 and 6.

In addition to the modal frequency distribution, the coupling between the loudspeakers and listener with the modal pressure is also important. The loudspeaker placement will accentuate or diminish coupling with the room modes. Similarly, a listener will hear different bass response depending on where he or she is seated. Figure 27-2 illustrates how the sound pressure is distributed along a room dimension. The room dimension is shown as a fraction ranging from 0 to 1; 0.5 would be in the center of the room and 1 would be against a wall. Examining Fig. 27-2 reveals that the fundamental first-order mode has no energy in the center of the room. Physically, this means that a listener seated in the center of the room would not hear this frequency. The second-order mode, however, is at a maximum. It can be inferred that in the center of the room all odd-order modal frequencies are absent and all even-order harmonics are at a maximum. Therefore, when we listen to music in a room, the music will be modified by the room's modal response, and this acoustic distortion will depend on where the speakers and the listener are located and how they couple with the room. Ideal room dimensions represent the acoustical search for the "Holy Grail." There are various suggested approaches.[7,8] While the distribution of the modal frequencies is important, it is equally important to consider the placement of the loudspeakers and the listeners with respect to the boundary surfaces to minimize acoustic distortion introduced by the room. Thus to minimize the modal coloration, we must optimize both the room dimensions and the locations for loudspeakers and listeners.

Speaker-Boundary Interference Response

In addition to modal pressure variations, the interaction of the direct sound from the loudspeakers with reflections from the walls can result in dips and peaks in the spectra due to interference effects. We refer to this as the speaker-boundary interference response. This issue has been examined by Allison,[9,10] Waterhouse,[11,12] and Waterhouse and Cook.[13] The interference occurs over the entire frequency range, with predominant effects at low frequency. The typical effect is a low-frequency emphasis followed by a notch. To illustrate the effect in Fig. 27.3, the speaker-boundary interference is averaged over listening positions[14] with the speaker located 1.22 m from one, two, and three

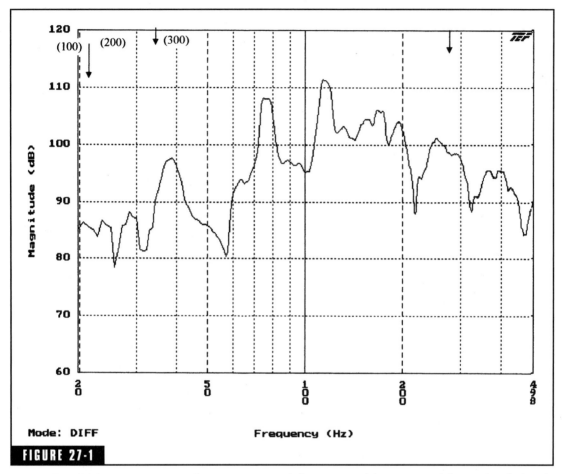

Mode: DIFF
Frequency (Hz)

FIGURE 27-1

Measured modal frequency response in a room 2.29 m (W) × 4.57 m (L) × 2.74 m (H). The (100), (200), and (300) modes are identified.

walls surrounding the loudspeaker. As each wall is added, the low-frequency response increases by 6 dB and the notch at roughly 100 Hz deepens. This demonstrates the fact that each time the solid angle into which a speaker can radiate is reduced by a factor of 2 (by adding a boundary surface, for example) the sound pressure at low frequencies is increased by a factor of 2 (6 dB). Thus by placing a loudspeaker on the floor near a corner (three boundaries) the full solid angle of 4π steradians is reduced to $\pi/2$ and the total low-frequency gain is increased by roughly 18 dB! Figure 27-3 also illustrates how the notch increases in frequency as the speaker spacing is decreased to 0.31 m

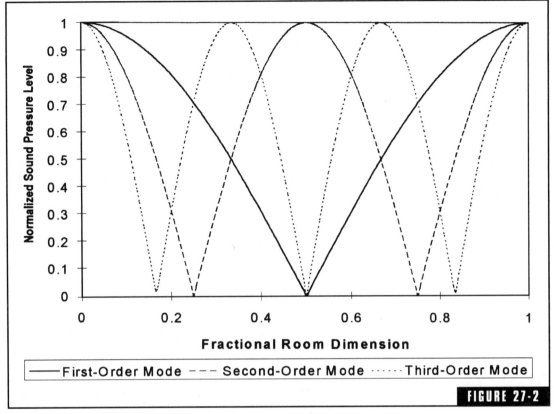

Normalized energy distribution of the first three modes in a room.

from each wall. By moving the loudspeakers and listeners to optimum positions in the room, the coloration produced by the room transfer function can be greatly reduced.

Optimization

Previous sections describe the complex interaction among the listening room, and the location of the listener and loudspeakers. Guidelines and procedures are already available that address these issues.[15–17] Modal frequencies for cuboid rooms and their pressure distribution are well known.[18] These can be used to aid listener and loudspeaker placement and room design. Positioning loudspeakers different distances from the nearest floor and walls can reduce the

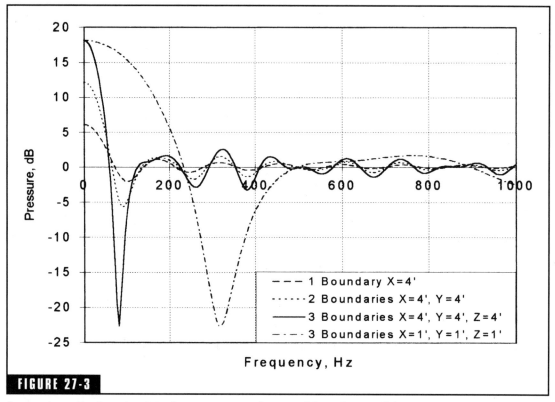

FIGURE 27-3

Averaged speaker-boundary interference response for several boundary conditions.

speaker-boundary interference. Simple computer programs that simulate the effect of loudspeaker and listener placement are also available. While these procedures are useful, they can never properly account for the complex sound field, which occurs in real listening rooms. Optimum placement of the loudspeakers and listener must be made taking all of these factors into consideration simultaneously, since the speaker-boundary interference and modal excitation are independent effects. That is, listener and loudspeaker locations that minimize the speaker-boundary interference do not necessarily lead to minimum modal excitation, and vice versa. To the authors' knowledge, such an algorithm has not been published. For this reason an iterative image method was developed to optimize the placement of listener and loudspeakers by monitoring the combined standard deviation of the speaker-boundary interference and modal response spectra.

In recent years there has been a great increase in knowledge concerning computer models to predict the acoustics of enclosed spaces.[19] In addition there has been a great increase in the computing power available on personal computers. This enables algorithms, which determine the best listener and loudspeaker positions within a space, to use complex calculation procedures based on more accurate predictions of the sound field received by the listener. These have many advantages over the simpler placement theories. For example, they take into account many more reflections from all surfaces in the room. This enables the examination of the subtle effects of many surfaces working in unison. Furthermore, by combining the room prediction models with optimization routines, the computer can determine the best positions for the loudspeakers and listener by processing the laborious trial-and-error optimization rather than the user.

In this chapter a program is described which combines an image source model to calculate the room transfer function with a simplex routine to carry out the optimization process. An appropriate cost function to characterize the quality of the spectra received by the listener has been developed. This parameter is based on the extensive subjective evaluations and listening tests of Toole and his colleagues.[20–23] They have confirmed that speakers which have flat on-axis frequency responses are preferred in standardized listening tests. In addition, similarly good off-axis response is also required, since the listener is hearing the combination of direct and reflected sound from the room's boundary surfaces. At low frequencies speakers are essentially omnidirectional. Since most rooms are not anechoic at low frequencies, we have chosen the flatness of the perceived spectra as a way to evaluate listener and loudspeaker placement and room dimensions. The cost parameter penalizes positions with uneven spectral responses. The optimization program we describe concentrates on lower frequencies (<300 Hz), where the individual room modes and the primary speaker-boundary interference are most influential and problematic. The goal of the program is to enable relative nonexperts to determine the best positions for listener, loudspeakers, and acoustical surface treatment in a cuboid room.

Theory

Prediction of Room Response

Considerable research has been carried out in recent decades concerning the predicted room acoustic responses using geometric models. Reference 19 details some of the model types available. The fastest and probably simplest prediction model for a cuboid room is based on the image source method. The image solution of a rectangular enclosure rapidly approaches an exact solution of the wave equation as the walls of the room become rigid. The image model provides a good time-domain transient description of the room response and is appropriate for listener and loudspeaker predictions because of its speed. While the small rooms considered here do not require extensive time to determine their impulse response, the program described will use an iterative optimization process, which will necessitate many hundreds of impulse-response calculations. The image method includes only those images contributing to the impulse-response and provides appropriate weighting of the modal frequencies. On the other hand, the alternate normal-mode solution of the enclosure[6,24] would require calculation of all modes within the frequency range of interest, plus corrections for those outside this range. Allen has derived the exact relationship between the normal-mode and the image solutions for a lossless room.[25] Since the impulse response can be equivalently viewed as a sum of normal modes, there must be repetitive patterns in the impulse response that form early in the response after a transient period. This has been demonstrated by Kovitz,[5] using the algorithm of Burrus and Parks. Kovitz[5] has shown that the full impulse response can be described by an IIR filter that is derived from the early time FIR impulse response. The equivalence between the impulse response and the modal frequency summation method will be demonstrated later in the chapter using the program.

The image source model algorithm constructs all possible image sources for the listener and loudspeaker pairs. Figure 27-4 shows a two-dimensional example, with only the first-order reflections shown. At the top the direct sound and a sidewall reflection are shown. In the lower illustration, each first-order reflection can be modeled by replacing the surface with an image source at the appropriate position. This is further illustrated in Fig. 27-4 showing a real source at coordi-

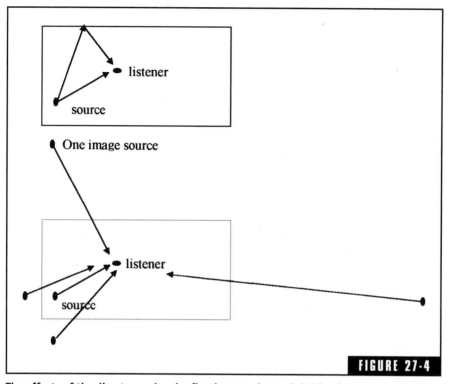

FIGURE 27-4

The effects of the direct sound and reflections can be modeled by the equivalent system of the original source, and all possible image sources.

nates (3,3,3) and its virtual image sources, which occur at an equivalent perpendicular distance on the opposite side of each boundary. The room is 10 units in height.

The sound at the listener is calculated by propagating the image source wave to the receiver, using the standard equation of a point source, and attenuating the wave according to the absorption coefficient of the boundary it "reflects" from. Obviously, the power of computing enables many orders of reflections to be accounted for, not just the first order indicated in Figs. 27-4 and 27-5. The prediction of the pressure received by the listener then reduces to a set of infinite series, which can be encoded and evaluated relatively easily.[19] The image source model as described is restricted to cuboid rooms. The model is also limited in that it does not account for any phase change on reflection. This is a common trait among geometric computer models and is also an implicit assumption in other procedures for calculating the

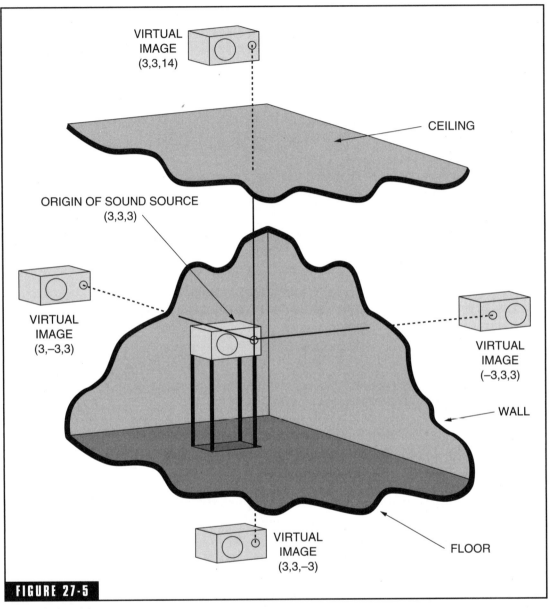

FIGURE 27-5

Real and virtual images construct.

position of loudspeaker and listeners. The lack of modeling of the phase change has arisen because it is not a simple process to treat the wall as an extended absorber. This means that the model is most accurate for rooms with relatively hard surfaces where the phase change effects are at their smallest. No account of diffusion caused by surface scattering is made. Fortunately, this is less crucial at the low frequencies being considered here.

The image source model produces an impulse response for the room, where the direct and reflected sounds are clearly distinguishable. An example is shown in Fig. 27-6. A Fourier transform of this gives the spectrum received by a listener if the sound source was producing a continuous tone. This long-term spectrum is shown in Fig. 27-7 and is similar to the modal response of the room (the modal response calculates the sound field at the listener by adding

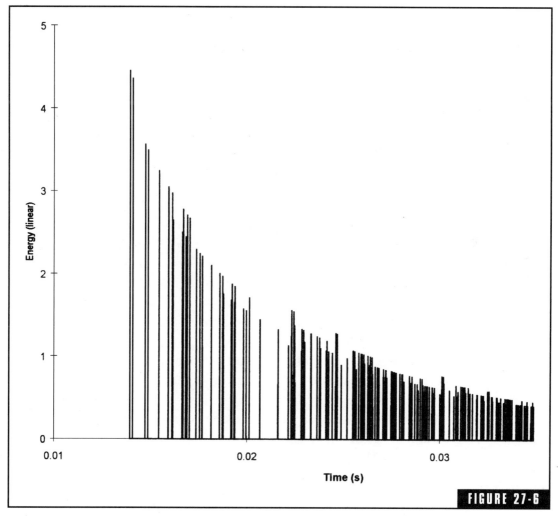

FIGURE 27-6

Typical energy impulse generated by image source model.

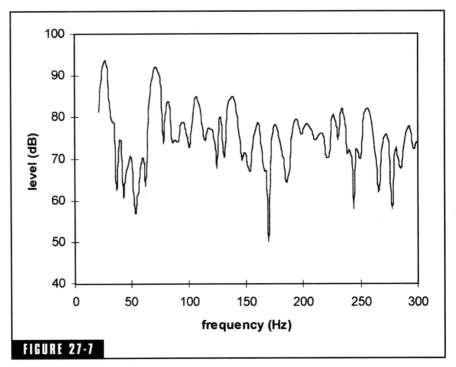

FIGURE 27-7

Example of long-term spectrum.

together the effects of individual modes in the frequency domain). In fact, the responses are equivalent, provided that the image source model takes into account an infinite number of sources and the modal calculation includes an infinite number of possible modes.

As an example we examine a 3 m × 3 m × 3 m room. The modes below 300 Hz are calculated with an exact frequency overlap algorithm, are shown in Table 27-1. Figure 27-8 compares this frequency-based calculation with the image model using 30 orders and a wall absorption of 0.12. The time- and frequency-based calculations give very similar answers. The discrepancies that do occur could be due to taking insufficient reflection orders and the fact that the impulse response is windowed with a cosine squared term for transforming and that the frequency-based calculation doesn't consider any modes above 300 Hz.

Music is naturally, however, a transient signal, and the ear can distinguish the effects of early arriving reflections. Furthermore, with many audio signals such as music, only the first few reflections are heard before the next musical note arrives and masks the later reflections. Consequently it is also necessary to investigate the response received by the listener for just the first few reflections—this will be referred to as the short-term spectrum. The short-term spectrum is a Fourier transform of the first 64 ms of the impulse response after the direct sound has arrived. The impulse response is windowed using a quarter period cosine squared window. The window starts at 32 ms after the direct sound and gradually weights the impulse response to zero at 64 ms. These times are motivated by the integration time of the ear, which is typically taken to be between 35 and 50 ms. The windowing is necessary to prevent the sudden cutoff of the impulse producing spurious effects in the spectrum. An example of the short-term spectrum is shown in Fig. 27-9. The two spectra shown in Figs. 27-7 and 27-9 are taken to represent the audible characteristics of the propagation from source to receiver, and it is the characteristics of the low-frequency spectra that are optimized to find the best locations for listener and loudspeakers within the room.

Table 27-1. Modal frequencies for a 3 m × 3 m × 3 m room.

n_x	n_y	n_z	f, Hz
0	0	0	0
0	0	1	56.67
0	1	1	80.14
1	1	1	98.15
2	0	0	113.33
2	1	0	126.71
2	1	1	138.8
0	2	2	160.28
2	1	2	170
3	0	1	179.2
3	1	0	179.2
3	1	1	187.94
2	2	2	196.3
3	2	0	204.32
3	2	1	212.03
4	0	0	226.67
4	1	0	233.64
4	1	1	240.42
3	3	1	247
4	2	0	253.42
4	2	1	259.68
3	3	2	265.79
4	2	2	277.61
5	0	0	283.33
5	1	0	288.94
5	1	1	294.45

Optimizing Procedure

The computer optimizing process finds the best position for the listener and loudspeakers by an iterative process illustrated in Fig. 27-10. The short- and long-term spectra are predicted by the image source method. From these spectra a cost parameter (described below) is derived which characterizes the quality of the sound produced.

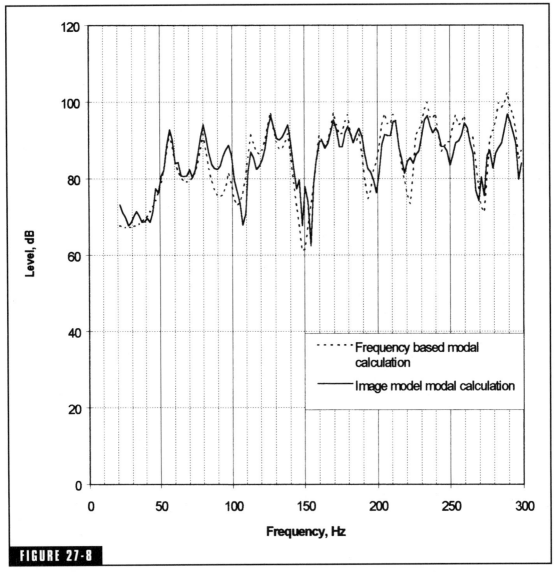

FIGURE 27-8

Comparison of image source calculation (solid line) and modal response (dotted line).

Then new listener and loudspeaker positions are repeatedly tried until a minimum in the cost parameter is found indicating that the best positions are found. The movement of the listener and the loudspeaker positions is carried out using a search engine following a standard minimization procedure.[26]

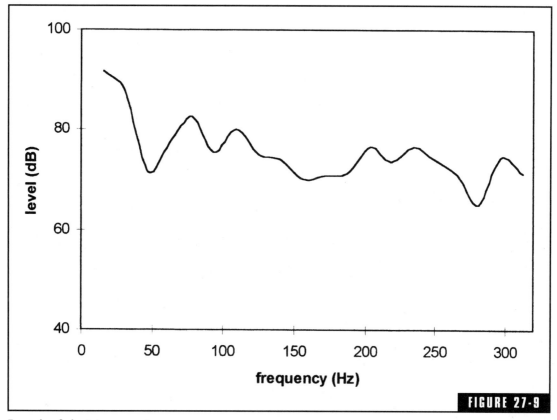

FIGURE 27-9

Example of short-term spectrum.

Cost Parameter

It is assumed that the best position within the room is represented by the position where the short-term and long-term spectra have the flattest frequency response. This then motivates the production of a cost parameter that measures how much the true frequency spectrum deviates from a flat response. Previous work by the authors has shown that a standard deviation function is a good measure for characterizing how even the pressure scattered from diffusers is.[27,28]

In reality, some smoothing over a few adjacent frequency bins is carried out, typically over 1 to 3 bins. This is done to simulate the effect of spatial averaging, which would naturally happen in actual listening rooms. Otherwise there is a risk that the optimization routine will find a solution which is overly sensitive to the exact solution position.

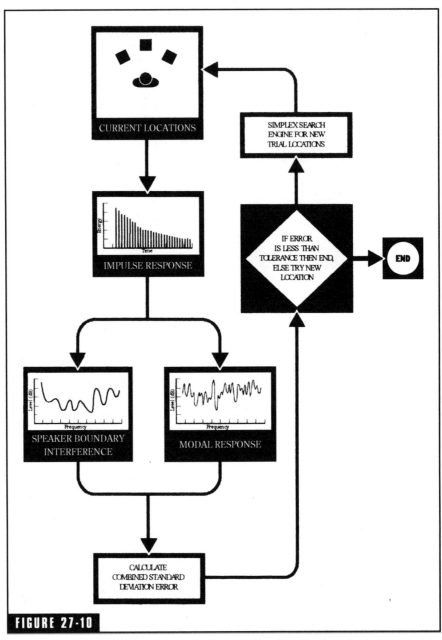

FIGURE 27-10

Iterative cycle to determine optimum listener and loudspeaker locations.

Figures 27-11 and 27-12 show an example of short-term speaker-boundary interference response and long-term modal response for an intermediate (nonoptimized) arrangement of listener and loudspeakers of a stereo pair configuration. The error parameters given in the captions indicate the ability of the standard deviation to measure the quality of the spectra.

Optimization Procedure

The optimization procedure used is a standard simplex routine. The simplex has a series of nodes, which are different points in the error

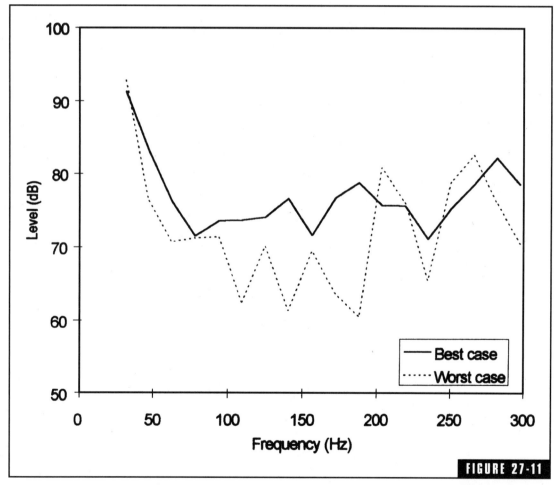

FIGURE 27-11

Short-term spectrum. The standard deviations are 4.67 and 8.13 dB for the best solution and the worst case, respectively.

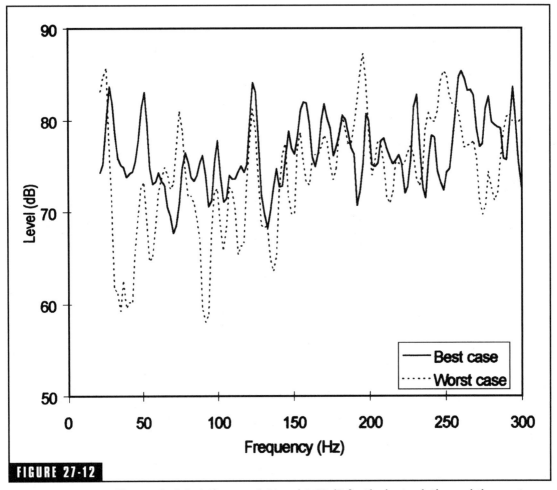

FIGURE 27-12

Long-term spectrum. The standard deviations are 3.81 and 6.28 dB for the best solution and the worst case, respectively.

space (representing different listener and loudspeaker positions). These nodes move around the space until either (*a*) the difference in the error parameter between the worst and best nodes is less than some tolerance variable or (*b*) a maximum number of iterations is exceeded (set at 500). Most of the time the program stops because of (*a*). The simplex routine has the advantage of being a robust system, which does not require first derivatives for calculations. Unfortunately, derivatives of the spectra with respect to the listener and loudspeaker positions are not immediately available from a numerical method such as the image source model. The penalty for using a function-only optimization routine is that the number of iterations used to find a solution is longer and so the procedure takes longer.

The user of the program inputs various parameters to define the optimization procedure. Table 27-2 lists the parameters that are under the user's control. These can be input via standard Windows dialog boxes, making the program user-friendly. The optimization routines cannot be used completely without some interpretation of the results from the user. For example, there is a tendency for the routines to want to place the source and receivers on the room boundaries as this will minimize the interference effects. Obviously, it is not always possible to build the loudspeakers into the walls and so this may not be a useful solution. Furthermore, there is a risk that the optimization routines will place the loudspeakers close enough to the surfaces to reduce the interference within the frequency band selected (say, 20 to 300 Hz), ignoring the fact that there may be audible interference effects just outside this frequency range. Also, the best solution found for the bass response will not always be optimized for stereo imaging, physical listener and loudspeaker placement, and other factors.

Table 27-2. Parameters presently under user control.

Room dimensions (width, length, height)

Rectangular volume defining the x, y, and z limits for the listener position

Rectangular volumes defining x, y, and z limits for the independent loudspeaker(s) position(s)

Number of independent and dependent loudspeakers

Displacement and symmetry constraints relating the dependent loudspeakers to the independent loudspeakers, e.g., simple mirror image symmetry or x, y, z displacement.

Stereo constraints

Minimum stereo pair separation

The weighting parameter w that determines the balance between the errors of the short- and long-term spectra

Frequency range of interest (default setting 20–300 Hz)

Number of solutions required

Therefore, the user is given the opportunity to limit the search range for the listener and loudspeakers. The limits of the listener and independent loudspeakers are defined in terms of rectangular volumes, determined by the minimum and maximum coordinates. The program allows the user to find the most appropriate solution within those imposed limits. The listener and loudspeakers can vary within the rectangular volume limits for an optimum practical solution. These limit constraints are applied to the simplex routine by brute force. For example, if the simplex routine asks for a prediction for a point outside the listener's rectangle, the program forces the coordinates of the point onto the nearest edge of the listener's constraint boundaries.

The loudspeakers can all be treated as independently varying. However, in most listening situations certain loudspeaker positions are determined by others. For example, in a simple stereo pair, both loudspeakers are related by mirror symmetry about the plane passing through the center of the room. As the number of loudspeakers increases in the 5.1 home theater and multichannel music surround formats, we can make the program more efficient by taking advantage of positional relationships between the speakers. It is usual to search the room and find several minima. The use of geometric constraints increases the chance of finding the global minimum.

To accomplish this, a system of independent and dependent loudspeakers is adopted with each dependent loudspeaker position being determined by an independent loudspeaker. For the stereo pair example, the left front loudspeaker can be considered the independent loudspeaker and the right front loudspeaker can be defined as the dependent loudspeaker with its position determined by a simple mirror image of the independent loudspeaker's position about the center of the room. The program allows mirror symmetry operations with respect to the x, y, z planes passing through the listener's position. Mirror symmetry about planes passing through the variable listener position allows constraints to be imposed on the rear surround speakers. For example, in a 5.1 multichannel music format with five matching speakers equispaced from the listener, we can set up constraint relationships with one independent speaker (left front) and four dependent speakers (center, right, front, left surround, and right surround).

To include the lessons we have learned about good stereo imaging, the program makes use of a stereo constraint. The stereo constraint refers to the normal angular constraints between a stereo pair and the listener, which are applied to give a good stereo image. In the program this constraint is applied by ensuring that the ratio of the distances between the listener to the center point of the speaker plane, and the distance between the stereo pair is within a specified range. The default is between 0.88 (equilateral triangle) and 1.33. Applying such a nonlinear constraint to the simplex routine can only be achieved by brute force. If a position which violates the constraint is required, the simplex routine moves both the listener and the loudspeaker positions to the nearest points in the room which comply with the constraint. The simplex routine can accommodate such abuses, but there is a risk that this will slow the procedure's finding of the optimum position. When optimizing independent loudspeakers, like subwoofers, for example, the constraint need not be applied. Table 27-3 lists the parameters that at the moment are not controlled by the user.

Results

Stereo Pair

The program was required to find the best solution for a stereo pair. The geometry and solution for the optimization are shown in Table 27-4. Figures 27-11 and 27-12 show example intermediate spectra for this configuration. It is apparent how poor the spectral responses can be, and also how they can be improved by the use of a positioning as outlined here. It is often found that the improvement for the short-term spectrum is more dramatic than for the long-term spectrum. The complexity and number of reflections in the long-term spectrum means that

---

Table 27-3. Parameters presently not under user control.

Minimum stereo pair separation (0.6 m)
Absorption coefficient of the surface (0.12)
Maximum order of reflections traced in image source model (15)
Number of frequency bins to smooth short-term and long-term spectra over
(1 and 3, respectively)

Table 27-4. Geometry and solution for a stereo pair configuration.

	X (m)	Y (m)	Z (m)
Room dimensions	7	4.5	2.8
Listener limits	2.0–6.0	2.25 (fixed)	1.14 (fixed)
Independent loudspeaker 1 (front left)	0.5–3	0.5–1.34	0.35–0.8
Dependent loudspeaker 1 (front right)	Mirror of left front at room center		
	$Y = 2.25$		
Stereo constraint	0.88–1.33		
Best error parameter	2.24		
Worst error parameter	3.84		
Positions of listener and loudspeakers for best solution			
Listener position	3.05	2.25	1.14
Independent (left front)	1.28	1.34	0.69
Dependent (right front)	1.28	3.16	0.69

there is less of a chance that there are positions in the room where large improvements of the standard deviation can be found. Even with this complexity, however, useful improvements are found by the optimization routines for the long-term spectrum. The short-term spectrum is much more sensitive to the positions of listener and loudspeaker.

Stereo Pair with Two Woofers Per Loudspeaker

We next determine the optimum arrangement of a stereo pair with two woofers vertically displaced by 0.343 m in each loudspeaker. The geometry and best solution are shown in Table 27-5. The optimization assumed constant-impedance loudspeaker floor mounting. The speaker-boundary interference response is shown in Fig. 27-13 and the modal response is shown in Fig. 27-14. The program makes use of mirror symmetry and displacement relationships to reduce the number of independent speakers that have to be optimized. In this case, even though there are four woofers, we actually only need to optimize one, which we will call the lower left front. The upper left front is constrained to follow the lower left front coordinates, while being displaced vertically in dimension z by 0.343 m. The lower and upper front right woofers are also dependent on the position of the lower left, because of the mirror sym-

Table 27-5. Geometry and solution for a stereo pair configuration with two woofers per loudspeaker.

	X (m)	Y (m)	Z (m)
Room dimensions	5.791	4.267	3.048
Listener limits	2.591–3.962	2.134 (fixed)	1.14 (fixed)
Independent loudspeaker 1 limits (lower left front)	0.61–1.829	0.457–1.067	0.381 (fixed)
Dependent loudspeaker 1 (upper left front)	Constrained to lower left front	Constrained to lower left front	Displaced 0.343
Dependent loudspeaker 2 (lower right front)	Mirror of lower left front at room center $Y = 2.134$		
Dependent loudspeaker 3 (upper right front)	Mirror of lower left front at room center and displaced in Z by 0.343		
Stereo constraint	0.88–1.33		
Best error parameter	2.1774		
Worst error parameter	4.0839		
Positions of listener and loudspeakers for the best solution			
Listener position	3.149	2.134	1.14
Independent lower left front	0.947	0.912	0.381
Dependent upper left front	0.947	0.912	0.724
Dependent lower right front	0.947	3.355	0.381
Dependent upper right front	0.947	3.355	0.724

metry of the stereo pair about the center plane of the room located at $Y = 2.134$ m. Thus in this optimization there is one independent loudspeaker, the lower left front, and three dependent loudspeakers, the upper left front, the lower right front, and the upper right front. Of particular note is the avoidance of the roughly 25 dB notch at about 180 Hz in the speaker-boundary interference of Fig. 27-12. The standard deviations for the best and worst solutions were 2.17 and 4.08 dB.

THX Home Theater

The emerging five-channel digital format, with low-frequency effects channel, commonly called 5.1, offers an exciting new aural experience. The program can be used to optimize the placement of any

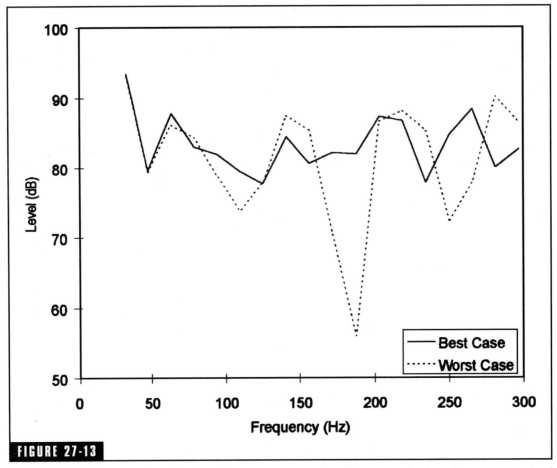

FIGURE 27-13

Comparison of the speaker-boundary interference response for two listener and loudspeaker arrangements of a stereo pair with two woofers per loudspeaker.

number of loudspeakers, so this surround configuration will provide a good example. We can illustrate this by examining the THX home theater surround configuration of a left/center/right (L/C/R) group of front speakers and a pair of dipoles used for the surround channels. We introduce a new type of constraint in this optimization. The center channel constraint assures that the center loudspeaker remains on the centerline of the room at a loudspeaker-listener distance equal to that of the left-front listener distance. In this way, all of the arrival times from the front speakers are maintained equal. Of course, if this is not desired, the constraint can simply not be applied. In this opti-

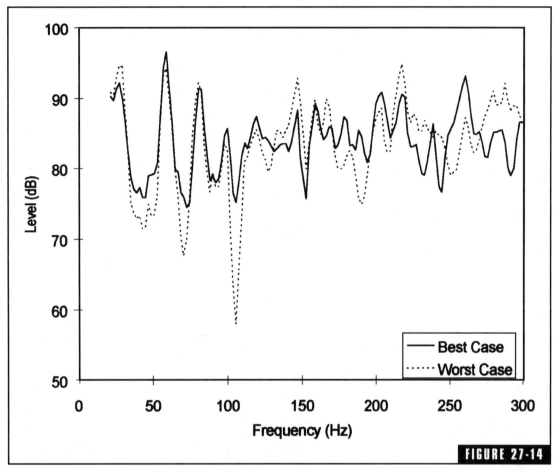

Comparison of the modal response for two listener and loudspeaker arrangements of a stereo pair with two woofers per loudspeaker.

mization the front loudspeakers are allowed to range in X, Y, and also Z. The Z search can be used to determine an appropriate elevation above the floor. In some instances this will be useful, while in others the location of the midrange and tweeter may take precedence for good imaging. The dipoles are omnidirectional below 300 Hz, so we will consider them as a point source, for the optimization. The geometry and solution for this configuration are listed in Table 27-6.

Another new constraint relation is added to this optimization to maintain that the dipole surrounds follow the listener's X coordinate, so that the listener will remain in the null. Thus, as the listener

Table 27-6. Geometry and solution for a THX surround sound configuration of L/C/R with dipole surrounds.

	X (m)	Y (m)	Z (m)
Room dimensions	5.791	4.267	3.048
Listener limits	2.286–3.962	2.134 (fixed)	1.14 (fixed)
Independent loudspeaker 1 limits (left front)	0.61–1.829	0.457–1.067	0.305–0.914
Independent loudspeaker 2 limits (left surround)	Constrained to listener	0.076–0.152	2.134–2.743
Dependent loudspeaker 1 (right front)	Mirror of left front at room center $Y = 2.134$		
Dependent loudspeaker 2 (center)	Constrained to left front/listener distance		
Dependent loudspeaker 3 (right surround)	Mirror of left surround at room center $Y = 2.134$		
Stereo constraint	0.88–1.33		
Best error parameter	1.9476		
Worst error parameter	3.7991		
Positions of listener and loudspeakers for best solution:			
Listener position	3.961	2.134	1.14
Independent left front	1.438	0.723	0.461
Dependent center	1.071	2.134	0.461
Dependent right front	1.438	3.544	0.461
Dependent left dipole surround	3.961	0.123	2.301
Dependent right dipole surround	3.961	4.144	2.301

moves forward and backward during the optimization the X coordinate of the dipoles follows this value. The Y coordinate, or spacing from the sidewalls, is optimized over a limited range, and the Z coordinate can assume any value within its range limits. A comparison of the speaker-boundary interference response and modal response of the best and worst solutions found for the THX configuration in a 5.791 m $\times$ 4.267 m $\times$ 3.048 m room are shown in Figs. 27-15 and 27-16. The standard deviations for the best and worst solutions were 1.95 and 3.80 dB.

Multichannel Music

In addition to the use of 5.1 in home theater, it is also being proposed as a music-only or multichannel music format. In addition to the previous dipole surround format, another configuration using five match-

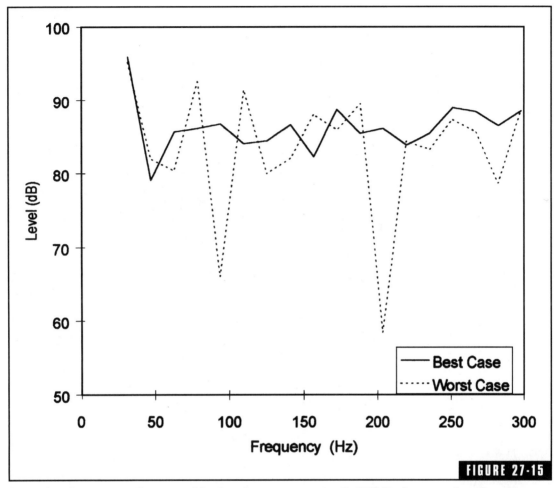

FIGURE 27-15

Comparison between two listener and loudspeaker arrangements of a THX dipole surround configuration.

ing loudspeakers is also being used. If the physical constraints of the room permit, all five loudspeakers would be equidistant from the listener.

To develop the constraints needed for this optimization, we would make use of the previous center channel constraint as well as a new rear channel constraint. The rear loudspeakers can be constrained to the front by the use of mirror planes about the listener. This is a dynamic constraint that follows the listener. The geometry and solution for this configuration is given in Table 27-7.

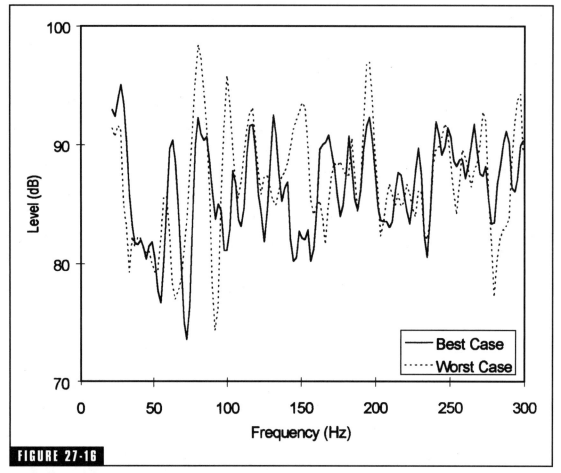

FIGURE 27-16

Comparison between the modal response for two listener and loudspeaker arrangements of a THX dipole configuration.

A comparison of the speaker-boundary interference response and modal response of the best and worst solutions found for the multichannel music configuration in a 5.791 m × 4.267 m × 3.048 m room are shown in Figs. 27-17 and 27-18. The standard deviations for the best and the worst solutions were 2.04 and 4.06.

Subwoofer

The use of subwoofers is growing in popularity, especially with the surround sound formats. The program can provide optimization of

Table 27-7. Geometry and solution for a 5.1 multichannel configuration.

	X (m)	Y (m)	Z (m)
Room dimensions	5.791	4.267	3.048
Listener limits	2.591–3.2	2.134 (fixed)	1.14 (fixed)
Independent loudspeaker 1 (left front)	0.914–1.829	0.457–1.067	0.305–0.914
Dependent loudspeaker 1 (right front)	Mirrors left front at room center $Y = 2.134$		
Dependent loudspeaker 2 (center)	Distance to listener constrained to equal left front-listener distance and constrained to lie on the $Y = 2.134$ line		
Dependent loudspeaker 3 (left rear)	Mirrors left front across X about the listener		
Dependent loudspeaker 4 (right rear)	Mirrors left front across X and Y about the listener		
Best error parameter	2.0399		
Worst error parameter	4.0633		
Positions of listener and loudspeakers for best solution:			
Listener position	3.099	2.134	1.14
Independent loudspeaker 1 (left front)	1.05	0.625	0.361
Dependent loudspeaker 1 (right front)	1.05	3.642	0.361
Dependent loudspeaker 2 (center)	0.554	2.134	0.361
Dependent loudspeaker 3 (left rear)	5.148	0.625	0.361
Dependent loudspeaker 4 (right rear)	5.148	3.642	0.361

subwoofers via a separate optimization over the 20- to 80-Hz frequency range. Once the listening position is determined, you can optimize any number of subwoofers. As an example, the geometry and solution for a single subwoofer in a 10 m × 6 m × 3 m room are listed in Table 27-8.

Figures 27-19 and 27-20 show the results of an optimization using a single subwoofer working in the range 20 to 80 Hz. The geometry for the optimization and solution is shown in Table 27-8. For the short-term spectrum the variation in the frequency response reduces from a range of about 30 dB in the worst case to around 10 dB for the solution found. The optimized solution provides a somewhat less

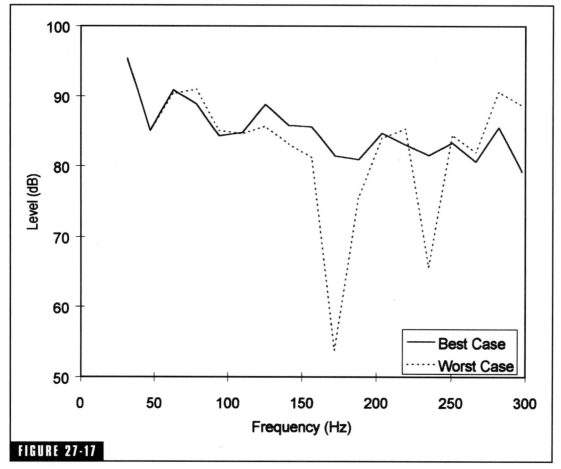

FIGURE 27-17

Comparison of the speaker-boundary interference response for two listener and loudspeaker arrangements of five equidistant matching loudspeakers.

dramatic improvement in the long-term spectrum. The standard deviations for the best and worst solutions were 2.7 and 5.6 dB, respectively.

Conclusion

A computer program has been developed that allows automated selection of positions for listeners and loudspeakers within listening rooms. The criterion for optimum listener and loudspeaker

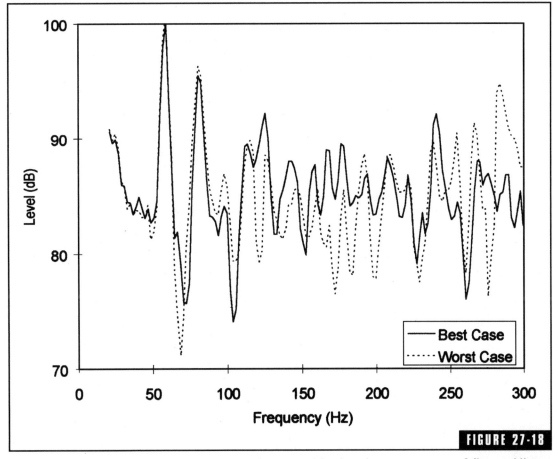

FIGURE 27-18

Comparison of the modal response for the two listener and loudspeaker arrangements of five equidistant matching loudspeakers.

positions within the room is the minimum standard deviation of the combined short- and long-term spectra. A cost parameter based on the standard deviation function was developed and used to monitor the quality of the short- and long-term spectra. Predictions of the spectra are carried out using an image source model. The optimization is carried out using a standard simplex routine. Some examples have been presented. All cases demonstrate the ability of the program to find the best positions for listener and loudspeakers within the room.

Table 27-8. Geometry and solution example for a subwoofer.

	X (m)	Y (m)	Z (m)
Room dimensions	10	6	3
Listener limits	5.5 (fixed)	3 (fixed)	1.2 (fixed)
Independent loudspeaker 1	5.6–9.9	0.1–2.9	0.1–2.9
Best error parameter	2.7051		
Worst error parameter	5.5535		
Positions of listener and loudspeakers for best solution:			
Listener position	5.5	3	1.2
Loudspeaker 1 (subwoofer)	7.398	0.231	1.227

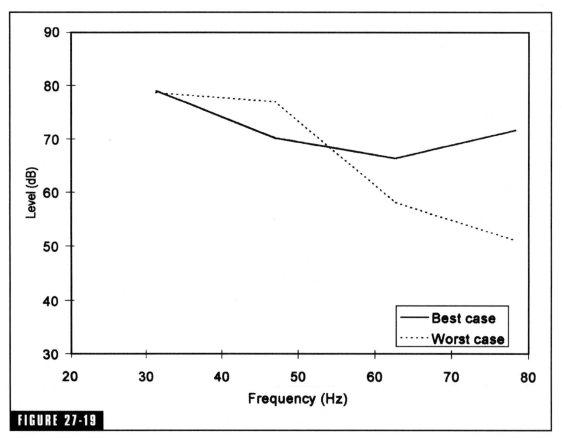

FIGURE 27-19

Comparison between the speaker-boundary interference response for two listener and loudspeaker positions of a subwoofer.

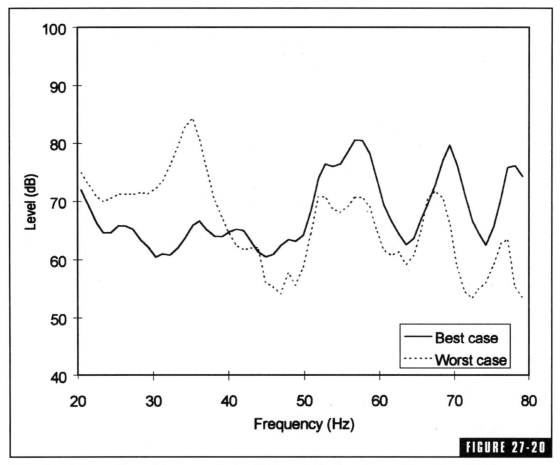

Comparison between the modal response for two listener and loudspeaker positions of a subwoofer.

Endnotes

[1]Schuck, P. L., S. Olive, J. Ryan, F.E. Toole, S. Sally, M. Bonneville, V. Verreault, and K. Momtohan, *Perception of Reproduced Sound in Rooms: Some Results from the Athena Project,* pp. 49–73, Proceedings of the 12th International Conference, Audio Engineering Soc. (June 1993).

[2]Olive, S. E., P. Schuck, S. Sally, and M. Bonneville, *The Effects of Loudspeaker Placement on Listener Preference Ratings,* J. Audio Eng. Soc., Vol. 42, pp. 651–669 (September 1994).

[3]Ballagh, K. P., *Optimum Loudspeaker Placement Near Reflecting Planes,* J. Audio Eng. Soc., Vol. 31, pp. 931–935 (December 1983); Letters to the Editor, J. Audio Eng. Soc., 31, p. 677 (September 1984).

[4]Groh, A. R., *High Fidelity Sound System Equalization by Analysis of Standing Waves,* J. Audio Eng. Soc., Vol. 22, pp. 795–799 (December 1974).

[5]Kovitz, P., *Extensions to the Image Method Model of Sound Propagation in a Room,* Pennsylvania State University Thesis (December 1994).

[6]Morse, P. M. and K. U. Ingard, *Theoretical Acoustics,* pp. 555–572, McGraw-Hill (1968).

[7]Louden, M. M., *Dimension-Ratios of Rectangular Rooms with Good Distribution of Eigentones,* Acustica, Vol. 24, pp. 101–104 (1971).

[8]Bonello, O. J., *A New Criterion for the Distribution of Normal Room Modes,* J. Audio Eng. Soc., Vol. 19, pp. 597–606 (September 1981).

[9]Allison, R. F., *The Influence of Room Boundaries on Loudspeaker Power Output,* J. Audio Eng. Soc., Vol. 22, pp. 314–319 (June 1974).

[10]Allison, R. F., *The Sound Field in Home Listening Rooms II,* J. Audio Eng. Soc., Vol. 24, pp. 14–19 (January/February 1976).

[11]Waterhouse, R. V., *Output of a Sound Source in a Reverberant Sound Field,* J. Acoust. Soc. Am., Vol. 27, pp. 247–258 (March 1958).

[12]Waterhouse, R. V., *Output of a Sound Source in a Reverberation Chamber and Other Reflecting Environments,* J. Acous. Soc. Am., Vol. 30 (March 1965).

[13]Waterhouse, R. V. and R. K. Cook, *Interference Patterns in Reverberant Sound Fields II,* J. Acous. Soc. Am., Vol. 37, pp. 424–428 (March 1965).

[14]D'Antonio, P. and J. H. Konnert, *The RFZ/RPG Approach to Control Room Monitoring,* 76th AES Convention Reprint No. 2157, New York (October 1984).

[15]Toole, F. E., *Subjective Evaluation,* in *Loudspeaker and Headphone Handbook,* 2nd Edition, chapter 11, J. Borwick, (Ed.), Focal Press (Oxford 1994).

[16]AES Recommended Practice for Professional Audio, J. Audio Eng. Soc., Vol. 44, No. 5, pp. 383–401 (May 1996).

[17]D'Antonio, P., C. Bilello, and D. Davis, *Optimizing Home Listening Rooms,* 85th AES Los Angeles Convention Preprint 2735 (November 1988).

[18]Everest. F. A., *The Master Handbook of Acoustics,* 3rd Edition, Tab Books.

[19]Stephenson, U., *Comparison of the Mirror Image Source Method and the Sound Particle Simulation Method,* Applied Acoustics, Vol. 29, pp. 35–72 (1990).

[20]Toole, F. E., *Loudspeaker Measurements and Their Relationship to Listener Preferences,* J. Audio Eng. Soc., Vol. 34, Pt. 1, pp. 227–235 (April 1986), Pt. 2, pp. 323–348 (May 1986).

[21]Toole, F. E. and S.E. Olive, *Hearing is Believing vs. Believing Is Hearing: Blind vs. Sighted Listening Tests and Other Interesting Things,* 97th Convention, Audio Eng. Soc. Preprint 3894 (November 1994).

[22]Olive, S. E. *A Method for Training of Listeners and Selecting Program Material for Listening Tests,* 97th Convention, Audio Eng. Soc., Preprint No. 3893 (November 1994).

[23]Toole, F. E. and S. E. Olive, *The Modification of Timbre by Resonances: Perception and Measurement,* J. Audio Eng. Soc., Vol. 36, pp. 122–142 (March 1988).

[24]Lam, Y. W. and D. C. Hodgson, *The Prediction of the Sound Field Due to an Arbitrary Vibrating Body in a Rectangular Enclosure,* J. Acous. Soc. Am., Vol. 88(4), pp. 1993–2000 (October 1990).

[25] Allen, J. B. and D. A. Berkeley, *Image Method for Efficiently Simulating Small-Room Acoustics,* J. Acous. Soc. Am., Vol. 65(4), pp. 943–950 (April 1979).

[26] W. H. Press, B. P. Flannery, S. A. Teukolsky, and W. T. Vetterling, *Numerical Recipes, The Art of Scientific Computing,* 2nd Edition, Cambridge University Press (1992).

[27] Cox, T. J., *Optimization of Profiled Diffusors,* J. Acous. Soc. Am., Vol. 97, pp. 2928–2936 (1995).

[28] Cox, T. J., *Designing Curved Diffusors for Performance Spaces,* J. Audio Eng. Soc., Vol. 44, pp. 354–364 (1996).

Desktop Auralization*

Introduction

In the past, acousticians have used a wide range of design techniques to provide good room acoustics, but acoustical options or the final result could not be tested prior to construction, except by evaluating scale models. Despite these limitations, acousticians have produced some remarkable performance and critical listening rooms. Wouldn't it be nice if we could listen to a room before it is built? In this way we might be able to avoid some of the acoustical problems and evaluate several design options and surface treatments. This process of acoustical rendering, analogous to visual rendering used by architects, has been called auralization. By way of definition, auralization is the process of rendering audible, by physical or mathematical modeling, the sound field of a source in a space, in such a way as to simulate the binaural listening experience at a given position in the modeled space.[1,2] Today, advances in computer processing power and digital signal processors are providing acousticians with new tools to predict and simulate the performance of critical listening rooms.

In the broadest of terms, to characterize a room it would be necessary to find a way to follow and catalog the reflection path history of the complex sound reflections as they proceed time through the room.

*Contributed by Peter D'Antonio, RPG Diffusor Systems, Inc., Upper Marlboro, Maryland 20774.

This reflection path history of sound level versus time as picked up at a location in the room can be called an echogram, and the process may be called "ray tracing." In this approach we would follow each ray through its reflection path history and catalog those rays that passed through a small volume at the listening position. This approach is easy to visualize, and if we could use audio slow motion and slow down sound from its 1,130 ft/s to 1 in/s, we could color-encode each sound ray and video tape the event.

Consequently, in the late 1960s, acousticians began using ray tracing to determine an echogram and estimate reverberation time. In the 1980s ray tracing became widely used. In ray tracing, the total energy emitted by a source is distributed according to the radiation characteristics of the source into a specified number of directions. In the simplest form, the energy of each ray is equal to the total energy divided by the number of rays. Depending on the type of surface, each ray from each boundary in its reflection history is either specularly reflected, in which the angle of incidence equals the angle of reflection, or diffusely reflected, in which the direction of the reflected ray is randomized. The reflected energy is diminished by absorption and also by the spherical attenuation due to propagation (in ray tracing this is achieved automatically because of the fixed receiver size). The number of rays passing through a receiver cell determines the sound-pressure level. An example of ray tracing is seen in Fig. 28-1, in which a ray emitted from source S is shown reflecting from three surfaces before it passes through the circular cross-section detection area of receiver $R1$. The energy contributions of the various rays to a certain receiver cell are added within prescribed time intervals, resulting in a histogram. Because of time averaging and the strongly random character of the ray arrivals, the histogram will only be an approximation of the true echogram. Ray tracing is straightforward, but its efficiency is achieved at the expense of limited time and spatial resolution in the echogram. The number of rays and the

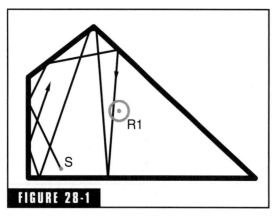

FIGURE 28-1

Ray tracing example of emitted ray from source S entering the circular cross-section detection area of receiver cell $R1$ after three specular reflections.

exact angles of emission from the source determine the accuracy of the sampling of room details and the reception of rays within the receiver volume.

In the late seventies, another approach, called the mirror image source method (MISM), was developed to determine the echogram. In this approach, a virtual image of the actual source is determined by reflecting the source perpendicularly across a room boundary. Thus the image source is located at a distance equal to twice the perpendicular distance d to the reflecting boundary. The distance between S1 and R1 in Fig. 28-2 is equal to the reflection path S, surface (1), R1. The idea is to reflect all real and virtual images across the room boundaries and develop a set of mirror images. The arrival times in the echogram are now simply determined by the distances between these images and the receiver. When applied to a rectangular room, all mirror sources are visible from every position in the room and the calculation is fast. In irregular rooms, however, this is not the case and "validity" tests have to be performed. For example, in Fig. 28-2, it can be seen that receiver R1 can be reached by a

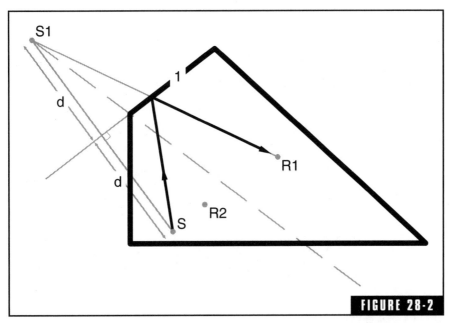

FIGURE 28-2

Mirror image source model (MISM) construction showing real source S, virtual source S1, boundary (1), and receiver R1.

first-order reflection from surface (1), but receiver $R2$ cannot. This means $R1$ is "visible" from $S1$ and $R2$ is not. Therefore, a fair amount of validation of each image is required. Since each further reflection order multiplies the number of image sources to be tested (by the number of walls minus one), the number of sources to validate grows very rapidly.

To minimize validation of sources, in the late eighties two approaches were developed. One was the hybrid method that used specular ray tracing to find potentially valid reflection paths and validated only those, while another used cones or triangular-based pyramidal "beams" instead of rays. First- and second-order cone tracing examples are shown in Figs. 28-3 and 28-4. If the receiver point lies within the projection of the beam, then a likely visible image source has been found and validation is not performed. Both of these methods can calculate a more detailed specular echogram than ray tracing and also faster than the MISM at the expense of leaving out reflection paths for the late part in the echogram (when the ray-cone separation is becoming larger than individual walls). However, it was only in ray tracing that diffuse reflection could be included in an efficient way, so the improved algorithms left out a very central acoustical phenome-

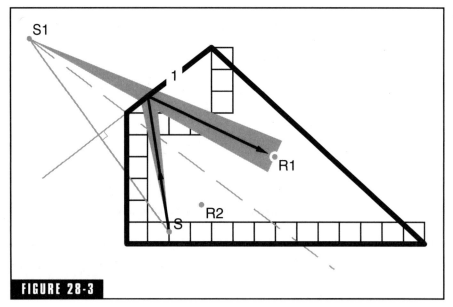

FIGURE 28-3

Equivalence of MISM and cone tracing for first-order reflection from surface (1).

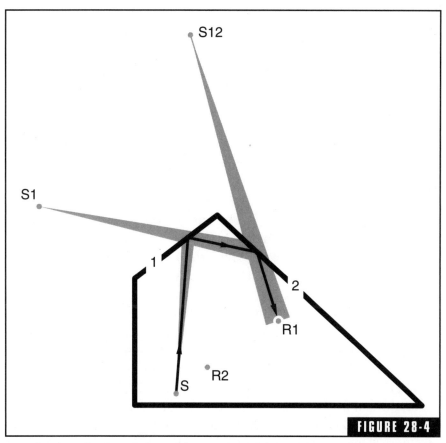

Second-order reflection of source *S* from surfaces (1) and (2) to receiver *R1* using cone tracing. *S1* and *S12* are virtual sources.

non. The current state of the art in geometrical room modeling programs all utilize a more complex combination of methods such as MISM for low-order reflections, and randomized cone tracing for higher-order reflections. Randomization is thus reintroduced to handle diffuse reflection.

The Auralization Process

The auralization process begins with predicting octave band echograms based on a 3D CAD model of a room, using geometrical acoustics. Frequency-dependent material properties (absorption and diffusion) are assigned to room surfaces and frequency-dependent source directivities are assigned to sound sources. From these echograms a great number of numerical objective measures, e.g., reverberation times, early and late energy ratios, lateral energy fraction, etc., can be estimated.

In Fig. 28-5 we illustrate the information needed to generate an echogram. Everything about the room needs to be defined, i.e., the sound sources, the room's geometry and surface treatments, and the listeners. Each source, be it a natural source or loudspeaker, is defined by its octave-band directivity at least at 125 Hz, 250 Hz, 500 Hz, 1 kHz, 2 kHz, and 4 kHz. The directivity balloons (a 3D version of the polar diagrams shown in Fig. 28-5) describe how the sound is directed in each octave band. The surface properties of the room's boundaries are described by their absorption and diffusion coefficients. Currently, there are working groups in the Audio Engineering Society and the ISO developing standards to measure diffusion coefficients. Each receiver is described by an appropriate response. If a binaural analysis is to be carried out, the listener is characterized by the head-related transfer functions (HRTFs as they are usually referred to). These frequency responses describe the difference in the response at each ear with and without a listener present. An example of HRTFs for sound arriving at 0° and at 20° is shown. Thus in the generation of the echogram each reflection contains information about its arrival time, its level, and its angle with respect to the receiver. The specular reflections are indicated as lines and the diffuse reflections as histograms.

In Fig. 28-6 we illustrate one method for converting the echogram into an impulse response by adding phase information. For each reflection the magnitude is determined by fitting the octave band levels (A through F), determined by a combination of the loudspeaker directivity, the absorption coefficient, and the diffusion coefficient. The phase for early specular reflections is often determined by minimum-phase techniques using the Hilbert transform. The impulse response for each reflection is then obtained by an inverse fast Fourier transform (IFFT).

There are several ways in which one can process the predicted echograms for final audition. These include binaural processing for headphones including headphone equalization or loudspeakers with crosstalk cancellation, mono processing, stereo processing, 5.1

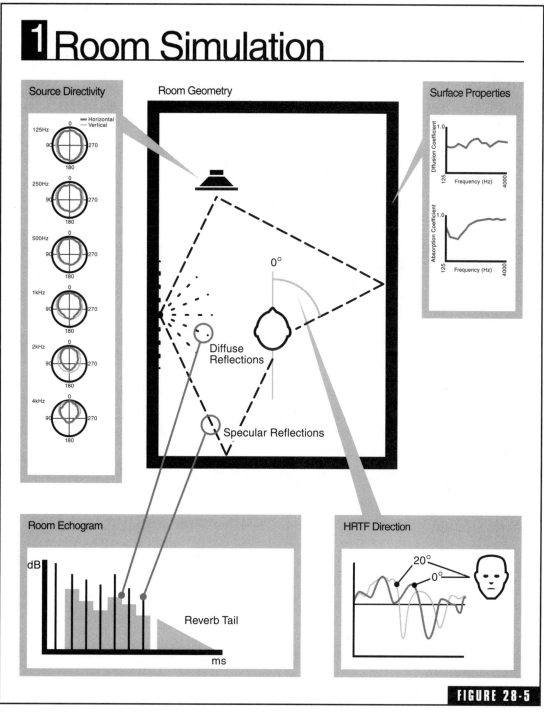

1 Room Simulation

Source Directivity

Horizontal
Vertical

125Hz
250Hz
500Hz
1kHz
2kHz
4kHz

Room Geometry

0°

Diffuse
Reflections

Specular Reflections

Surface Properties

Diffusion Coefficient
1.0
125   Frequency (Hz)   4000

Absorption Coefficient
1.0
125   Frequency (Hz)   4000

Room Echogram

dB

Reverb Tail

ms

HRTF Direction

20°
0°

FIGURE 28-5

Simulation of a room echogram using geometrical acoustics and descriptions of the room geometry, the surface properties, the source directivity, and the receiver's HRTFs.

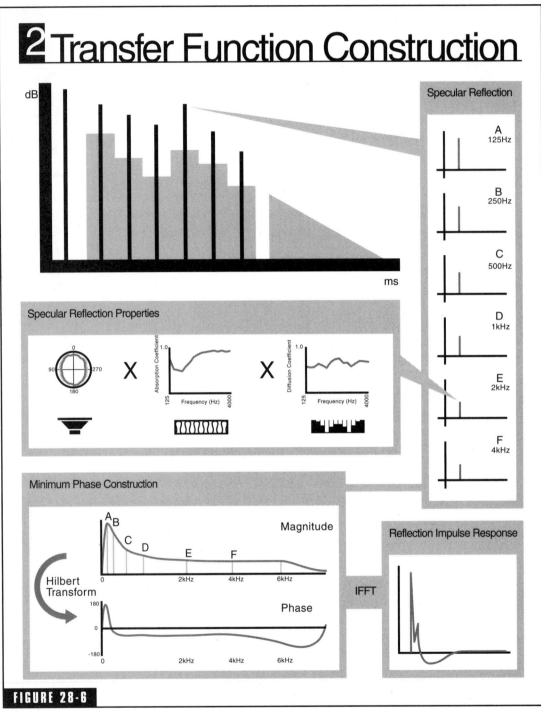

2 Transfer Function Construction

FIGURE 28-6

Transfer function construction is used to convert an echogram reflection into an impulse response.

processing, or B-format processing (for ambisonic replay). In Fig. 28-7 we illustrate the path for binaural processing. The impulse response of each reflection in turn is modified by the HRTFs for the appropriate angle of arrival and transformed into the time domain (IFFT). Thus each reflection is transformed into a binaural impulse response, one for the left and one for the right ear. This is done for each of the reflections from 1 to N and the resultant binaural impulse responses are added to form the total left and right room impulse responses. We are now able to auralize the room. This is achieved by convolving the binaural impulse responses with anechoic or nonanechoic music as is illustrated in Fig. 28-8. This can be done in software or in "real time" using DSP convolution platforms by companies like Lake Technologies, Australia (*www.laketechnology.com*).

In Fig. 28-9 we show a 3D model for a critical listening room created with a program called CATT-Acoustic (www.catt.se). 3D models are either created in AutoCAD or with a text program language, which allows the user to use variables, such as length, width, height, etc. Changing these variables allows one to easily change the model. Objects can also be created with variable orientations, to easily model the effect of different designs. In Fig. 28-9 we see the right (A1), center (A0), left (A2), right surround (A3), and left surround (A4) speakers. Also the room is divided vertically into thirds and horizontally in half to allow for different acoustical surface treatment. Adjusting the appropriate variable can easily change the size of these areas. The corners can be 90° or contain a bass trap the size of which is variable. The ceiling soffit is also variable. The receiver is indicated by location 01.

In Fig. 28-10 we see the early part of the echogram showing the specular and diffuse reflections, the emitted angles of the sound from the source, the arrival angles at the receiver, and an isometric view of the room. By moving the cursor to each reflection, the reflection path from source to receiver is visible in the room illustration. In this way one can determine the boundary responsible for a possibly problematic reflection.

Figure 28-11 illustrates the objective parameters that can be determined directly from the echogram after backward integration. EDT is the time it takes to reduce the sound-pressure level (SPL) from its steady-state level by 10 dB. T_{15} and T_{30} refer to the time is takes to decrease the SPL from -5 to -20 and from -5 to -35 dB, respectively.

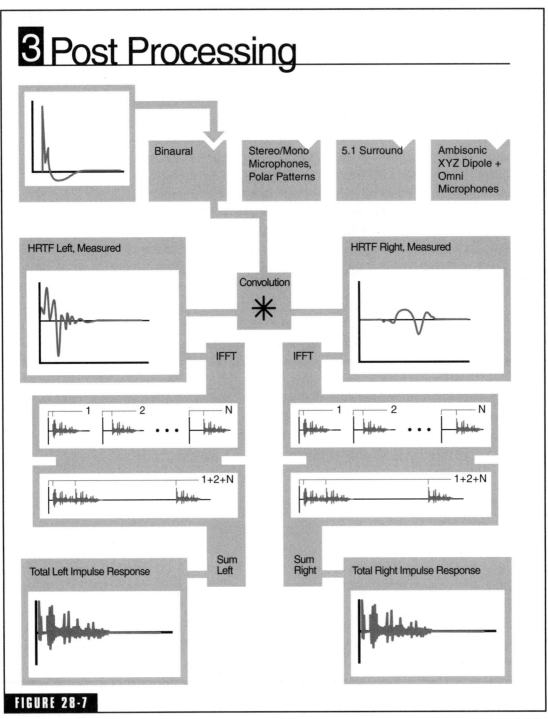

3 Post Processing

FIGURE 28-7

For binaural auralization, the postprocessing step converts the room impulse response into a left and right ear binaural impulse response through convolution with the HRTFs.

4 Auralization

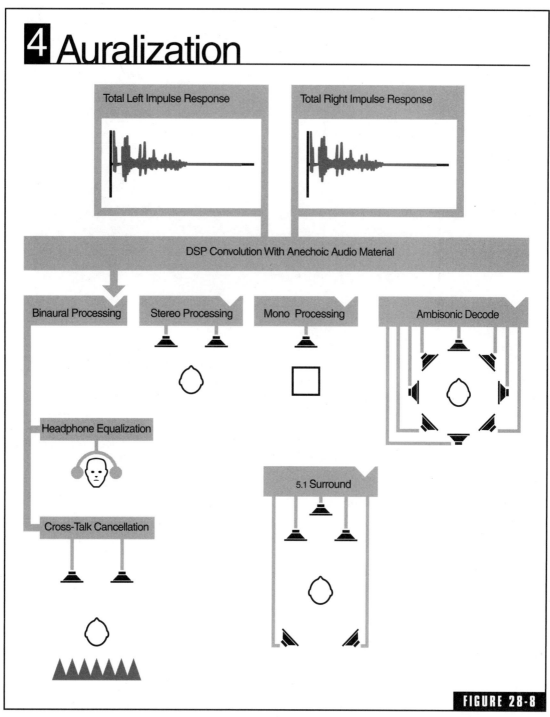

FIGURE 28-8

Auralization of the room is achieved by convolving the binaural impulse responses with music.

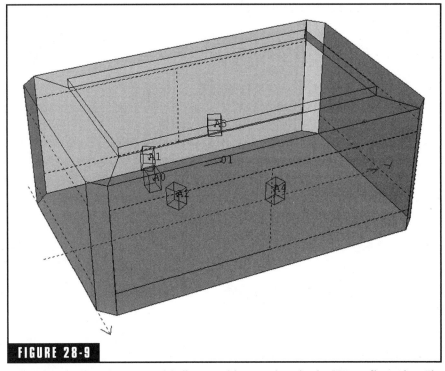

FIGURE 28-9

3D model of a listening room with five matching speakers in the ITU configuration. The model illustrates how the chosen variables describe the corner detail, soffit, and allocation of acoustical treatment to various sections on the room's boundaries.

There are several objective parameters based on ratios of early- to late-arriving sound. D50 is the definition criterion and is a measure of speech intelligibility. It is a percent ratio of the sound arriving in the first 50 ms following the direct sound to the total sound. The music clarity index C80 is based on a ratio of the first 80 ms of sound to the rest of the sound. C-80 is expressed in dB. Lateral energy fractions LFC and LF (referred to as LEF1 and LEF2 in CATT-Acoustic) are measures of the impression of spaciousness. They are based on the ratio of laterally arriving reflections between 5 and 80 ms to all of the omnidirectionally arriving reflections before 80 ms.

The center-of-gravity time T_s is a measure constructed to describe where the sound is concentrated in the echogram. T_s will have a low value if the arriving sound is concentrated in the early part of the echogram and a high value if the early reflections are weak or if the

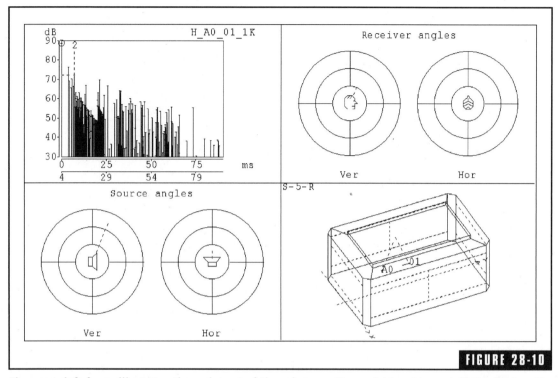

The upper left frame illustrates the early part of the echogram. Upper right and lower left panels show the directionality of the receiving and emitted sound rays. Lower right shows selected reflection from A0 to receiver 01 bouncing off the ceiling.

decay is slow. Sound-pressure level SPL is 10 times the 10-based log of the ratio of the pressure squared over a reference pressure squared. G10 is the sound pressure at a location in a room as compared to what the direct sound from an omnidirectional source would give at 10 m distance. G10 is a measure that is used for characterizing how loud the sound will be and can be compared between halls.

The upper left panel shows the impulse response with the backward integration. The reverberation time from a 15-dB interval T_{15} is shown along with the reverberation time from a 30-dB interval T_{30}.

Another useful function of the geometrical modeling programs is the ability to map the objective parameters, so that we can see at a glance how a parameter varies over the listening area. (These maps are normally in color giving a much better visual resolution than the gray-scale versions presented here. Color maps may be viewed on the

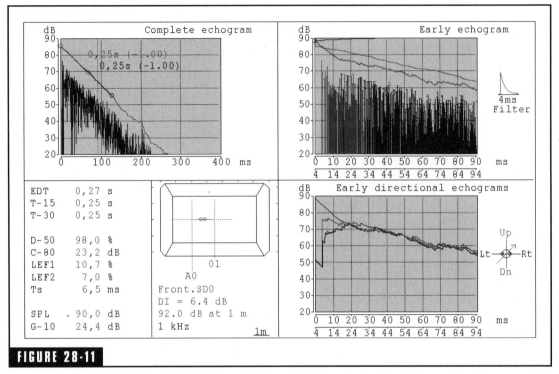

FIGURE 28-11

The upper left shows the complete echogram with T_{15} and T_{30} from the backward integration. The upper right shows the early echogram with backward and forward integration. Lower left panel lists the objective parameters and the lower right compares the directional echograms (up/down, left/right, and front/back).

R.P.G. website: *http://www.rpginc.com.*) In Fig. 28-12 we show a mapping of RT′ (which is related to the EDT), the LEF2, SPL, and D50 for five matching monopole speakers in the International Telecommunications Union (ITU) surround format.

Although these parameters are mostly useful for large-room analysis, they also give some information in the case of a listening room. They will, for example, give an opportunity to study how the effective listening area is affected by loudspeaker placement and surface treatment. Note that the RT′ value in Fig. 28-12 is very low in between the speakers (= a distinct sound) while for the THX case in Fig. 28-13, the situation is different since the listener is not reached by much direct sound from the surround dipoles. This is, of course, one of the design goals of using dipoles, giving a more diffuse sound from the surround speakers, but it shows the ability of the mapping technique to illustrate the behavior of different setups.

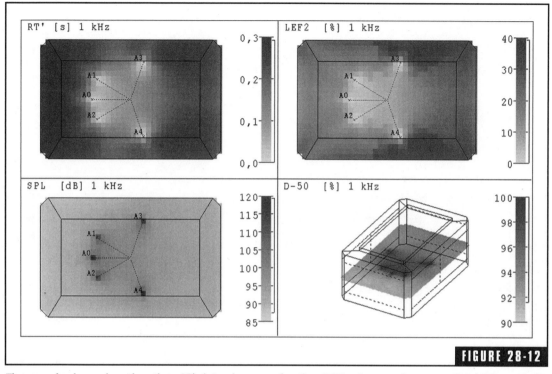

The perceived reverberation time RT', lateral energy fraction LEF2, the sound-pressure level SPL, and the energy ratio D50 are mapped at 1 kHz across the room for five matching speakers in the ITU surround configuration, left/center/right with monopole surrounds.

Another informative mapping plot shows how a given parameter changes with time. In Fig. 28-14 the SPL is plotted in four time intervals so that we can view how the sounds from the various loudspeakers over time combine to energize the room. If an RFZ™* (reflection-free zone) is created, it will show up in the upper right map as low SPL values around the listener position, and in the same map the effect of absorptive surfaces behind the front speakers can be seen as a lower SPL. By selecting the time windows well the efficiency of a design can be studied.

In addition to analyzing the acoustical parameters, we can also listen to what the room might sound like with the current acoustical design and speaker configuration. To do this we need to combine the binaural impulse responses of the left and right ear, shown in Fig. 28-15, with music. For binaural auralization the binaural impulse responses

*RFZ™ is a registered trademark of RPG Diffusor Systems, Inc.

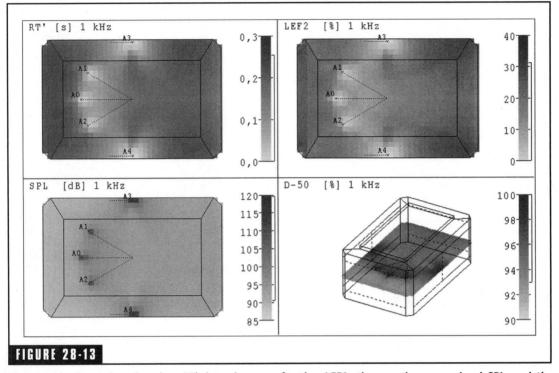

FIGURE 28-13

The perceived reverberation time RT′, lateral energy fraction LEF2, the sound-pressure level SPL, and the energy ratio D50 are mapped at 1 kHz across the room with a THX configuration of left/center/right and dipole surrounds.

are derived from the room impulse response by convolution of each reflection with HRTFs, which are stored in the program from a mannequin and also from a real person. The next step is to select an anechoic or nonanechoic music sample and convolve it with the binaural impulse responses. An anechoic music sample can be used to auralize the sound in a music production room, such as a concert hall. Since the music is anechoic, it does not contain any room signature. Therefore, convolving with the room response allows us to audition the room at various source and listening positions. A nonanechoic source might be used to determine the extent to which a sound reproduction room colors or influences what is heard by adding its own room signature to the sound. Thus by comparing the nonanechoic music sample with that heard in the room, we can make an evaluation of the coloration introduced.

Figure 28-16 illustrates the left and right ear wave files that can be

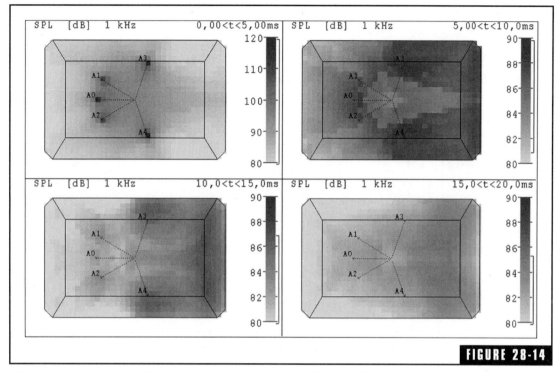

FIGURE 28-14

Illustrates how the sound fields from the five speakers combine over time to energize the room.

listened to by headphones, but as previously indicated several other auralization formats can be selected during the postprocessing step.

Summary

Thus we have described an approach using geometrical acoustics that generates the room impulse response of a virtual room. This impulse response can be postprocessed for several types of auralization including mono, stereo, 5.1, binaural, and ambisonic. We can also auralize real rooms if we measure their impulse response.

Architectural drawings have evolved from plan and section, to 3D, to fully rendered image, to walk-throughs. Now we have the opportunity to also auralize the visually rendered environment. At this time high-quality auralizations are possible for fixed locations and real-time auralizations can be used primarily for virtual reality applications. We might end with a bit of caution and reiterate that

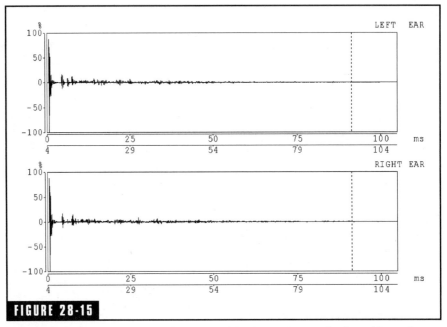

FIGURE 28-15

Convolution of the echogram reflections with the HRTFs results in a binaural room impulse response shown for the left ear (top) and right ear (bottom).

the geometrical acoustics methods presently used are generally more accurate at the higher frequencies and consequently the lower octaves cannot be accurately represented. However, despite the approximations involved, auralization is still a very valid and useful tool, especially in the hands of experienced acousticians.

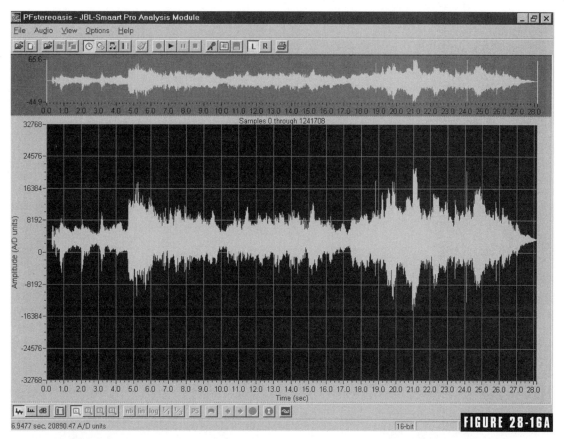

The convolution of the binaural room impulse responses with music generates the left (A) and right (B) wave files which can be played back for auralization.

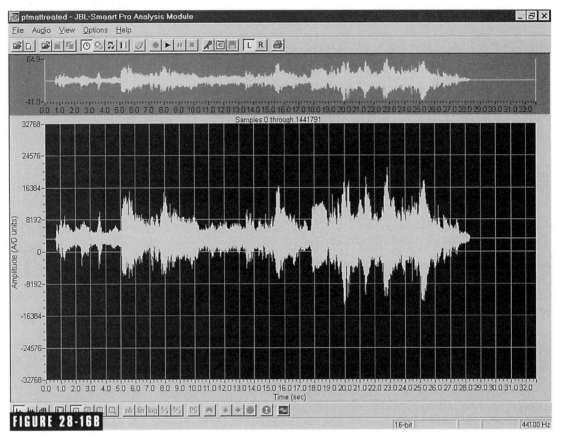

FIGURE 28-16B

The convolution of the binaural room impulse responses with music generates the left (A) and right (B) wave files which can be played back for auralization.

Endnotes

[1]*Auralization—an Overview,* Kleiner, M., Dalenbäck, B.-I., and P. Svensson, J. Audio Eng. Soc., Vol. 41, No. 11, pp. 861–875 (1993).

[2]*A Macroscopic View of Diffuse Reflection,* Dalenbäck, B.-I., M. Kleiner, and P. Svensson, J. Audio Eng. Soc., Vol. 42, pp. 973–807 (1994).

SELECTED
ABSORPTION COEFFICIENTS

Material	125 Hz	250 Hz	500 Hz	1 kHz	2 kHz	4 kHz	Reference (Chapter)
POROUS TYPE							
Drapes: cotton 14 oz/sq yd							
draped to 7/8 area	0.03	0.12	0.15	0.27	0.37	0.42	Mankovsky, ref 9-4
draped to 3/4 area	0.04	0.23	0.40	0.57	0.53	0.40	Mankovsky, ref 9-4
draped to 1/2 area	0.07	0.37	0.49	0.81	0.65	0.54	Mankovsky, ref 9-4
Drapes: medium velour, 14 oz/sq yd							
draped to 1/2 area	0.07	0.31	0.49	0.75	0.70	0.60	Mankovsky, ref 9-4
Drapes: heavy velour, 18 oz/sq yd							
draped to 1/2 area	0.14	0.35	0.55	0.72	0.70	0.65	Compendium, ref 9-1
Carpet: heavy on concrete	0.02	0.06	0.14	0.37	0.60	0.65	Compendium, ref 9-1
Carpet: heavy on 40 oz hair felt	0.08	0.24	0.57	0.69	0.71	0.73	Compendium, ref 9-1
Carpet: heavy with latex backing on foam or 40 oz hair felt	0.08	0.27	0.39	0.34	0.48	0.63	Compendium, ref 9-1
Carpet: indoor/ outdoor	0.01	0.05	0.10	0.20	0.45	0.65	Seikman, ref 9-17
Acoustical tile, ave, 1/2" thick	0.07	0.21	0.66	0.75	0.62	0.49	—
Acoustical tile, ave, 3/4" thick	0.09	0.28	0.78	0.84	0.73	0.64	—

Material	125 Hz	250 Hz	500 Hz	1 kHz	2 kHz	4 kHz	Reference (Chapter)
MISC. BUILDING MATERIALS							
Concrete block, coarse	0.36	0.44	0.31	0.29	0.39	0.25	Compendium, ref 9-1
Concrete block, painted	0.10	0.05	0.06	0.07	0.09	0.08	Compendium, ref 9-1
Concrete floor	0.01	0.01	0.015	0.02	0.02	0.02	Compendium, ref 9-1
Floor: linoleum, Asphalt-tile, or cork tile on concrete	0.02	0.03	0.03	0.03	0.03	0.02	Compendium, ref 9-1
Floor: wood	0.15	0.11	0.10	0.07	0.06	0.07	Compendium, ref 9-1
Glass: large panes, heavy glass	0.18	0.06	0.04	0.03	0.02	0.02	Compendium, ref 9-1
Glass, ordinary window	0.35	0.25	0.18	0.12	0.07	0.04	Compendium, ref 9-1
<u>Drop Ceiling</u>							
Owens-Corning Frescor, painted, 5/8" thick, Mounting 7	0.69	0.86	0.68	0.87	0.90	0.81	Compendium, ref 9-1
Plaster, gypsum or lime, smooth finish on tile or brick	0.013	0.015	0.02	0.03	0.04	0.05	Compendium, ref 9-1
Plaster: gypsum or lime, smooth finish on lath	0.14	0.10	0.06	0.05	0.04	0.03	Compendium, ref 9-1
Gypsum board: 1/2" on 2 x 4s, 16" on centers	0.29	0.10	0.05	0.04	0.07	0.09	Compendium, ref 9-1
RESONANT ABSORBERS							
Plywood panel: 3/8" thick	0.28	0.22	0.17	0.09	0.10	0.11	Compendium, ref 9-1
<u>Polycylindrical:</u>							
chord 45" height 16" empty	0.41	0.40	0.33	0.25	0.20	0.22	Mankovsky, ref 9-4
chord 35" height 12" empty	0.37	0.35	0.32	0.28	0.22	0.22	Mankovsky, ref 9-4
chord 28" height 10"empty	0.32	0.35	0.3	0.25	0.2	0.23	Mankovsky, ref 9-4
chord 28" height 10" filled	0.35	0.5	0.38	0.3	0.22	0.18	Mankovsky, ref 9-4

Material	125 Hz	250 Hz	500 Hz	1 kHz	2 kHz	4 kHz	Reference (Chapter)
RESONANT ABSORBERS							
Polycylindrical *(Continued)*:							
chord 20″ height 8″ empty	0.25	0.3	0.33	0.22	0.2	0.21	Mankovsky, ref 9-4
chord 20″ height 8″ filled	0.3	0.42	0.35	0.23	0.19	0.2	Mankovsky, ref 9-4
Perforated Panel							
5/32″ thick, 4″ depth, 2″ glass fiber							
Perf: 0.18%	0.4	0.7	0.3	0.12	0.1	0.05	Mankovsky, ref 9-4
Perf: 0.79%	0.4	0.84	0.4	0.16	0.14	0.12	Mankovsky, ref 9-4
Perf. 1.4%	0.25	0.96	0.66	0.26	0.16	0.1	Mankovsky, ref 9-4
Perf:. 8.7%	0.27	0.84	0.96	0.36	0.32	0.26	Mankovsky, ref 9-4
8″ depth, 4″ glass fiber							
Perf: 0.18%	0.8	0.58	0.27	0.14	0.12	0.1	Mankovsky, ref 9-4
Perf: 0.79%	0.98	0.88	0.52	0.21	0.16	0.14	Mankovsky, ref 9-4
Perf: 1.4%	0.78	0.98	0.68	0.27	0.16	0.12	Mankovsky, ref 9-4
Perf: 8.7%	0.78	0.98	0.95	0.53	0.32	0.27	Mankovsky, ref 9-4
With 7″ air space plus 1″ mineral fiber of 9-10 16 cu ft/lb density, 1/4″ cover							
Wideband, 25% perf or more	0.67	1.09	0.98	0.93	0.98	0.96	BBC, ref 9-18
Midpeak, 5% perf	0.60	0.98	0.82	0.90	0.49	0.30	BBC, ref 9-18
Lo-peak, 0.5% perf	0.74	0.53	0.40	0.30	0.14	0.16	BBC, ref 9-18
With 2″ air space filled with mineral fiber, 9-10 lb/cu ft density							
Perf: 0.5%	0.48	0.78	0.60	0.38	0.32	0.16	BBC, ref 9-18

GLOSSARY

Abffusor A proprietary panel offering both absorption and diffusion of sound.

absorption In acoustics, the changing of sound energy to heat.

absorption coefficient The fraction of sound energy that is absorbed at any surface. It has a value between 0 and 1 and varies with the frequency and angle of incidence of the sound.

acoustics The science of sound. It can also refer to the effect a given environment has on sound.

AES Audio Engineering Society.

algorithm Procedure for solving a mathematical problem.

ambience The distinctive acoustical characteristics of a given space.

amplifier, line An amplifier designed to operate at intermediate levels. Its output is usually on the order of one volt.

amplifier, output A power amplifier designed to drive a loudspeaker or other load.

amplitude The instantaneous magnitude of an oscillating quantity such as sound pressure. The peak amplitude is the maximum value.

amplitude distortion A distortion of the wave shape of a signal.

analog An electrical signal whose frequency and level vary continuously in direct relationship to the original electrical or acoustical signal.

anechoic Without echo.

anechoic chamber A room designed to suppress internal sound reflections. Used for acoustical measurements.

articulation A quantitative measure of the intelligibility of speech; the percentage of speech items correctly perceived and recorded.

artificial reverberation Reverberation generated by electrical or acoustical means to simulate that of concert halls, etc., added to a signal to make it sound more lifelike.

antinodic Point of maximum vibration in a vibrating body.

ASA Acoustical Society of America.

ASHRAE American Society of Heating, Refrigerating, and Air-Conditioning Engineers.

attack The beginning of a sound, the initial transient of a musical note.

attenuate To reduce the level of an electrical or acoustical signal.

attenuator A device, usually a variable resistance, used to control the level of an electrical signal.

audio frequency An acoustical or electrical signal of a frequency that falls within the audible range of the human ear, usually taken as 20 Hz to 20 kHz.

audio spectrum *See* audio frequency

auditory area The sensory area lying between the threshold of hearing and the threshold of feeling or pain.

auditory cortex The region of the brain receiving nerve impulses from the ear.

auditory system The human hearing system made up of the external ear, the middle ear, the inner ear, the nerve pathways, and the brain.

aural Having to do with the auditory mechanism.

A-weighting A frequency response adjustment of a sound-level meter that makes its reading conform, very roughly, to human response.

axial mode The room resonances associated with each pair of parallel walls.

baffle A movable barrier used in the recording studio to achieve separation of signals from different sources. The surface or board upon which a loudspeaker is mounted.

bandpass filter A filter that attenuates signals both below and above the desired passband.

bandwidth The frequency range passed by a given device or structure.

basilar membrane A membrane inside the cochlea that vibrates in response to sound, exciting the hair cells.

bass The lower range of audible frequencies.

bass boost The increase in level of the lower range of frequencies, usually achieved by electrical circuits.

beats Periodic fluctuations that are heard when sounds of slightly different frequencies are superimposed.

binaural A situation involving listening with two ears.

boomy Colloquial expression for excessive bass response in a recording, playback, or sound-reinforcing system.

byte A term used in digital systems. One byte is equal to 8 bits of data. A bit is the elemental "high" or "low" state of a binary system.

capacitor An electrical component that passes alternating currents but blocks direct currents. Also called a *condenser*, it is capable of storing electrical energy.

clipping An electrical signal is clipped if the signal level exceeds the capabilities of the amplifier. It is a distortion of the signal.

cochlea The portion of the inner ear that changes the mechanical vibrations of the cochlear fluid into electrical signals. It is the frequency-analyzing portion of the auditory system.

coloration The distortion of a signal detectable by the ear.

comb filter A distortion produced by combining an electrical or acoustical signal with a delayed replica of itself. The result is constructive and destructive interference that results in peaks and nulls being introduced into the frequency response. When plotted to a linear frequency scale, the response resembles a comb, hence the name.

compression Reducing the dynamic range of a signal by electrical circuits that reduce the level of loud passages.

condenser *See* capacitor.

correlogram A graph showing the correlation of one signal with another.

cortex *See* auditory cortex.

crest factor Peak value divided by rms value.

critical band   In human hearing, only those frequency components within a narrow band, called the *critical band*, with mask a given tone. Critical bandwidth varies with frequency but is usually between 1/6 and 1/3 octave.

crossover frequency   In a loudspeaker with multiple radiators, the crossover frequency is the −3-dB point of the network dividing the signal energy.

crosstalk   The signal of one channel, track, or circuit interfering with another.

cycles per second   The frequency of an electrical signal or sound wave. Measured in hertz (Hz).

dB   *See* decibel.

dB(A)   A sound-level meter reading with an A-weighting network simulating the human-ear response at a loudness level of 40 phons.

dB(B)   A sound-level meter reading with a B-weighting network simulating the human-ear response at a loudness level of 70 phons.

dB(C)   A sound-level meter reading with no weighting network in the circuit, i.e., flat. The reference level is 20 μPa.

decade   Ten times any quantity or frequency range. The range of the human ear is about 3 decades.

decay rate   A measure of the decay of acoustical signals, expressed as a slope in dB/second.

decibel   The human ear responds logarithmically and it is convenient to deal in logarithmic units in audio systems. The bel is the logarithm of the ratio of two powers, and decibel is 1/10 bel.

delay line   A digital, analog, or mechanical device employed to delay one audio signal with respect to another.

diaphragm   Any surface that vibrates in response to sound or is vibrated to emit sound, such as in microphones and loudspeakers. Also applied to wall and floor surfaces vibrating in response to sound or in transmitting sound.

dielectric   An insulating material. The material between the plates of a capacitor.

diffraction   The distortion of a wavefront caused by the presence of an obstacle in the sound field.

diffusor   A proprietary device for the diffusion of sound through reflection-phase-grating means.

digital   A numerical representation of an analog signal. Pertaining to the application of digital techniques to common tasks.

distance double law   In pure spherical divergence of sound from a point source in free space, the sound pressure level decreases 6 dB for each doubling of the distance. This condition is rarely encountered in practice, but it is a handy rule to remember in estimating sound changes with distance.

distortion   Any change in the waveform or harmonic content of an original signal as it passes through a device. The result of nonlinearity within the device.

distortion, harmonic   Changing the harmonic content of a signal by passing it through a nonlinear device.

DSZ   Diffused sound zone.

dynamic range   All audio systems are limited by inherent noise at low levels and by overload distortion at high levels. The usable region between these two exptremes is the dynamic range of the system. Expressed in dB.

dyne   The force that will accelerate a 1-gram mass at the rate of 1 cm/sec. The old standard reference level for sound pressure was 0.0002 dyne/cm^2. The same level today is expressed as 20 micropascals, or 20 μPa.

ear canal   The external auditory meatus; the canal between the pinna and the eardrum.

eardrum   The tympanic membrane located at the end of the ear canal that is attached to the ossicles of the middle ear.

echo   A delayed return of sound that is perceived ty the ear as a discrete sound image.

echogram   A record of the very early reverberatory decay of sound in a room.

EES   Early, early sound. Structure-borne sound may reach the microphone in a room before the air-borne sound because sound travels faster through the denser materials.

EFC   Energy-frequency curve.

EFTC   Energy-frequency-time curve.

eigen function   A defining constant in the wave equation.

ensemble   Musicians must hear each other to function properly; in other words, ensemble must prevail. Diffusing elements surrounding the stage area contribute greatly to ensemble.

equal loudness contour   A contour representing a constant loudness for all audible frequencies. The contour having a sound pressure level of 40 dB at 1,000 Hz is arbitrarily defined as the 40-phon contour.

equalization   The process of adjusting the frequency response of a device or system to achieve a flat or other desired response.

equalizer   A device for adjusting the frequency response of a device or system.

ETC   Energy-time curve.

ETF   AcoustiSoft's loudspeaker and room acoustic analysis program.

Eustachian tube   The tube running from the middle ear into the pharynx that equalizes middle-ear and atmospheric pressure.

external meatus   The ear canal terminated by the eardrum.

feedback, acoustic   Unwanted interaction between the output and input of an acoustical system, e.g., between the loudspeaker and the microphone of a system.

FFT   Fast Fourier transform. An iterative program that computes the Fourier transform in a shorter time.

fidelity   As applied to sound quality, the faithfulness to the original.

filter, high pass   A filter that passes all frequencies above a cutoff frequency.

filter, low pass   A filter that passes all frequencies below a certain cutoff frequency.

filter, bandpass   A filter that passes all frequencies between a low-frequency cutoff point and a high-frequency cutoff point.

filter IIR   Digital filter of special type.

filter FIR   Digital filter of special type.

flanging   The term applied to the use of comb filters to obtain special sound effects.

flutter   A repetitive echo set up by parallel reflecting surfaces.

Fourier analysis   Application of the Fourier transform to a signal to determine its spectrum.

frequency   The measure of the rapidity of alterations of a periodic signal, expressed in cycles per second or Hz.

frequency response   The changes in the sensitivity of a circuit or device with frequency.

FTC   Frequency-time curve.

fundamental   The basic pitch of a musical note.

fusion zone   All reflections arriving at the observer's ear within 20 to 40 msec of the direct sound are integrated, or fused together, with a resulting apparent increase in level and a pleasant change of character. This is the *Haas effect*.

gain   The increase in power level of a signal produced by an amplifier.

graphic-level recorder   A device for recording signal level in dB vs. time on a tape. The level in dB vs. angle can be recorded also for directivity patterns.

grating, diffraction   An optical grating consists of minute, parallel lines used to break light down into its component colors. The principle is now used to achieve diffraction of acoustical waves.

grating, reflection phase   An acoustical diffraction grating to produce diffusion of sound.

Haas effect   *See* fusion zone. Also called the *precedence effect*. Delayed sounds are integrated by the auditory apparatus if they fall on the ear within 20 to 40 msec of the direct sound. The level of the delayed components contributes to the apparent level of the sound, and it is accompanied by a pleasant change in character.

hair cell   The sensory elements of the cochlea that transduce the mechanical vibrations of the basilar membrane to nerve impulses that are sent to the brain.

harmonic distortion   *See* distortion, harmonic.

harmonics   Integral multiples of the fundamental frequency. The first harmonic is the fundamental, and the second is twice the frequency of the fundamental, etc.

hearing loss   The loss of sensitivity of the auditory system, measured in dB below a standard level. Some hearing loss is age-related; some related to exposure to high-level sound.

Helmholtz resonator   A reactive, tuned, sound absorber. A bottle is such a resonator. It can employ a perforated cover or slats over a cavity.

henry   The unit of inductance.

hertz   The unit of frequency, abbreviated Hz. The same as cycles per second.

high-pass filter   *See* filter, high pass.

HRTF   Head-related transformation.

IEEE   Institute of Electrical and Electronic Engineers.

image source   A loudspeaker located at an image point.

impedance   The opposition to the flow of electric or acoustic energy measured in ohms.

impedance matching   Maximum power is transferred from one circuit to another when the output impedance of the one is matched to the input impedance of the other. Maximum power transfer may be less important in many electronic circuits than low noise or voltage gain.

impulse   A very short, transient, electric or acoustic signal.

in phase   Two periodic waves reaching peaks and going through zero at the same instant are said to be "in phase."

initial time-delay gap   The time gap between the arrival of the direct sound and the first sound reflected from the surfaces of the room.

inductance   An electrical characteristic of circuits, especially of coils, that introduces inertial lag because of the presence of a magnetic field. Measured in henrys.

intensity   Acoustic intensity is sound energy flux per unit area. The average rate of sound energy transmitted through a unit area normal to the direction of sound transmission.

interference   The combining of two or more signals results in an interaction called interference. This may be constructive or destructive. Another use of the term is to refer to undesired signals.

intermodulation distortion   Distortion produced by the interaction of two or more signals. The distortion components are not harmonically related to the original signals.

inverse-square law   *See* spherical divergence.

ITD   Initial time-delay gap.

JASA   Journal of the Acoustical Society of America.

JAES   Journal of the Audio Engineering Society.

Korner Killer   A proprietary sound absorbing/diffusing unit for use in corners of rooms.

kHz   1,000 Hz.

law of the first wavefront   The first wavefront falling on the ear determines the perceived direction of the sound.

level   A sound pressure level in dB means that it is calculated with respect to the standard reference level of 20 μPa. The word "level" associates that figure with the appropriate standard reference level.

LEDE   Live end dead end.

LFD   Low-frequency diffusion.

linear   A device or circuit with a linear characteristic means that a signal passing through it is not distorted.

live end dead end   An acoustical treatment plan for rooms in which one end is highly absorbent and the other end reflective and diffusive.

logarithm   An exponent of 10 in the common logarithms to the base 10. For example, 10 to the exponent 2 is 100; the log of 100 is 2.

loudness   A subjective term for the sensation of the magnitude of sound.

loudspeaker   An electroacoustical transducer that changes electrical energy to acoustical energy.

masking   The amount (or the process) by which the threshold of audibility for one sound is raised by the presence of another (masking) sound.

mean free path   For sound waves in an enclosure, it is the average distance traveled between successive reflections.

microphone   An acoustical-electrical transducer by which sound waves in air may be converted to electrical signals.

middle ear   The cavity between the eardrum and the cochlea housing the ossibles connecting the eardrum to the oval window of the cochlea.

millisecond   One-thousandth of a second, abbreviated ms or msec.

MLS   Maximum length sequence.

mixer   A resistive device, sometimes veery elaborate, that is used for combining signals from many sources.

modal resonance   *See* mode.

mode   A room resonance. Axial modes are associated with pairs of parallel walls. Tangential modes involve four room surfaces and oblique modes all six surfaces. Their effect is greatest at low frequencies and for small rooms.

monaural   *See* monophonic.

monitor   Loudspeaker used in the control room of a recording studio.

monophonic   Single-channel sound.

multitrack   A system of recording multiple tracks on magnetic tape or other media. The signals recorded on the various tracks are then "mixed down" to obtain the final recording.

NAB   National Association of Broadcasters.

node   Point of no vibration.

noise   Interference of an electrical or acoustical nature. Random noise is a desirable signal used in acoustical measurements. Pink noise is random noise whose spectrum falls at 3 dB per octave; it is useful for sound analyzers with constant percentage bandwidths.

noise criteria   Standard spectrum curves by which a given measured noise may be described by a single NC number.

nonlinear   A device or circuit is nonlinear if a signal passing through it is distorted.

normal mode   A room resonance. *See* mode.

nul   A low or minimum point on a graph. A minimum pressure region in a room.

oblique mode   *See* mode.

octave   The interval between two frequencies having a ratio of 2:1.

oscilloscope   A cathode-ray type of indicating instrument.

ossicles   A linkage of three tiny bones providing the mechanical coupling between the eardrum and the oval window of the cochlea consisting of the hammer, anvil, and stirrup.

out of phase   The offset in time of two related signals.

oval window   A tiny membranous window on the cochlea to which the foot plate of the stirrup ossicle is attached. The sound from the eardrum is transmitted to the fluid of the inner ear through the oval window.

overtone   A component of a complex tone having a frequency higher than the fundamental.

partial   One of a group of frequencies, not necessarily harmonically related to the fundamental, which appear in a complex tone. Bells, xylophone blocks, and many other percussion instruments produce harmonically unrelated partials.

passive absorber   A sound absorber that dissipates sound energy as heat.

PFC   Phase-frequency curve.

phase   The time relationship between two signals.

phase shift   The time or angular difference between two signals.

phon   The unit of loudness level of a tone.

pink noise   A noise signal whose spectrum level decreases at a 3-dB-per-octave rate. This gives the noise equal energy per octave.

pinna   The exterior ear.

pitch   A subjective term for the perceived frequency of a tone.

place effect   The theory that pitch perception is related to the pattern of excitation on the basilar membrane of the cochlea.

plenum   An absorbent-lined cavity through which conditioned air is routed to reduce noise.

polar pattern   A graph of the directional characteristics of a microphone or loud-speaker.

polarity   The relative position of the high (+) and the low (−) signal leads in an audio system.

preamplifier   An amplifier designed to optimize the amplification of weak signals, such as from a microphone.

precedence effect   For delay time, less than 50 msec, echoes are no longer annoying even if the echo is stronger than the primary sound. This is called the precedence (or Haas) effect.

pressure zone   As sound waves strike a solid surface, the particle velocity is zero at the surface and the pressure is high, thus creating a high-pressure layer near the surface.

PRD   Primitive root diffusor.

psychoacoustics   The study of the interaction of the auditory system and acoustics.

pure tone   A tone with no harmonics. All energy is concentrated at a single frequency.

Q-factor   Quality-factor. A measure of the losses in a resonsnce system. The sharper the tuning curve, the higher the Q.

QRD   Quadratic-residue diffusor.

random noise   A noise signal, commonly used in measurements, which has constantly shifting amplitude, phase, and frequency and a uniform spectral distribution of energy.

ray   At higher audio frequencies, sound may be considered to travel in straight lines, in a direction normal to the wavefront.

reactance   The opposition to the flow of electricity posed by capacitors and inductors.

reactive absorber   A sound absorber, such as the Helmholtz resonator which involves the effects of mass and compliance as well as resistance.

reactive silencer   A silencer in air-conditioning systems that uses reflection effects for its action.

reflection   For surfaces large compared to the wavelength of impinging sound, sound is reflected much as light is reflected, with the angle of incidence equaling the angle of reflection.

RFZ   Reflection-free zone.

RPG   Reflection-phase grating. A diffusor of sound energy using the principle of the diffraction grating.

refraction   The bending of sound waves traveling through layered media with different sound velocities.

resistance   That quality of electrical or acoustical circuits that results in dissipation of energy through heat.

resonance   A resonant system vibrates at maximum amplitude when tuned to its natural frequency.

resonator silencer   An air-conditioning silencer employing tuned stubs and their resonating effect for its action.

response   *See* frequency response.

RT60   Reverberation time.

reverberation   The tailing off of sound in an enclosure because of multiple reflections from the boundaries.

reverberation chamber   A room with hard boundaries used for measuring sound absorption coefficients.

reverberation time   The time required for the sound in an enclosure to decay 60 dB.

RFZ   Reflection-free zone.

ringing   High-Q electrical circuits and acoustical devices have a tendency to oscillate (or ring) when excited by a suddenly applied signal.

room mode   The normal modes of vibration of an enclosed space. *See* mode.

round window   The tiny membrane of the cochlea that opens into the middle ear that serves as a "pressure release" for the cochlear fluid.

RPG Cloud   Reflection-phase gratings arranged for overhead mounting.

sabin   The unit of sound absorption. One square foot of open window has an absorption of 1 sabin.

Sabine   The originator of the Sabine reverberation equation.

Schroeder plot   A reverberation decay computed by the mathematical process defined by Manfred Schroeder.

semicircular canals   The three sensory organs for balance that are a part of the cochlear structure.

sequence, maximum length   A mathematical sequence used in determining the well depth of diffusors.

sequence, primitive root   A mathematical sequence used in determining the well depth of diffusors.

sequence, quadratic residue   A mathematical sequence used in determining the well depth of diffusors.

signal-to-noise ratio   The difference between the nominal or maximum operating level and the noise floor in dB.

sine wave   A periodic wave related to simple harmonic motion.

slap back   A discrete reflection from a nearby surface.

solid state   The branch of physics having to do with transistors, etc.

sone   The unit of measurement for subjective loudness.

sound absorption coefficient   The practical unit between 0 and 1 expressing the absorbing efficiency of a material. It is determined experimentally.

sound power level   A power expressed in dB above the standard reference level of 1 picowatt.

sound pressure level   A sound pressure expressed in dB above the standard sound pressure of 20 micropascals.

sound spectrograph   An instrument that displays the time, level, and frequency of a signal.

spherical divergence   Sound diverges spherically from a point source in free space.

SBIR   Speaker boundary interference response.

spectrum   The distribution of the energy of a signal with frequency.

spectrum analyzer   An instrument for measuring, and usually recording, the spectrum of a signal.

specularity   A new term devised to express the efficiency of diffraction-grating types of diffusors.

splaying   Walls are splayed when they are constructed somewhat "off square," i.e., a few degrees from the normal rectilinear form.

standing wave   A resonance condition in an enclosed space in which sound waves traveling in one direction interact with those traveling in the opposite direction, resulting in a stable condition.

steady-state   A condition devoid of transient effects.

stereo   A sterophonic system with two channels.

superposition   Many sound waves may traverse the same point in space, the air molecules responding to the vector sum of the demands of the different waves.

synthesizer   An electronic musical instrument.

T60   *See* RT60.

tangential mode   A room mode produced by reflections off four of the six surfaces of the room.

TDS   Time-delay spectrometry.

TEF   Time, energy, frequency

threshold of feeling (pain)   The sound pressure level that makes the ears tickle, located about 120 dB above the threshold of hearing.

threshold of hearing   The lowest level sound that can be perceived by the human auditory system. This is close to the standard reference level of sound pressure, 20 μPa.

timbre   The quality of a sound related to its harmonic structure.

time-delay spectrometry   A sophisticated method for obtaining anechoic results in echoic spaces.

tone   A tone results in an auditory sensation of pitch.

tone burst   A short signal used in acoustical measurements to make possible differentiating desired signals from spurious reflections.

tone control   An electrical circuit to allow adjustment of frequency response.

transducer   A device for changing electrical signals to acoustical or vice versa, such as a microphone or loudspeaker.

transient   A short-lived aspect of a signal, such as the attack and decay of musical tones.

treble   The higher frequencies of the audible spectrum.

Tube Traps   Proprietary sound-absorbing units.

volume   Colloquial equivalent of sound level.

watt   The unit of electrical or acoustical power.

wave   A regular variation of an electrical signal or acoustical pressure.

wavelength   The distance a sound wave travels in the time it takes to complete one cycle.

weighting   Adjustment of sound-level meter response to achieve a desired measurement.

white noise   Random noise having uniform distribution of energy with frequency.

woofer   A low-frequency loudspeaker used with the main loudspeaker.

***Boldface** numbers indicate illustrations; *italic* indicates a table.

ABOUT THE AUTHOR

F. Alton Everest is a legend in the world of sound. The creator of numerous technical innovations, and the author of scores of books and scholarly papers, he has been a leader in television engineering, sound recording, motion pictures, radio, and multimedia. A co-founder and director of the Science Film Production division of the Moody Institute of Science, he was also a section chief of the Subsea Sound Research section of the University of California. An educator who has taught at several leading institutions, he has consulted on acoustics to numerous industries for nearly 30 years. Having touched many of the technical highlights of the 20th century, he celebrated his 90th birthday in 1999. He and his wife live in Santa Barbara, California.